D0085088

EVOLUTION

OF

GENETIC SYSTEMS

EVOLUTION OF GENETIC SYSTEMS

Editor

H. H. SMITH

Editorial Committee

H. J. PRICE
A. H. SPARROW
F. W. STUDIER
J. D. YOURNO

BIOLOGY DEPARTMENT
BROOKHAVEN NATIONAL LABORATORY
UPTON, NEW YORK

QH366.2
.E86

GORDON AND BREACH
New York • London • Paris

INDIANA
UNIVERSITY
LIBRARY

NORTHWEST

Copyright©1972 GORDON AND BREACH, Science Publishers, Inc.
440 Park Avenue South, New York, N.Y. 10016

Library of Congress Catalog Card Number: 74-18433

Editorial office for Great Britain
   Gordon and Breach, Science Publishers Ltd.
   12 Bloomsbury Way
   London W.C. 1 England

Editorial office for France
   Gordon and Breach
   7-9 rue Emile Dubois
   Paris 14^e, France

"The papers that make up the chapters of this book
were delivered at a Symposium held at Brookhaven
National Laboratory under sponsorship of the United
States Atomic Energy Commission."

ISBN 0 677 1223 0 (cloth); 0 677 1223 5 (paper): All rights
reserved. No part of this book may be reproduced or utilized
in any form or by any means, electronic or mechanical, including
photocopying, recording, or by any information storage and
retrieval system, without permission from the publishers.

Printed in the United States of America

PREFACE

The development of new tools for examining the structures that transmit hereditary information is opening up new and significant directions for the study of the evolution of genetic systems. The purpose of this symposium is to bring together recent research on genetic mechanisms, from the molecular to chromosome level, and from lower to higher forms, in order to present a picture of current activities and ideas in this field of biology.

Evolutionary changes in the kind and amount of DNA, in macromolecular complexity, in the grouping of genes into operon and chromosome, and the significance of extrachromosomal systems are main topics of discussion.

The Committee wishes to express appreciation to those who gave advice during the planning stages, to the speakers and chairmen, to the Brookhaven staff who handled the arrangements, and to Mrs. Helen Kondratuk for special assistance throughout.

THE SYMPOSIUM COMMITTEE

H. H. SMITH, Chairman

H. J. PRICE

A. H. SPARROW

F. W. STUDIER

J. D. YOURNO

PROGRAM AND TABLE OF CONTENTS

*No Manuscript Received

PROGRAM AND TABLE OF CONTENTS

THE RATE OF CHANGE OF DNA IN EVOLUTION

Brian J. McCarthy and Margaret N. Farquhar
Departments of Biochemistry and Genetics
University of Washington
Seattle, Washington 98191

Abstract

Introduction

DNA annealing as an approach to the quantitation of evolutionary
divergence
    Repeated and unique sequences
    Estimation of base substitution
    Effect of base substitution on renaturation rate

Total DNA sequence divergence
    Base substitutions in rodent DNA's
    Comparison of the rate of evolutionary change in DNA and various
    proteins

Divergence within specific DNA sequences
    Ribosomal DNA
    Nuclear and cytoplasmic RNA
    Histone genes

Conclusions

Abstract

     Estimates of base sequence divergence in total DNA and specific
parts of the genome as determined by DNA annealing techniques are
discussed. Redundant and unique DNA sequences are shown to be
defined by the experimental conditions rather than to exist as two
distinct classes. Hybridization of sea urchin histone messenger RNA
to DNA from diverse organisms indicates that these sequences are
highly conserved in evolution and have not rapidly accumulated
neutral third position codon changes. The fact that histone genes
and other genes occur as repeats of identical sequences suggests
the probability of a correction mechanism which functions in
evolution to maintain intragenomic homogeneity.

INTRODUCTION

The changes in nucleic acid base sequences resulting from the
evolutionary divergence of two related species can be measured by the
formation of DNA-DNA duplexes or DNA-RNA hybrids. Duplex formation
with total DNA has been used extensively to determine the average
rates of base sequence change in bacteria, plants and animals.
Nucleotide substitution in the genes for ribosomal RNA (rRNA) has been
determined in many of the major phyla (Sinclair and Brown, 1971;
Bendich and McCarthy, 1970), and divergence in tRNA genes has been
measured in bacteria (Goodman and Rich, 1962; Brenner et al., 1970)
but studies of rates of divergence of other specific genes or groups
of genes have been limited by the difficulty of purification of such
genes or their messenger RNA products. Direct determination of base
substitution has been made by comparison of nucleotide sequence data
for a few tRNA molecules (see Dayhoff and McLaughlin, 1969 for review).
However, most information for total DNA or specific parts of the
genome derives from the indirect approach offered by annealing of
nucleic acids. Although this method provides only an indirect
determination of specific base changes, analyses of the reaction
rates and the properties of heteroduplexes, DNA-DNA or DNA-RNA, can
be used to quantitate the number of base substitutions which exist
between two partially complementary nucleic acid molecules.

DNA ANNEALING AS AN APPROACH TO THE QUANTITATION OF EVOLUTIONARY
DIVERGENCE

Repeated and unique sequences. DNA renaturation kinetics are often
displayed as "$C_o t$ curves" as described by Britten and Kohne (1968).
The kinetics of renaturation have been used to calculate genome size,
the relative proportions of repeated and unique sequences and the

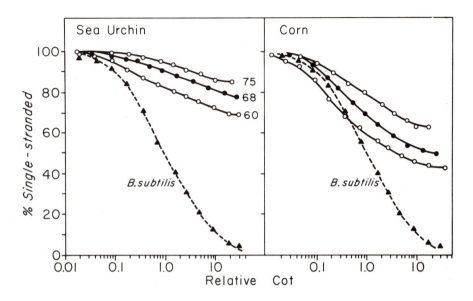

Figure 1. Renaturation of sea urchin and corn DNA as a function of
temperature. The rate of renaturation of sea urchin and corn DNA's
with B. subtilus DNA as an internal standard was followed in a recording spectrophotometer at 270 mμ. All DNA samples were at concentrations of 60 μg/ml in 44% formamide, 4XSSC. Renaturation was
allowed to proceed for approximately 64 hrs at each temperature. The
initial reaction was at 37° C followed by denaturation and renaturation
at 45° C, and a third denaturation and renaturation at 52° C. These
conditions are equivalent in stringency to renaturation in 1XSSC at
60° C, 68° C and 75° C, respectively (McConaughy et al., 1969). The
fraction reassociated as determined by hyperchromicity is plotted as a
function of relative $C_o t$ ($C_o t$ = initial concentration in moles P/liter
multiplied by time in seconds). Relative $C_o t$ values are used to correct
for the rate differences which exist at the different temperatures.
This correction was made by superimposing the B. subtilus renaturation
curves and plotting the sea urchin and corn kinetics relative to this
standard.

homogeneity of the repeat frequency. However, it is our contention
that repeated and unique sequences can be defined only operationally
in terms of the criteria involved, especially the conditions of renaturation (McCarthy and Church, 1970). It is apparent that most of
the sequences which cross-react and therefore appear to be repeated
are by no means completely complementary (Britten and Kohne, 1968).
Their classification as repeated results from their tendency to form
imperfect partially complementary duplexes. Under conditions of
higher stringency, such as higher temperatures or lower salt concentrations,
less DNA appears to be repeated. Figure 1 illustrates an experiment
in which sea urchin and corn DNA were renatured under different
conditions. The proportion of repeated DNA is directly dependent
on the reaction conditions. An increase in reaction temperature of 15° C
decreases the proportion of repeated sequences in sea urchin RNA from
30% to 15%, and in corn from 56% to 35%. Similar results have been
presented for mouse DNA (McCarthy and Duerksen, 1970). Together
these results suggest that no sharp distinction can be made between
unique and repetitive DNA.

.At the present time, it appears that a continuous gradation of
complex interrelationships exists among the large set of sequences
comprising a typical eucaryotic genome. Since many sequences have
arisen from duplication and divergence of other sequences, eucaryotic
genomes contain families of quite similar sequences as well as
apparently unrelated sequences depending on the amount of divergence
since the original duplication event. By varying the reaction
conditions more or less of these variously related sequences can be
made to react. At conditions of low stringency quite distant intragenomic evolutionary relationships may reveal themselves through the

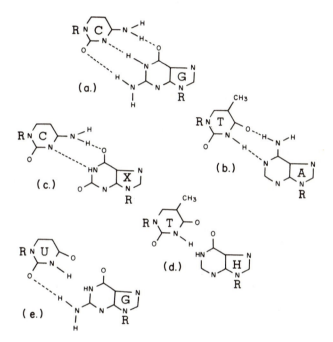

Figure 2. The structures of the standard Watson-Crick base pairs and those formed after deamination. The normal base pairs guanine-cytosine (a) and adenine-thymine base pairs (b) are compared to cytosine-xanthine (c), thymine-hypoxanthine (d) and guanine-uracil (e) base pairs formed after the deamination of guanine, adenine and cytosine, respectively (from Ullman and McCarthy, 1971).

formation of mismatched duplexes. Alternatively, under the most
stringent conditions only those sequences which are nearly identical
show renaturation kinetics typical of repeated sequences. At
conditions of moderate stringency (60° C, 1 X SSC) 15-20% of mouse
DNA appears redundant (McConaughy and McCarthy, 1970); however,
renaturation experiments at very high stringency (80° C, 1 X SSC) with
a purified redundant fraction of mouse DNA suggest that less than 3%
of the total DNA (M-D Chilton, personal communication) is comprised of
identical sequences typified by tRNA, rRNA and histone genes. The
contention that redundant and unique DNA should not be considered as
two distinct classes is further supported by the finding that the
rates of evolutionary divergence of these fractions from primates are
the same (Kohne,1970). Conversely, large differences in evolutionary
rates are seen when one compares highly reiterated satellite DNA
sequences and the rest of the genome.

Estimation of Base Substitution. The effect of mismatched bases on the
properties of DNA duplexes has been studied by Ullman and McCarthy
(1971) by producing duplexes which contain various non-Watson-Crick
base pairs. These altered base pairs were obtained by deamination of
various bases as illustrated in Figure 2. Deamination of DNA under
alkaline conditions specifically converts deoxycytosine to deoxyuracil
(Marrian et al., 1950). Thus renatured alkali deaminated DNA will
contain dG-dU pairs in place of the normal dG-dC pairs. Other
conditions can be used to produce different alterations. Deoxyadenine, deoxyguanine and deoxycytosine are deaminated by nitrous
acid (Schuster, 1960; Kotaka and Baldwin,1964). Elimination of bases
by depurination (McConaughy and McCarthy, 1967) also affects the
stability of renatured DNA duplexes (Ullman and McCarthy, 1971).

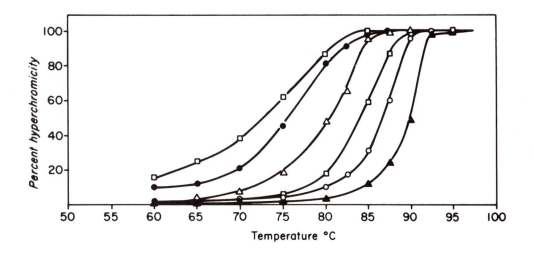

Figure 3. Thermal dissociation profiles of alkali deaminated E. coli

DNA. Renatured DNA duplexes (50 µg/ml in 1XSSC) were dissociated by

heating in a recording spectrophotometer. Absorbance changes were

monitored at 260 mµ. The percent altered base pairs of the duplexes

i.e. % dG to dU deamination is as follows: unaltered ▲ ; 1.0, ◐; 2.3,

□ ; 3.4, △ ; 5.6, ● ; and 6.8, □ . (from Ullman and McCarthy, 1971)

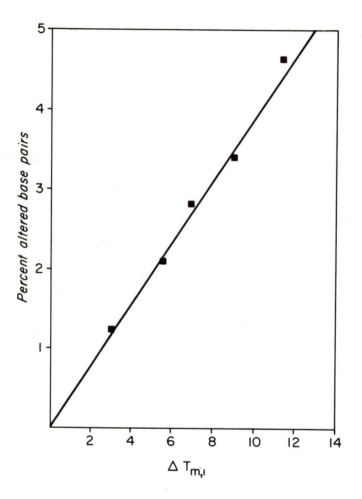

Figure 4. Thermal stability changes as a function of base pair alterations. The $T_{m,i}$ values were determined by melting duplexes formed from alkali deaminated E. coli DNA and unaltered filter bound DNA as described by McConaughy and McCarthy (1968) (from Ullman and McCarthy, 1971)

The effect of alkali deamination on the thermal stability
of renatured DNA is shown in Figure 3. Increased deamination leading
to more dG-dU pairs decreases the T_m of the duplexes. A linear
relationship between the ΔT_m, the decrease in mean thermal stability,
and the percent alteration in alkali deaminated DNA is seen in
Figure 4. The effect of deoxycytosine deamination on thermal
stability is 2.2^O C for each one percent altered base pairs.
Deamination of deoxyadenine and deoxyguanine reduces the T_m by 0.7^O C
and 1.6^O C for each one percent alteration, respectively. The largest
effect on the T_m is produced by depurination, which causes a ΔT_m of
3.2^O C for each one percent eliminated bases (Ullman and McCarthy, 1971).

    Although data exists for only a few of the possible mismatched
base pairs, this preliminary study suggests that the range of effects
on thermal stability is between 0.7 and 3.2^O C for each one percent
mismatched bases. After considering the relative frequency of
transitions, transversions and deletions occurring in evolution,
Ullman and McCarthy (1971) suggest that the overall effect of mismatched base pairs on the thermal stability is roughly 1.6^O C per
one percent altered base pairs. This estimation attempts to take
into account all the possible kinds of changes and, though tentative,
is probably accurate to within 25%. This value can be used to
estimate the base pair differences in DNA-DNA heteroduplexes or RNA-DNA
hybrids formed from nucleic acids of different organisms.

Effect of base substitution on renaturation rate. In considering
heterologous reactions occurring between two partially complementary
DNA strands representative of two related genomes, several effects
of the resultant base pair mismatching must be taken into account.
The rate of renaturation of the partially complementary duplex and its

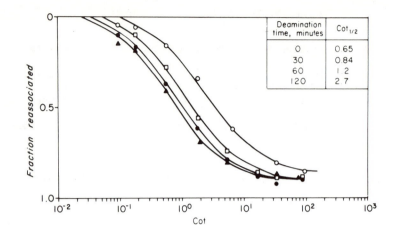

Deamination time, minutes	$Cot_{1/2}$
0	0.65
30	0.84
60	1.2
120	2.7

Figure 5. Rate of renaturation of alkali deaminated B. subtilus DNA's.
H^3-thymidine labeled B. subtilus DNA's (2500 cpm/μg) were deaminated
for various times then sheared at 12,000 psi in a French pressure
cell. The DNA's (1 mg/ml) were incubated in 48% formamide and 5XSSC
at 37° C for times indicated by the C_ot value. 100 μl samples were
diluted into 5 ml of 0.12 M phosphate buffer, pH 6.8 (PB) and applied
to columns containing 1 ml packed volume of Bio Rad HTP hydroxylapatite.
Column packing, equilibration with 0.12 M PB and all elutions were
carried out at 60° C. The 5 ml sample eluant and subsequent 5 X 1 ml
0.12 M PB washes were collected as a single fraction. Double stranded
DNA was eluted from the column by 5 X 1 ml washes with 0.5 M PB which
was also collected as a single fraction. The 0.5 M PB fraction was
diluted to a final concentration of 0.12 M PB and each fraction was
precipitated with TCA, collected on Whatman GFC glass filters and
counted. The fraction reassociated is the percent of radioactivity in
the 0.5 M PB fraction of each sample. Renaturation kinetics are shown
for unaltered DNA (▲), 30 minute deaminated DNA (●), 60 minute deaminated
DNA (□) and 120 minute deaminated DNA (O). Theoretical curves are
fitted to the data points for a 90% reaction for the first 3 reactions
and an 85% reaction for the 120 minute deaminated DNA reaction. $C_ot_{1/2}$
refers to the midpoint of the renaturation curve (Marsh, Gregg and
McCarthy, unpublished results).

thermal stability are both directly influenced by the extent of mismatching. An example of the effect on reaction rate is shown in
Figure 5 which illustrates an experiment in which partially deaminated
B. subtilis DNA was renatured. Different extents of deamination of
deoxycytosine residues to deoxyuridine were effected by alkali
treatment. As the deamination time was increased the reaction rate
declined, a direct consequence of the presence of mismatched dG-dU
base pairs. The number of mismatched base pairs and the reaction
rate (k) can be calculated from the data in Figure 5 by the following
equations: rate of alkali deamination = 4.5% base pair alteration/
hour (Ullman and McCarthy, 1971) and $1/k = (C_0 t_{1/2})$ (Britten and
Kohne, 1968). An exponential decline in reaction rate as the number
of mismatched base pairs is increased is apparent in Figure 6. Well-
matched sequences react approximately four times as fast as those
which contain 9% altered base pairs.

The explanation for this effect probably derives from the
existence of a rate limiting nucleation step in DNA renaturation
(Wetmur and Davidson, 1968). The exact number of base pairs
necessary for a stable nucleation will depend upon the reaction
conditions such as temperature and salt concentration. At a given
set of conditions the probability of a stable effective collision
will obviously be diminished if mismatched base pairs are formed
since longer stretches of bases in DNA must interact to compensate
for those which do not form hydrogen-bonded pairs. The magnitude
of the effect of mismatched bases will likely increase when
conditions of higher stringency are used for renaturation reactions.
More base pairs are necessary for stable duplexes at higher
temperatures or lower salt concentrations; therefore even longer
sequences are needed for stable collisions when these sequences

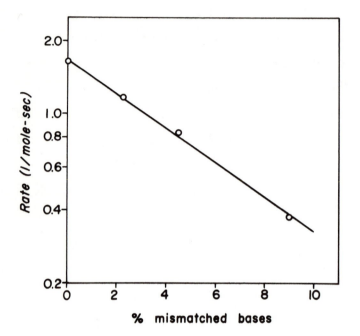

Figure 6. Renaturation rate as a function of mismatched nucleotide pairs.
Reaction rates and percent mismatched bases are calculated from data
in Figure 5 as described in the text. (from Marsh, Gregg and McCarthy,
unpublished results)

contain mismatched base pairs. Preliminary experiments with alkali deaminated B. subtilus DNA at high stringency conditions suggest that, in fact, renaturation rates do decrease as stringency is increased (Marsh, Gregg and McCarthy, unpublished results).

This effect has several important implications. In the first place, calculation of the number of interacting sequences from the rate of renaturation is justifiable only if perfectly matched duplexes are formed exclusively. This consideration brings into question the practice of calculating family size, or number of redundant inter- acting sequences from renaturation kinetics (Britten and Kohne, 1968; Davidson, et al., 1971). Certainly with mammalian nucleic acids, the redundant DNA is so mismatched that a very sizeable effect on renaturation kinetics must be expected. Since the rate of reaction will be severely reduced, the family size or number of interacting partially complementary sequences will be severely underestimated (Sutton and McCallum, 1971).

TOTAL DNA SEQUENCE DIVERGENCE

Base substitutions in rodent DNA's. Determinations of the rates of nucleotide substitutions in eucaryotes are usually made with that fraction of DNA which appears to be unique under the reaction conditions used. Families of related base sequence are absent so that the ΔT_m of the various heterologous duplexes will reflect only interspecies evolutionary divergence. However, the ΔT_m is a measure of the divergence only in those sequences which actually formed duplexes. McConaughy and McCarthy (1970) have determined the extent of base substitutions among the unique DNA fractions of various rodents. The thermal stabilities of the duplexes indicate 9, 11 and 12% mismatched bases among mouse-rat, mouse-hamster and mouse-guinea

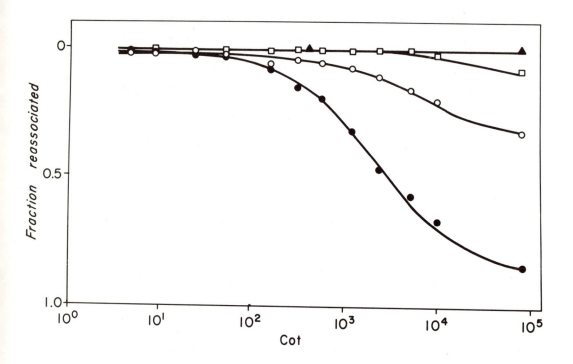

Figure 7. Renaturation kinetics of mouse L-cell unique DNA with DNA's of other rodents. H³-thymidine labeled mouse L-cell unique sequences were incubated at 0.5 µg/ml alone and with unlabeled DNA's from mouse, rat and guinea pig (5 mg/ml) in 48% formamide and 5XSSC at 37° C. The fraction of H³-L-cell DNA associated with the various rodent DNA's was determined by hydroxylapatite chromatography as described in Figure 5. Kinetics are shown for the renaturation of unique mouse L-cell DNA alone (▲) and with unlabeled mouse (●), rat (○) and guinea pig (□) DNA's.

pig respectively. Any consideration of the meaning of this relative difference must take into account the fact that very different degrees of cross-reaction were obtained in these three cases.

Since the rate of reaction of mismatched sequences is slower than that of well-matched ones, a high degree of mismatching carries the implication that the reaction may not have attained completion. For example, Figure 7 shows the rates of reaction of unique mouse H^3-DNA with other unlabeled rodent DNA's. At $C_ot = 10,000$ the homologous reaction is almost complete although the heterologous reactions apparently are not. Even at $C_ot = 90,000$ the heterologous reactions may not be complete; however, at this point all of the unlabeled DNA has been renatured and further heterologous reaction is impossible. This consideration points out a major misinterpretation (McConaughy and McCarthy, 1970) of earlier experimental data. The fact that heterologous reactions among rodent DNA's are not quantitative does not necessarily prove that a major fraction of the DNA base sequences have diverged so far as to be unrecognizable. Even if cross-reactions cannot be demonstrated one should not assume that such DNA base sequences have diverged beyond recognition. Comparison of the rate of evolutionary change in DNA and various proteins. It is of obvious interest to compare quantitative estimates of the rate of base substitution in DNA during evolution with estimates of amino acid substitution in various proteins. Figure 8 presents such a comparison. Rates of sequence change in artiodactyl DNA obtained from experiments analogous to those described in previous paragraphs are plotted together with rates of change in the messengers for several proteins (Laird et al., 1970).

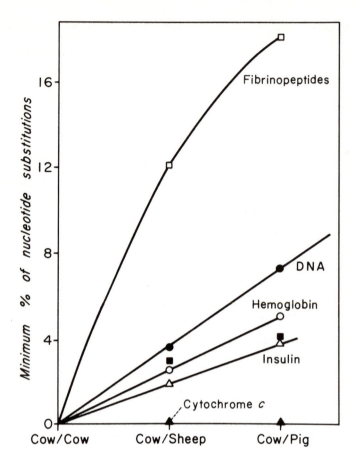

Figure 8. Nucleotide sequence divergence among artiodactyls. Estimates
of gene divergence are recalculated from the data of Laird et al. (1970)
based on the relationship of 1.6° C ΔT_m per 1% altered base pairs (Ullman
and McCarthy, 1971). Inferred percentage of nucleotide substitution
for the various protein genes is from amino acid sequence data compiled
by Dayhoff and Eck (1967-8). Data are included for fibrinopeptides (□),
hemoglobin α (■), hemoglobin β (O), insulin (Δ) and cytochrome C (▲).
(from Laird et al., 1970)

Since the number of base substitutions in a messenger is calculated as
the minimum number of changes in the genetic code necessary for an
amino acid replacement, the actual change in the mRNA sequence may
well be greater as a result of third position substitutions which
do not lead to amino acid changes (King and Jukes, 1969). In much
the same way the values for the rate of change in DNA could be
underestimated since incomplete cross-reactions were obtained and
it is assumed that the T_m is a valid measurement for the total DNA.
As is well known, the rate of substitution in proteins is extremely
variable. Cytochrome C shows extreme conservatism, fibrinopeptides
accumulate amino acid substitutions rapidly and intermediate rates of
change are seen in other proteins (Dayhoff and Eck, 1967-8). The
apparent rate of change in DNA is higher than that for all of the
proteins except fibrinopeptides. The reason for this is far from
clear at the present time. It could be explained by a high
frequency of third position changes or even more simply by the fact
that the majority of the DNA does not code for structural genes
and could evolve at a faster rate.

DIVERGENCE WITHIN SPECIFIC DNA SEQUENCES

Ribosomal DNA. In view of the difficulties inherent in comparing
rates of change in total DNA with those for specific proteins, it is
of importance to determine rates of change in individual genes
or families of genes in DNA. This is most readily accomplished by
isolating specific classes of RNA molecules or individual messengers.
Ribosomal RNA can be easily isolated and purified and therefore
has been used extensively to study divergence of a restricted
portion of the genome. These genes occur as tandem repeats of
identical sequences (Birnstiel et al., 1971); consequently

Table 1. Divergence of ribosomal RNA in yeast and bacteria

Yeast species	DNA	25 S rRNA
Saccharomyces lactis	100	100
S. dobzhanskii	32	98
S. wickerhamii	13	99
S. globosus	13	83
S. fragilis	5	98

Bacterial species	2 Min Pulse labeled RNA	23 S rRNA
Escherichia coli	100	100
Aerobacter aerogenes	30	63
A. cloacae	17	89
Proteus vulgaris	1	105
Bacillus subtilus	0	56

S. lactis DNA filters (25 µg DNA) and H^3 labeled S. lactis DNA (0.5 µg)
were incubated in 0.3 ml 2XSSC for 12 hrs at 60° C in the presence of
the indicated unlabeled DNA's. Yeast rRNA homologies were assayed
by direct binding of H^3 labeled S. lactis rRNA (0.1 µg) to the
various DNA filters (20 µg DNA/filter). Hybridization conditions
were 70° C, 2XSSC for 12 hr (from Bicknell and Douglas, 1970).
Bacterial RNA hybridization reactions were carried out by binding
C^{14} pulse labeled E. coli RNA (33 µg) or E. coli P^{32} labeled rRNA
(3.9 µg) to bacterial DNA filters (10 µg DNA/filter). All incubations were in 2XSSC at 66° C for 16 hrs (from Moore and
McCarthy, 1967). Cross reactions are recorded as the percent binding
relative to the homologous reaction.

Table 2. Interaction of pea ribosomal RNA with various DNA's.

Filter bound DNA	Bound Relative to Pea (%)		T_m (°C)		ΔT_m (°C)		T_m 25S minus T_m 16S
	25S	16S	25S	16S	25S	16S	
Pea	100	100	86.4*	87.3*			
Cucumber	211	184	81.7	83.0	4.8	4.5	0.3
Barley	46	53	80.2	81.8	6.3	5.7	0.6
Oats	17	26	79.8	80.0	7.3	6.8	0.5
Wheat	24	26	75.6	78.3	10.7	9.0	1.7
Daffodil	12	16	76.8	80.7	8.8	7.0	1.8
Cedar	20	31	76.0	78.4	10.5	9.1	1.4
Yeast (S. fragilis)	105	125	75.0	78.5	12.1	8.3	3.8
Sea urchin	18	18	71.7	75.0	14.6	12.3	2.3
Tetrahymena	51	44	72.3	73.2	14.2	14.3	-0.1
Euglena	9	17	63.0	69.9	22.6	18.8	3.8
B. subtilis	20	41		71.6		15.2	
E. coli	18	43	64.3	68.1	22.8	18.7	4.1
Clam	6	11					
Mouse	3	6					
Guinea pig	3	5					
φ80	1	1					

*T_m values for the homologous pea hybrids were 86.3, 86.5, 87.1, and 85.6 degrees for 25S rRNA and 87.3, 87.5, 86.8, and 87.7 degrees for 16S rRNA; the average value is presented in the table.

Five-tenths ug ^{3}H-labeled pea 25S or 16S rRNA (2100 or 1900 cpm) was incubated 17 hr at 60° C in 2 X SSC with 19-23 µg filter-bound DNA. After direct binding was assayed, thermal stability profiles and T_m values were determined using DNA filters from duplicate experiments. 8.1 and 6.6% of the input ^{3}H-25S and 16S rRNA, respectively, were the means bound to pea DNA-filters. The ΔT_m values in the table were derived from thermal elution profiles run in parallel rather than from the average pea T_m.

annealing reactions proceed quite rapidly with little or no
degradation of RNA.

Several sets of experiments have demonstrated that rRNA genes
are highly conserved throughout evolution and hybridization of rRNA
has been extensively used to measure relatedness among widely
divergent species. Table 1 compares the hybridization of rRNA or
pulse labeled RNA with total DNA for several species of bacteria
and yeast. In many cases the rRNA of closely related species
appears to be quite similar where there are sizeable differences
between the two DNAs or between pulse labeled RNA and DNA. For
example only 1% of Escherichia coli pulse labeled RNA can form stable
duplexes with Proteus vulgaris DNA, but E. coli rRNA reacts equally
well with P. vulgaris and homologous DNA. Similar situations exist
among the yeasts, e.g. Saccharomyces lactis and S. fragilis show
only 5% homology when DNA duplexes are compared, but heterologous
rRNA/DNA hybrids are formed with a 98% efficiency. However, small
differences are detectable even between closely related species
if the T_m of the heterologous hybrids is used as the criterion
(Moore and McCarthy, 1967; Bicknell and Douglas, 1970). The very
high degree of conservation of base sequence in rRNA is illustrated
by the ability of a plant rRNA to form hybrids with animal or
bacterial DNA's. The reaction of pea rRNA with DNA of several
organisms is shown in Table 2. Cross reaction occurs not only with
other members of the plant kingdom but also with invertebrates,
ciliates and bacteria. Comparison of the thermal stabilities of the
28S and 18S rRNA hybrids suggests that the large rRNA molecule is
evolving at a faster rate than the smaller one. Implications of
these findings have been discussed elsewhere (Bendich and McCarthy, 1970).

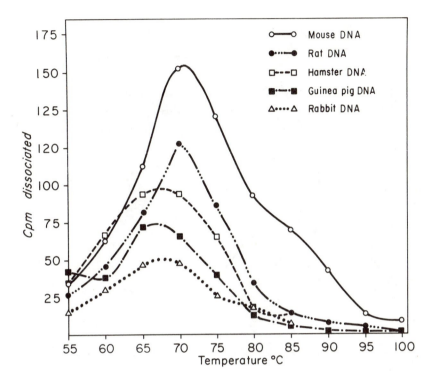

Figure 9. Thermal dissociation profiles of hybrids formed between mouse

L-cell cytoplasmic RNA and mouse DNA or DNA of related organisms.

Forty-five μg of 100 minute H^3-uridine pulse-labeled RNA was incubated

with 15 μg of filter bound DNA at 50^0 C for 18 hrs in 0.5 ml of 2XSSC

buffered with 10^{-3} M TES, pH7. Filter bound hybrids were dissociated

by incubation in 3 ml SSC for 8 minutes at the indicated temperatures.

The RNA eluted from the filters was monitored by TCA precipitation and

scintillation counting. (from Shearer and McCarthy, 1970)

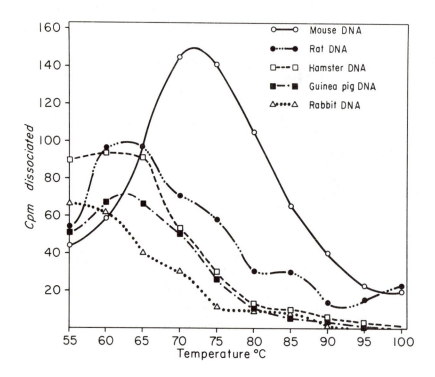

Figure 10. Thermal dissociation profiles of the hybrids formed between mouse L-cell nuclear restricted RNA and mouse DNA or DNA of related organisms. Hybridization conditions and thermal dissociation procedures are the same as in Figure 9. Nuclear restricted RNA is total nuclear RNA incubated in the presence of 1.2 mg of unlabeled mouse L-cell cytoplasmic RNA. (from Shearer and McCarthy, 1970)

In summary, ribosomal RNA genes contain some segments which are so highly conserved as to permit the formation of heteroduplexes between two very distantly related species. Nevertheless, the fact that small differences are discernible even between closely related species emphasizes that some parts of the ribosomal RNA molecule are much less resistant to evolutionary change. This situation is analogous to that established for evolutionary change in protein molecules: some positions are invariant while others are extremely variable between species (Fitch and Margoliash, 1967).

Nuclear and cytoplasmic RNA. Eucaryotic cells transcribe large amounts of RNA which is restricted to the nucleus and therefore unavailable for translation (Harris, 1959; Attardi et al., 1966; Scherrer et al., 1966; Soeiro et. al., 1968). If natural selection operates mainly at the level of the structure and function of protein molecules, then the nuclear restricted RNA might accumulate more base substitutions than the cytoplasmic messenger RNA (cRNA). This rather elementary expectation appears to be validated by some preliminary experimental data (Shearer and McCarthy, 1970). The reactions of mouse cRNA and mouse nuclear restricted RNA with various rodent and rabbit DNA's are shown in Figures 9 and 10. Since total RNA isolated from nuclei contains nuclear restricted RNA as well as cRNA sequences, the annealing reactions in Figure 10 were carried out in the presence of high concentrations of unlabeled cRNA. This competition precludes the reaction of labeled cRNA molecules also present in the nucleus but permits reaction of the labeled nuclear restricted RNA. Comparison of the hybrids formed with cRNA and nuclear restricted RNA reveal that divergence does occur more rapidly in the nuclear restricted sequences. More of the

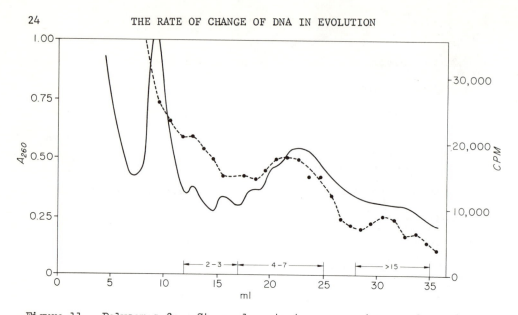

Figure 11. Polysomes from <u>Strongylocentrotus</u> <u>purpuratus</u> morulae. Sea urchin embryos were cultured at 9-10° C as described by Whiteley et al. (1966). Morulae were harvested and resuspended in filtered sea water (1 ml packed embryos: 10 ml sea water) for pulse labeling with H^3-uridine (5 μc/ml) for 1 hour. Cultures were placed on a reciprocal shaker and aerated during the labeling period. Cell homogenates were prepared by the method of Iverson and Cohen (1969) with the following modifications: cells were ruptured by passing the embryos twice through 54 μ Nitex cloth, then twice through 25 μ Nitex cloth without detergent; the homogenate was centrifuged at 10,000xg for 10 minutes to remove cell debris and mitochondria. Approximately 3 ml of the post-mitochondrial supernatant (cell homogenate from 3 ml of packed embryos) was applied to 15-40% sucrose gradients containing the higher salt buffer of Iverson and Cohen (1969) and was centrifuged at 26,000 rpm in an SW 27 rotor at 4° C for 2.5 hours. The A260 was monitored by pumping through an Isco Model S Density Gradient Fractionator. Radioactivity in each 1 ml fraction was determined by liquid scintillation counting of a 50 μl aliquot in 10 ml of toluene scintillant : Triton-X-100 (2:1). Polysome fractions were collected as indicated by precipitation with 2 volumes of ethanol at -20° C overnight.

cRNA reacts with the various heterologous DNA's, as compared to the homologous reaction, and the higher thermal stabilities of the products indicate fewer mismatched bases in the hybrids. For example, in the most divergent pair of animals, mouse and rabbit, the heterologous cRNA reaction is 21% of the homologous reaction, while the heterologous nuclear restricted reaction is 11%. The ΔT_m of the mouse-rabbit cRNA hybrid is 4° C; the nuclear restricted ΔT_m is 12° C. This indicates about 3% mismatching in the cRNA hybrids and 8% in the nuclear restricted hybrids.

Histone genes. Histones, the predominant chromosomal proteins of most eucaryotes, show fewer amino acid substitutions among divergent species than any other protein for which sequence data is available. Smith et al. (1970) have shown that there are only two amino acid differences between the histones IV of peas and cattle. Although species specific differences have been found in histone I (Zweidler and Gotchel, 1971) the other histones also appear to be highly conserved (Stellwagen and Cole, 1969). Clearly the genes for histones should reflect this same conservation of sequence, showing only those nucleotide substitutions leading to synonomous codons. Indeed the readily hybridizable RNA synthesized in early echinoderm cleavage has already been shown to be conserved in the evolution of this phylum (Whiteley et al., 1970).

The mRNA for histones is carried on small polysomes which predominate during early cleavage in sea urchin embryos (Nemer and Lindsay, 1968; Kedes and Gross, 1969). Figure 11 shows the polysome profile of Strongylocentrotus purpuratus morulae. Most of the polysomes at this stage are small (4-7 ribosomes) and readily separable from the larger ones. The H^3-uridine pulse-labeled mRNA

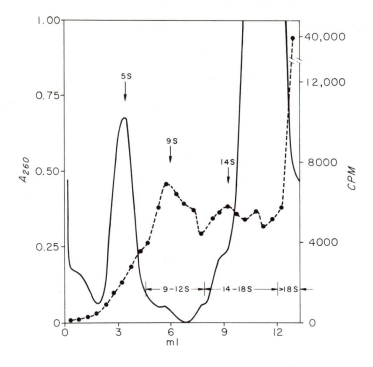

Figure 12. RNA from small polysomes of sea urchin morulae. Alcohol
precipitated RNA from the small polysome region (4-7 ribosomes) of
15-40% sucrose gradients was resuspended in SD buffer (0.02 M sodium
acetate, pH 5.2, 0.2 M LiCl, 0.5% SDS), extracted twice with cold SD
saturated phenol and once with cold chloroform : octanol (24:1). The
RNA was precipitated with 2 volumes ethanol at -20° C, then resuspended
in distilled water. One-tenth volume of 10 X SDS buffer (Chamberlain
and Metz, 1971) was added, the RNA was layered on 5-20% sucrose
gradients containing the same buffer and centrifuged in an SW 40 rotor
at 38,000 rpm at 20° C for 13 hours. A260 and radioactivity were
monitored as described in Figure 11. RNA fractions from the 9-12S
region of the gradient were pooled and dialyzed, then concentrated
by evaporation. Precipitation with 2% potassium acetate removed the
SDS. The 9-12S fractions from 12-15 ml of packed morulae contained
approximately 100 μg RNA with specific activities of 600-900 cpm/μg.

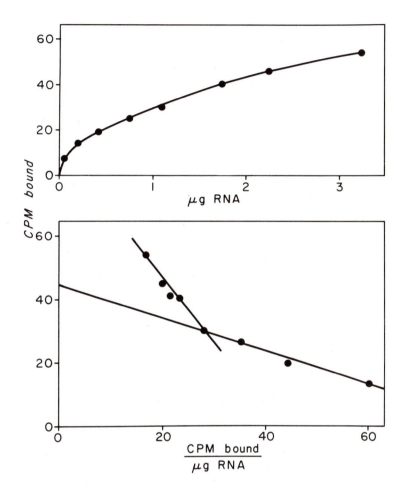

Figure 13. Saturation hybridization of 9-12S mRNA with sea urchin

DNA. DNA filters containing 18 µg of sea urchin sperm DNA per filter

were prepared as described by McCarthy and McConaughy (1968) and

incubated with the indicated amounts of H[3]-uridine labeled 9-12S mRNA

(900 cpm/µg). All incubations were in 0.2 ml 2XSSC at 67° C for

16 hours. Filters were washed 3 times in 2 ml of the incubation

buffer at 67° C. Radioactivity was monitored by liquid scintillation

counting.

extracted from these polysomes appears on a sucrose gradient
as a 9S peak with a 12S shoulder (Figure 12). The 9-12S RNA was
hybridized with DNA filters in order to obtain an estimate of the
fraction of the genome complementary to this group of RNA molecules.
The saturation hybridization curve is shown in Figure 13. The
reciprocal plot (lower curve) suggests that this curve is made up of
at least two components. Since hybridization with DNA filters
monitors only those sequences which appear repeated under the
conditions employed (McCarthy and Church, 1970), a substantial
number of redundant or highly mismatched hybrids are formed in
experiments with total RNA. With RNA preparations containing
identical as well as partially complementary sequences, the identical
sequences would react more rapidly and form hybrids with higher
thermal stabilities. Indeed the most rapidly reacting fraction
proves to have a higher thermal dissociation temperature. The best
estimate from the saturation curve yields a value of 2000 complementary DNA sites per genome, based on a molecular weight of
200,000 for a 9S molecule and 0.8 pg of DNA for the genome size
of S. purpuratus (Hinegardner, personal communication). In vitro
translation of the small polysomes suggests the presence of mRNA's
for histone fractions II, III and IV, probably four or five
different proteins (Moav and Nemer, 1970). This suggests a few
hundred sites for each histone species. An estimate of 400 for
the average repeat frequency of genes coding for histones was
obtained by independent methods by Kedes and Birnstiel (1971).

The 9-12S mRNA forms stable hybrids with DNA of very divergent
species, which further substantiates the contention that this RNA
fraction contains histone mRNA. The results of hybridization with

Table 3. Reaction of Histone mRNA with various DNA's (60° C, 2XSSC)

DNA	%cpm in hybrid	T_m
Echinoderms:		
Sea Urchin	16.9	77.0
Starfish	3.2	70.5
Other Invertebrates:		
Jellyfish	3.5	69.0
Oyster	3.4	69.0
Crab	7.9	70.0
Acorn Worm	4.5	69.5
Vertebrates:		
Salmon	3.1	69.0
Frog	1.1	69.0
Chicken	1.8	67.5
Mouse	1.1	67.0
Plant:		
Pea	2.0	67.5
Bacterium:		
B. subtilus	0.3	

DNA filters were prepared as described by McConaughy and McCarthy (1968). Filters containing 80-100 μg DNA were incubated with 1.6 μg 9-12S H^3-RNA (1100 cpm) in 0.2 ml 2XSSC at 60° C for 16 hrs. The thermal stability of the duplexes was assayed by washing the filters in 2 ml of 1XSSC at increasing temperatures. The T_m is the temperature at which 50% of the filter bound radioactivity was removed.

various DNAs are shown in Table 3. The amounts of hybrid formed
show large variation due to the fact that the genome size and
probably the numbers of genes devoted to histones vary widely among
the organisms tested. However, the thermal stabilities of the hybrids
are remarkably similar; slightly lower T_ms are seen with the DNA of
those species which are the most distantly related to the sea urchin
i.e. the higher vertebrates and a higher plant.

Although the lower stringency conditions (60° C, 2 X SSC)
allow higher amounts of reaction with the divergent species and are,
therefore, quite useful in testing a wide variety of organisms,
these conditions are not optimal for well-paired hybrids of high G+C
content. The T_m of the homologous hybrid probably reflects mismatching of contaminating heterogenous RNA molecules rather than an
accurate measure of the thermal stability of the histone mRNA hybrids.

Kedes and Birnstiel (1971) have estimated that histone mRNA is
51-53% G+C. The theoretical T_m for a 52% G+C RNA-DNA hybrid is 85° C
(Marmur and Doty, 1962; Church and McCarthy, 1970). Higher
temperatures are necessary for optimal hybridization of high
G+C nucleic acids. Increased temperatures also minimize the reaction
of the heterogeneous RNA since sequences with a high degree of mismatching will be unstable under these conditions. For these reasons,
the T_m of the homologous hybrids formed at 67° C in **2** X SSC is more
likely to reflect the thermal stability of histone messenger-DNA
hybrids than is the T_m of hybrids formed at 60° C, 2 X SSC. The T_m
of homologous hybrids formed at 67° C is 83.5° C (Table 4). Although
the hybridization conditions favor the formation of high G+C DNA/RNA
hybrids, the T_m is probably a minimum value due to some reaction of
RNA molecules contaminating the histone mRNA, even at higher

Table 4. Reaction of Histone mRNA with various DNA's (67° C, 2XSSC)

DNA	%cpm in hybrid	T_m
Echinoderms:		
Sea Urchin	13.6	83.5
Starfish	1.9	80.0
Sea Cucumber	4.1	77.0
Other Invertebrates:		
Jellyfish	2.6	75.0
Acorn Worm	2.7	74.5
Vertebrate:		
Mouse	0.7	72.5
Bacterium:		
B. subtilus	0.02	

Experimental conditions were the same as in Table 3 except that all incubations were at 67° C.

temperatures. Therefore, the exact number of base substitutions within a family of histone genes cannot be estimated but the small T_m difference between the theoretical (85° C) and experimental (83.5° C) values suggests that the repeated genes for a particular histone must be very nearly identical.

The higher stringency conditions tend to amplify the differences in heterologous reactions. Less hybridization under these conditions is probably due to two factors: 1) the most highly mismatched sequences are not stable at the higher temperature, and 2) the rate of reaction of sequences containing mispaired bases is substantially decreased. Mispaired bases appear to decrease the reaction rate more as stringency is increased (Marsh, Gregg and McCarthy, unpublished results). Hybridization at 60° C (Table 3) suggests that sea urchin histone genes are equally divergent from those of other echinoderms, other invertebrates and lower vertebrates. However, the 67° C reactions (Table 4) indicate that the number of base substitutions increases with evolutionary distance. Detectable divergence has occurred within the echinoderms and between echinoderms and other invertebrate phyla. The most distantly related pair tested at 67° C was sea urchin-mouse which has a ΔT_m of 11° C compared to the homologous reaction. This indicates about 8% mismatched bases in these hybrids.

CONCLUSIONS

Since the pioneering study of Britten and Kohne (1968), it has been customary to consider the eucaryotic genome as consisting of two distinct kinds of DNA base sequence, termed redundant (or repetitive or repeated) and unique. Since that time it has become clear that a special type of redundant DNA having the highest degree

of repetition does have special characteristics. The most highly
repetitive DNA, which in many species appears as a density satellite,
seems to play a structural role, and is located in heterochromatic
regions (Pardue and Gall, 1970; Jones, 1970; Botchan et al., 1971).
These sequences are seldom, if ever, transcribed (Flamm et al.) and
appear to be rapidly evolving. With respect to the so-called
intermediate or less repetitive fraction, several authors have
suggested a specific regulatory role (Britten and Davidson, 1970;
Georgiev, 1970). However, as we have stressed in the preceding
paragraphs, no clear distinction between redundant and unique
sequences can be made on the basis of renaturation kinetics or rate of
evolution. The assignment of any given sequence to one or another
category is arbitrary and completely dependent upon the reaction
conditions chosen by the experimenter. This, together with the fact
that the proportion of apparently redundant and unique sequences is
so variable from species to species, leads one to conclude that models
of regulation based upon a sharp distinction between rendundant and
unique sequences are speculative and premature at best. Nevertheless,
it is perfectly conceivable that some stretches of DNA base sequence
performing specialized roles can be distinguished by renaturation
kinetics. The polyadenylic acid sequences existing in high molecular
weight nuclear RNA (Edmonds et al., 1971, Lee et al., 1971 and Darnell
et al., 1971), which may act as recognition signals for the cleavage
mechanism, are one obvious category which would appear highly
rendundant under appropriate DNA renaturation conditions.

DNA which is not transcribed, or transcribed but not translated,
might be expected to evolve rapidly. This may well be a predominant
effect since several lines of evidence suggest that very little of the

DNA of higher organisms is used for structural genes. Functions
other than coding for proteins may require a certain number of bases
with little or no requirement for a specific sequence. Nucleotide
substitutions would then accumulate at the maximum rate, although
few if any deletions or insertions would be tolerated. Genome sizes
would be relatively constant in closely related species but base
sequences would be different. DNA which serves a structural function,
such as that supposedly involved in chromosomal bridges (De Praw, 1971)
might be rapidly evolving.

An example of rapid evolution of non-transcribed DNA is provided
by the spacer regions between transcriptional units in ribosomal DNA
of Xenopus. Regions separating the 45S initial transcriptional
product differ between X. laevis and X. mulleri (Brown, personal
communication), although the mature gene products, 18S and 28S rRNA
are virtually indistinguishable. Within the 45S precursor itself,
those regions which are transcribed but not preserved i.e., the
spacer regions between the 18S and 28S rRNA molecules, are also
rapidly evolving (Loening, 1970). Presumably the selection is mostly
determined by the specificity of the cleavage enzymes, and most of the
spacer sequence is free to accumulate non-deleterious substitutions.

A similar situation appears to exist for base sequences within
the giant nuclear RNA of mammalian cells. The rate of accumulation
of base substitutions is much greater for those parts of the precursor
which are not transported to the cytoplasm but degraded within the
nucleus, than is the case for cytoplasmic messenger RNAs.

The fact that base substitutions in DNA appear to occur more
rapidly than amino acid substitutions in proteins (Figure 8) has been
attributed to 1) an accumulation of third position changes in

synonomous codons, 2) more rapid base substitution in non-
transcribed or non-translated DNA sequences than in those sequences
which code for structural genes and 3) the addition of DNA sequences
in speciation (Laird et al., 1970). All of these mechanisms probably
contribute to the increased rate of base substitutions, but the effect
of third position changes is perhaps less than expected if such
changes are strictly neutral.

Kimura (1968) has calculated that out of 549 possible codon
changes, 134, or 25% could lead to synonomous codons which would not
result in amino acid changes. Assuming complete conservation of
amino acid sequence, corresponding genes of two distantly related
organisms could differ by as much as 17% if all possible synonomous
condon changes accumulate (Kohne, 1970). In the case of histone IV,
the change in the mRNA base sequence expected from the amino acid
substitutions would be approximately 0.5% for the pair cattle-peas.
According to our results the divergence in histone mRNA for the
distantly related pair of species, sea urchin and mouse, amounts to
some 8%. Presumably most of the base sequence divergence is
attributable to third position substitutions and could approach the
expected 17% if an even more distantly related pair of species were
considered. However, it should be emphasized that such a rate of
accumulation of base substitutions is not as rapid as would be
expected if these third position changes are strictly neutral. Thus
within the phylum Echinodermata itself, a much more limited number
of changes are apparent despite the fact that the age of the phylum
is estimated at 600×10^6 years (Moore, 1966), so that the most
divergent species within this group must be separated by a comparable
period of evolutionary time.

It is now well-established that the multiple genes coding for ribosomal RNA are made up of tandemly arranged sequences which are identical or nearly so (Birnstiel et al., 1970). Although these sets of genes do show some base sequence divergence even between two closely related organisms, intra-genome sequence divergence appears to be precluded. This phenomenon raises the fundamental question as to how identical sequences are preserved and what correction mechanism is involved. An analogous situation exists for the multiple 5S genes; here actual sequence analysis reveals little or no heterogeneity (Forget and Weissman, 1967; Brownlee et al., 1968). In these two cases, it is possible to account for lack of heterogeneity simply in terms of natural selection. If the base sequence of each ribosomal RNA molecule is uniquely adapted to the intracellular environment in a particular organism, all mutations would perhaps be disadvantageous. This is feasible since the RNA molecule is the final gene product and natural selection can operate directly at the level of base sequence. Nevertheless it is difficult to accept the possibility that one mutant ribosomal RNA molecule in the presence of hundreds of normal copies could be so disadvantageous. It seems more likely that mechanisms other than selection also play a role.

The results presented recently by Kedes and Birnstiel (1971) together with those summarized above suggest that the group of genes coding for histones has similar properties. Again interspecies divergence is evident, but intra-genome heterogeneity appears to be absent although this conclusion must be regarded as preliminary. Nevertheless it is apparent that nucleotide substitutions, pre-dominantly in the third position of a codon do occur in evolution, yet are not apparent within the family of sequences within a species.

Since the group of histone genes itself must be at least as old in evolutionary terms as the time since the divergence of many of these pairs of organisms in question, it follows that some mechanism selects against intra-genome heterogeneity. It is worth emphasizing that the process involving selection alone cannot readily be invoked for the histone genes. Unlike the ribosomal genes, the final gene product is a protein and natural selection does not operate directly at the level of base sequence. If it is admitted that synonym codons can be translated with equal or nearly equal efficiency, it would seem that mutations occurring in the third position of a codon would be permissible. Since this apparently occurs between species but not within a genome, one is again led to suggest a correction mechanism.

The simplest scheme which would serve as a model for a correction mechanism would involve classical unequal crossover mechanisms. In a tandemly arranged set of genes this is likely to occur with high frequency. This would lead to individuals with much lower numbers of genes. Together with a mechanism for regenerating the optimal number of genes from this limited number, this scheme could serve to maintain homogeneity within the set of genes. Obviously the most effective mechanism would involve the occurrence of an individual with only one such gene, whose immediate progeny would again develop the optimum number. However, it is implausible that such an individual would survive and possess a selective advantage. It is more likely that a very frequent reversion to a limited but viable number of genes would be sufficient. Support for this general mechanism derives from studies of the ribosomal genes of amphibia and Drosophila. For example in Bufo marinus and D. melanogaster the number of genes is variable from individual to individual (Miller and Brown, 1970; Ritossa, 1970). Moreover, in Drosophila at least, the number of ribosomal genes does change with extreme rapidity (Tartof, 1971).

ACKNOWLEDGEMENTS

We are indebted to John S. Ullman, Mary-Dell Chilton, Larry Marsh and Laura Gregg for the use of their unpublished data. We also wish to thank Marion Namenwirth and Byron Gallis for reading and discussing the manuscript. This research was supported by a grant from the National Science Foundation (GB6099). One of us (M.N.F.) was supported by a predoctoral Training Grant fellowship USPHS GM 5214.

REFERENCES

1. G. Attardi, H. Parnas, M.I.H. Hwang and B. Attardi, J. Mol. Biol. 20, 145 (1966)

2. A.J. Bendich and B.J. McCarthy, Proc. Nat. Acad. Sci. 65, 349 (1970)

3. J. Bicknell and H.C. Douglas, J. Bact. 101, 505 (1970)

4. M.L. Birnstiel, M. Chipchase and J. Speirs, Prog. Nuc. Acid. Res. and Mol. Biol. 11, 357 (1971)

5. M. Botchan, R. Kram, C.W. Schmid and J.E. Hearst, Proc. Nat. Acad. Sci. 68, 1125 (1971)

6. D.J. Brenner, M.J. Fournier and B.P. Doctor, Nature 227, 448 (1970)

7. R.J. Britten and E.H. Davidson, Science 165, 349 (1970)

8. R.J. Britten and D.E. Kohne, Science 161, 529 (1968)

9. G.G. Brownlee, F. Sanger and B.G. Barrell, J. Mol. Biol. 34, 379 (1968)

10. J.R. Chamberlain and C.B. Metz, submitted to J. Mol. Biol. (1971)

11. J.E. Darnell, R. Wall and R.J. Tushinski, Proc. Nat. Acad. Sci. 68, 1321 (1971)

12. E.H. Davidson, B.R. Hough, M.E. Chamberlain and R.J. Britten, Dev. Biol. 25, 445 (1971)

13. M.O. Dayhoff and R.V. Eck, Atlas of Protein Sequence and Structure 3, 89 (1967-68)

14. M.O. Dayhoff and P.J. McLaughlin, Atlas of Protein Sequence and Structure 4, 89 (1969)

15. E.J. Du Praw, this symposium (1971)

16. M. Edmonds, M.H. Vaughan and H. Nakazato, Proc. Nat. Acad. Sci. 68, 1336 (1971)

17. W.M. Fitch and E. Margoliash, Biochem. Genet. 1, 65 (1967)

18. W.G. Flamm, M. McCallum and P.M.B. Walker, J. Mol. Biol. 42, 441 (1969)

19. B.G. Forget and S.M. Weissman, Science 158, 1695 (1967)

20. G.P. Georgiev, J. Theoret. Biol. 25, 473 (1969)

21. H.M. Goodman and A. Rich, Proc. Nat. Acad. Sci. 48, 2101 (1962)

22. H. Harris, Biochem. J. 73, 362 (1959)

23. R.M. Iverson, and G.H. Cohen, The Cell Cycle, (Academic Press, 1969) pp. 299

24. K. Jones, Nature 255, 912 (1970)

25. L.H. Kedes and M.L. Birnstiel, Nature 230, 165 (1971)

26. L.H. Kedes and P.R. Gross, Nature 223, 1335 (1969)

27. M. Kimura, Genet. Res. 11, 247 (1968)

28. J.L. King and T.H. Jukes, Science 164, 788 (1969)

29. D.E. Kohne, Quart. Rev. Biophys. 3, 327 (1970)

30. T. Kotaka and R.L. Baldwin, J. Mol. Biol. 9, 323 (1966)

31. C.D. Laird, B.L. McConaughy and B.J. McCarthy, Nature 224, 149 (1970)

32. S.Y. Lee, J. Mendecki and G. Brawerman, Proc. Nat. Acad. Sci. 68, 1331 (1971)

33. U. Loening, Symp. Soc. Gen. Microbiol. 20, 77 (1970)

34. J. Marmur and P.J. Doty, J. Mol. Biol. 5, (1962)

35. D.H. Marrian, V.C. Spicer, M.C. Balis and G.B. Brown, J. Biol. Chem. 189, 533 (1950)

36. B.J. McCarthy and R.B. Church, Ann. Rev. Biochem. 39, 131 (1970)

37. B.J. McCarthy and J.D. Duerksen, Cold Spring Harbor Symp. 35, 621 (1970)

38. B.L. McConaughy, C.D. Laird and B.J. McCarthy, Biochemistry 8, 3289 (1969)

39. B.L. McConaughy and B.J. McCarthy, Biochim. Biophys. Acta, 149, 180 (1967)

40. B.L. McConaughy and B.J. McCarthy, Biochem. Genetics, 2, 37 (1968)

41. B.L. McConaughy and B.J. McCarthy, Biochem. Genetics, 4, 446 (1970)

42. L. Miller and D.D. Brown, Chromosoma 28, 430 (1969)

43. B. Moav and M. Nemer, Biochemistry 10, 881 (1971)

44. R.C. Moore, Treatise on Invertebrate Palentology (Univ. of Kansas Press, 1966) p. U282

45. R.L. Moore and B.J. McCarthy, J. Bact. 94, 1066 (1967)

46. M. Nemer and D.T. Lindsay, Biochem. Biophys. Res. Comm. 35, 156 (1969)

47. M.L. Pardue and J.G. Gall, Science 168, 1356 (1970)

48. F. Ritossa, Proc. Nat. Acad. Sci. 60, 509 (1968)

49. K. Scherrer, L. Marcaud, F. Zajdela, I. London and F. Gros, Proc. Nat. Acad. Sci. 56, 1571 (1966)

50. H. Schuster, Biochem. Biophys. Res. Comm. 2, 320 (1960)

51. R.W. Shearer and B.J. McCarthy, Biochem. Genetics 4, 395 (1970)

52. J.H. Sinclair and D.D. Brown, Biochemistry 10, 2761 (1971)

53. E.L. Smith, R.J. De Lange and J. Bonner, Physiol. Rev. 50, 159 (1970)

54. R. Soeiro, M.H. Vaughan, J.R. Warner and J.E. Darnell, J. Cell Biol. 39, 112 (1968)

55. W.D. Sutton and M. McCallum, Nature 232, 83 (1971)

56. R.H. Stellwagen and R.D. Cole, Ann. Rev. Biochem. 38, 951 (1969)

57. K.D. Tartof, Science 171, 294 (1971)

58. J.S. Ullman and B.J. McCarthy, in preparation (1971)

59. J. Wetmur and N. Davidson, J. Mol. Biol. 31, 349 (1968)

60. H.R. Whiteley, B.J. McCarthy and A.H. Whiteley, Dev. Biol. 21, 216 (1970)

61. A. Zweidler and B.V. Gotchel, Fed. Proc. Abs. 30, 187 (1971)

DISCUSSION

FITCH: I do not believe that it makes sense to claim that perhaps 90% of the DNA is useless and simultaneously maintain that every nucleotide replacement was necessarily selected for.

MC CARTHY: Certainly both of these are extreme views and no one would maintain at the present time that 90% of the DNA is useless in the absolute sense. Even spacer sequences may serve specific purposes.

STUDIER: Do you see a selective advantage in making ten times as much nuclear RNA as is translated?

MC CARTHY: Certainly not if ten times as much of each sequence is made as translated. However, if cleavage and transport are selective so that some sequences are translated with high efficiency, the explanation could rest on the details of nuclear RNA processing and the advantages of large precursors to messenger RNAs.

MELTON: Would you also consider as an alternative explanation that there is some sort of functional restriction of tRNA species, in the eggs for example, or in different tissues, such that the various third position nucleotides are not equivalent after all, for functional reasons, even though they may be equivalent codons in the genetic dictionary.

MC CARTHY: Yes this point is well taken. There is no convincing evidence to show that synonym codons are translated with equal efficiency in vivo.

MELTON: Returning to an earlier part of your talk on deamination of C to U residues and its effect on DNA duplex stability, I would like to hear your comments on the recent work of Uhlenbeck and colleagues (Nature 230, 362; J. Mol. Biol. 57, 217) indicating that the G-U pair has no destabilizing effect on the RNA helix.

MC CARTHY: I do not understand the difference unless it derives from the fact that the double-stranded regions in tRNA are so much shorter than in the DNA duplexes we consider.

MILKMAN: You've suggested that DNA heterogeneity in a species might be reduced by the reduction of 400 genes to one and subsequent regeneration of the 400 from that one. This would certainly reduce heterogeneity within a single organism, but not within the species. What if the chosen gene is a newly arisen (neutral) mutant?

MC CARTHY: I should point out that this is an extreme and artificial model. Obviously the organism in which this occurred would require an extreme selective advantage in order to disseminate this change through the species. A more realistic view would involve frequent reduction of the number of copies by unequal crossing-over and regeneration of the optimum number. If this occurs frequently enough, it could serve as a means of controlling intragenome heterogeneity.

LEE: Have you looked at other kinds of mRNA besides histone mRNA?

MC CARTHY: No, we haven't looked at any other mRNA. However, Falkow and co-workers have done some work with β-galactosidase mRNA of enterobacteria.

CHANGE IN REPEATED DNA IN EVOLUTION

Nancy Reed Rice
Carnegie Institution of Washington
Department of Terrestrial Magnetism
Washington, D. C.

Abstract. Studies of repeated nucleotide sequences in various rodent DNAs have led to the following conclusions: (1) DNA families held in common by two or more of the rodents tested reassociate to form products of low to moderate thermal stability; (2) the greater the evolutionary separation of two DNAs, the fewer the families held in common and the lower their reassociated thermal stability; (3) a substantial fraction of each DNA tested is capable of reassociating with high thermal stability with homologous DNA. Among pairs of rats, mice, and hamster DNAs, however, heterologous duplexes of high thermal stability are not observed; (4) it is argued that the reassociated material of high thermal stability which can be detected in homologous but not in heterologous DNA has presumably appeared in the former since the divergence of the two species; (5) the same conclusions can be drawn from studies of primate DNAs, though it appears that change in DNA is slower (per unit time) among the primates than among the rodents.

Introduction

DNA sequences which are repeated a few or many times have been identified in the DNA of every eukaryotic organism examined, with the possible exception of the fungi. The examined species now number well over 60, include protozoa, both mono- and dicotyledonous plants, and animals ranging from Porifera to mammals.[1] Not only is the function of the repeated sequences an intriguing puzzle, for with a few exceptions it is unknown, but so is their evolutionary origin and history.

Sufficient evidence has accumulated over the past few years to warrant

some inferences on this latter topic, and these are the subject of this

paper. To begin, it may be appropriate to offer some general com-

ments on the nature of repetitious DNA.

Characteristics of Repetitious DNA

Repeated sequences in DNA are recognized by their reassociation

rate, [1-3] and comparison of the reassociation kinetics of various DNAs

reveals very considerable differences among them. For example,

estimates of the fraction of total DNA composed of repeated elements

cover a wide spectrum. When reassociation is carried out at 60°C, in

0.14M phosphate buffer, with a fragment size of about 400-500 nucleo-

tides, it is found that about 90% of Necturus DNA consists of repeated

elements. [4] Under the same conditions, repeated sequences constitute

about 78% of Rana clamitans DNA, [4] about 45% of Xenopus DNA, [5] about

40% of S. purpuratus [6] or calf DNA, [1] and about 15% of Drosophila

melanogaster DNA. [7]

The degree of repetition also varies. Some DNAs have extremely

highly repetitive elements. Ten percent of mouse DNA, for example,

consists of a sequence repeated about 10^6 times;[2] guinea pig DNA

appears to contain an even more highly repeated element. [8] In other

DNAs, those of Microtus agrestis[9] and Scaphiopus couchi, [4] for example,

the maximum degree of repetition appears to be very considerably lower.

The reassociation profiles of most DNAs are very broad and suggest at

least several classes of repeated sequences with differing degrees of

repetition. [1]

From studies of the thermal stability of reassociated molecules,
members of a single class or <u>family</u> of repeated DNA sequences have
been judged to be similar but not identical in base sequence. If there is
no mispairing in a reassociated double helix, its thermal stability can
approach that of native DNA; such is the case with reassociated E. coli
DNA, for example.[1,10] Double-stranded DNA containing some non-
complementary base pairs, on the other hand, has a thermal stability
lower than that of native DNA. Laird, et al.[11] have estimated this de-
pression in T_m to be about 1°C per 1.5% mismatched base pairs. Now,
one observes that the stability of reassociated repeated sequences is
significantly lower than that of native DNA. It is presumed that this is
attributable in large part to the presence of non-complementary base
pairs in the reassociated molecules; and it is concluded that members
of a family, while sufficiently similar to reassociate with one another,
are not identical.

The repeated sequences of sea urchin DNA provide a good
example of the range of thermal stabilities which may be observed following reassociation. As shown in Fig. 1, there is considerable material
whose stability approaches that of native DNA and which presumably
results from the reassociation of very similar sequences. On the other
hand, a substantial fraction of sea urchin repeated DNA has a reassociated thermal stability more than 20°C below that of native DNA.
This wide range presumably reflects at least several classes of repeated
sequences which differ considerably in the degree of similarity among

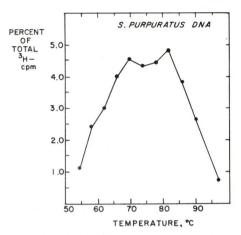

Figure 1. Thermal stability of sea urchin repeated DNA. ^{3}H-S. purpuratus DNA was sheared at 50,000 psi, denatured at 100°C, and incubated at 100 μg/ml for 22 hours at 50°C in 0.13M phosphate buffer. Reassociated material was bound to hydroxyapatite at 50°C in the same buffer and eluted in 4°C steps.

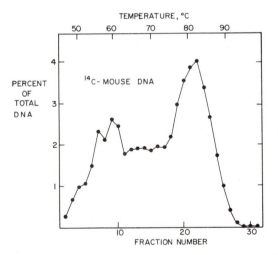

Figure 2. Thermal stability of mouse repeated DNA. Sheared, denatured ^{14}C-C3H DNA was incubated at 300 μg/ml for 18 hours at 50°C in 0.13M phosphate buffer. 53% of the total DNA bound to hydroxyapatite at 50°C and was eluted during a linear temperature gradient.

and thermal stability of reassociated sequences. Further, comparison
of kinetic or thermal stability profiles of DNA of various species reveals
very considerable differences between the DNAs of different organisms.

For the most part, functions of these sequences are unknown.
Some repeated elements do appear to be transcribed, but with unknown
effect; and still others may not be transcribed at all. The ribosomal
RNA,[12] tRNA,[13, 14] 5S RNA,[14] and apparently the histone genes[15]
exist in multiple copies, but generally these comprise a small percentage of the total repeated DNA of a somatic cell.

Some Families Are Found in Only One or a Few Species

There are some families of repeated sequences which appear to be
shared by the DNAs of many species. It has been known for some time,
for example, that at least some of the repeated DNA of one species can
reassociate with that of another, and that the closer the DNA donors taxonomically, the greater the cross-reaction will be.[16, 17] On the other hand,
repeated sequences which appear to be restricted to one or perhaps a few
closely related species have also been found to be quite common. The
mouse satellite, the best known example of this phenomenon, comprises
some 10% or more of total mouse DNA and was discovered because it
differs significantly from the bulk of nuclear DNA in average buoyant
density.[18] This distinctive element is of chromosomal origin,[19-22]
and is present in apparently constant amount in the DNA from all mouse
tissues and cell lines examined.[18] Measurements of its very rapid reassociation rate led Waring and Britten[2] to estimate that mouse satellite

about 10^6 times. Since thermal stability of the reassociated product is
very high, about 5°C below that of native satellite, [23] it can be concluded
that these copies are very similar to one another.

There appear to be few or no sequences capable of reassociating
with mouse satellite DNA in the DNAs of other species. Neither DNA
from Apodemus, Rattus, Peromyscus, Cricetus, nor Cavia shows any
appreciable binding to mouse satellite DNA at 60°C in 0.12M phosphate
buffer; from the limits of their experiment, Flamm, et al. [24] have been
able to calculate that there can be no more than 20 mouse satellite
sequences in the rat genome. It appears, then, that this million copy
10% component of Mus musculus DNA is essentially without counterpart
in other DNAs. One must entertain the notion, therefore, that the
mouse satellite has arisen, either by large-scale amplification of an
original sequence or by introduction of new sequences, in the mouse
genome since the mouse-rat divergence. In this sense, a very substantial amount of DNA has been added to the mouse genome in this period.

Nuclear satellites have been described in DNAs of well over 60
species. [25, 26] In those cases which have been examined, these elements have properties similar to those of the mouse satellite: they
consist of repeated elements which reassociate to rather high thermal
stability, and are without known function. [8, 27-32] The extent to which
the same satellite may be shared by several species is not yet clear.
Crab dAT is the only satellite which has been found to be quite widespread; it has been described in several different genera. [33-36] In
other cases, a satellite present in one species appears absent from
another species of the same genus. [9, 25] However, conclusions as to

whether satellites are usually species-specific await further study.

What can be said at this point is that a satellite present in the DNA of

one organism is usually absent from that of some fairly close relative.

Just as with the mouse, therefore, one must consider that a satellite

has been added to its DNA since the divergence of even close relatives

from a common ancestor.

Repeated DNA of High Thermal Stability after Reassociation

Satellites may be a special case of a far larger class of DNA

elements. There are DNA components which share many of the proper-

ties of satellite DNA - repetition, apparent similarity of copies, and

presence in one member of a closely related pair but not the other - but

whose density may not differ appreciably from the bulk of nuclear DNA.

Among the rodents, at least, such components appear to be very common

and, as I will demonstrate, they are easily identified. Coupled with the

evidence from satellite DNAs, they indicate that more-or-less species-

specific elements of high repetition may be a rather general phenomenon.

As already pointed out, the reassociated mouse satellite has quite

high thermal stability, indicating a considerable degree of similarity

among its many copies. This satellite material is clearly visible in

the thermal stability profile of total mouse repeated DNA, as shown in

Fig. 2. Here mouse DNA fragments of length about 400-500 nucleotides

were denatured and then incubated at 50°C in 0.14M phosphate buffer

for sufficient time to allow reassociation of most of the repeated

elements (C_0t 65). The sample was applied to a water-jacketed hydroxy-

apatite column in the same buffer, also at 50°C - conditions under

which double-stranded molecules are retained by the column, but **single-**

stranded ones are not[37] - and ~53% of the total DNA was found to bind.
The temperature of the column was then slowly raised, as washing with
buffer continued. When bound molecules denature, they elute from the
column and thus the amount of DNA in each fraction provides a measure
of the thermal stability of the reassociated repeated molecules.   The
general shape of the elution profile and the two large peaks, one at
about 60°C and the other at over 80°C, are typical of the repeated
sequences of mouse DNA and are reproducible from experiment to
experiment; smaller secondary fluctuations within this general pattern
may vary with the experiment.   The reassociated mouse satellite is a
major part of the high stability material. For comparison purposes,
native DNA of similar size would dissociate at about 85°C under these
conditions.

Rat DNA also has repeated sequences which reassociate to high
thermal stability, as shown in Fig. 3.   Here, 48% of total rat DNA
bound to hydroxyapatite after incubation to $C_o t$ 49 at 50°C.   As with
mouse DNA, the range of stabilities is broad, and while there is less
high thermal stability material than in mouse DNA, about 16% of the
total DNA is found to elute above 75°C.   Incidentally, rat DNA apparently
has no satellite[38, 39] and shows no appreciable homology to mouse
satellite.[24]

Rat and mouse DNAs can be more directly compared by measuring
the degree and nature of cross-reaction between them.   This is accomplished by reassociating rat and mouse DNAs together, ^{3}H-mouse DNA
at high concentration and ^{14}C-rat DNA at very low concentration.

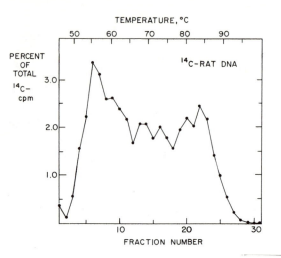

Figure 3. Thermal stability of rat repeated DNA. Sheared, denatured ^{14}C-R. norvegicus DNA was incubated at 210 µg/ml for
19 hours at 51°C in 0.13M phosphate buffer. 48% of the total
DNA bound to hydroxyapatite at 50°C and was eluted during a
linear temperature gradient.

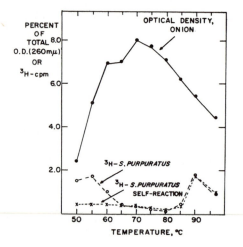

Figure 4. Thermal stability of onion repeated DNA. A mixture
of sheared, denatured onion DNA (100 µg/ml) and ^{3}H-S. purpuratus DNA (0.04 µg/ml) was incubated at 45°C for 24 hours in
0.14M phosphate buffer, 0.02% sodium dodecyl sulphate. 60% of
the onion DNA and 8.3% of the sea urchin DNA bound to hydroxyapatite at 45°C and were eluted in the same buffer in 5°C
steps. Self-reaction of the ^{3}H-sea urchin DNA was measured
under the same conditions, except that onion DNA was omitted
from the incubation mixture.

Because of its low level, self-reaction of the latter is essentially prevented, and it reassociates, if at all, with
the heterologous mouse DNA. After incubation sufficient to
allow reassociation of most of the repeated sequences of the
mouse DNA (C_0t 63 at 50°C, 0.14M phosphate buffer), the
sample is applied to hydroxyapatite and the bound material
is eluted at increasing temperatures. If the repeated sequences of the two DNAs are identical, then rat-mouse duplexes
should be as stable as mouse-mouse duplexes, and the elution
profiles of ^{14}C-rat DNA and ^{3}H-mouse DNA should be identical.
If rat and mouse repeated sequences were very very dissimilar,
then rat DNA might be unable to pair with mouse DNA; in that
case one expects that no ^{14}C-rat DNA would be bound to the
column at all.

To demonstrate that an experiment of this sort is feasible, we take two DNAs which are expected to be quite dissimilar, say those of a plant and an animal, and show that
there is very little association between them, even when incubation is performed at 45°C in 0.14M phosphate buffer. Annealing a very small amount of sea urchin DNA with excess
onion DNA under these conditions results in 60% of the onion
DNA but only 8% of sea urchin DNA binding to hydroxyapatite,
as shown in Fig. 4. Further, at least half of the sea urchin
DNA binding results not from association with onion DNA, but
from self-reaction.

In contrast, there is appreciable pairing of ^{14}C-rat

DNA with excess ^{3}H-mouse DNA, indicating considerable similarity between the repeated elements of these two species. As shown in Fig. 5, however, they are clearly not identical. Thermal stability of the homologous (mouse) duplex covers the broad range indicated earlier; but the heterologous duplex elutes over a more restricted range, 50° to about 80°C (the small amount of ^{14}C-rat DNA eluting above 80°C is attributable to self-reaction). Thus, there is very little rat DNA capable of forming high thermal stability reaction products with mouse DNA. There are two possible explanations for this: 1) the high thermal stability mouse sequences are represented, if at all, only at a very low level in the rat genome 2) rat DNA does react with high stability mouse sequences, but this heterologous product has a reduced stability. Though the issue is not settled, we have two indications that the former explanation may be the correct one: 1) the mouse satellite comprises a substantial fraction of the mouse high thermal stability material, and Flamm, et al.[24] found few if any sequences in rat DNA able to pair with mouse satellite at 60°C 2) when a heterologous duplex is formed it appears to be very similar in thermal stability to the homologous - as Fig. 5 shows, the ^{14}C and ^{3}H elution profiles are indistinguishable between 50° and about 70°C. Thus, the high thermal stability mouse material, amounting to about 20% of the total DNA, appears to be without measurable counterpart in rat DNA.

What of the rat DNA which is capable of reacting with

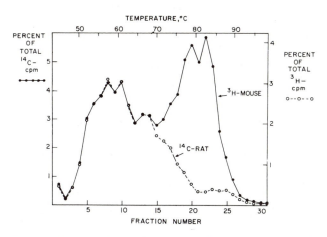

Figure 5. Comparison of the repeated sequences of rat and
mouse DNAs. A mixture of sheared, denatured ^{3}H-mouse DNA
(290 µg/ml) and ^{14}C-rat DNA (0.086 µg/ml) in 0.14M phosphate
buffer was incubated at 51°C for 18.3 hours. 58.3% of the
total mouse DNA and 51.6% of the rat DNA bound to hydroxyapatite at 50°C and were eluted during a linear temperature
gradient. To emphasize the constant mouse DNA:rat DNA ratio
which is observed during the first half of the gradient, the
ordinates have been selected so as to produce coincident
elution profiles in this region.

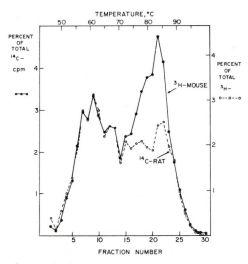

Figure 6. Thermal stability of rat and mouse repeated DNAs.
Sheared, denatured ^{3}H-C3H DNA, incubated at 295 µg/ml for 18.5
hours in 0.14M phosphate buffer at 51°C, was applied to hydroxyapatite; the column was thoroughly washed to eliminate
all non-reassociated material. 56% of the total mouse DNA was
bound. Sheared, denatured ^{14}C-rat DNA, incubated at 225 µg/ml
for 21.5 hours, was then applied, 51.7% of the total was bound,
the column was again washed, and bound material was eluted during a linear temperature gradient.

rat DNA to high thermal stability: are similar sequences present in mouse DNA? Apparently not, for a substantial fraction of these rat sequences remain single-stranded after a lengthy incubation with mouse DNA (C_0t 60) and can be located in the material which does not bind to hydroxyapatite.

If this interpretation is correct, the repeated sequences of rat or mouse DNA can be arbitrarily divided into two classes, those common to both rat and mouse, and those apparently found only in mouse or in rat. The former reassociate with rather low thermal stability, the latter with quite high stability. As one test of this interpretation, we anneal ^{3}H-mouse DNA and ^{14}C-rat DNAs separately, each to a C_0t sufficient to allow reassociation of most of the repeated sequences, apply them to the same column and elute them together. Thus, the elution profile of each isotope represents dissociation of a homologous duplex only. On the basis of the previous results, we expect the two profiles to be similar between 50° and 70°C and to differ markedly in the higher temperature region where the rat-specific or mouse-specific material is expected to elute. As shown in Fig. 6, these expectations are fulfilled. The elution profiles differ in the high temperature region both in the amount (about 20% of total rat DNA, 27% of mouse DNA) and pattern of material that elutes above 75°C. Much of this high stability material, detectable in the homologous genome only, has presumably appeared in each since the mouse-rat divergence.

Similar experiments with a variety of rodent DNAs have yielded very similar findings. A sizable fraction of the repeated sequences of each DNA both reassociates to high thermal stability and appears absent from a fairly close relative. Further, whatever the pair of DNAs compared, some amount of low thermal stability material appears to be held in common; the greater the evolutionary separation of the two organisms, the smaller is the amount of this material and the lower its T_m. Results of experiments demonstrating these points follow.

Annealing trace amounts of ^{14}C-Chinese hamster DNA with excess ^{3}H-Syrian hamster DNA to a C_0t sufficient to allow reassociation of most of the repeated sequences of the Syrian hamster DNA, followed by chromatography on hydroxyapatite, yields the results shown in Fig. 7. Both homologous and heterologous duplexes of low thermal stability are found, and these appear quite similar. Further, a substantial fraction of Syrian hamster DNA reassociates to high thermal stability – about 11% elutes above 75°C, for example. Yet there are few if any heterologous duplexes of high thermal stability, and it can be estimated from the two elution profiles that about 17% of Syrian hamster DNA appears to be without counterpart in the DNA of the Chinese hamster. The reciprocal experiment, annealing a very small amount of Syrian hamster DNA with a large amount of Chinese hamster DNA, demonstrates that Chinese hamster DNA also contains repeated sequences able to reassociate with high thermal sta-

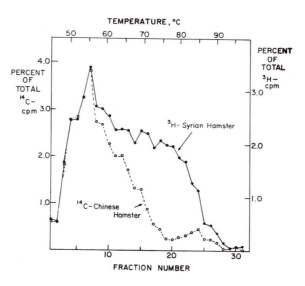

Figure 7. Comparison of repeated sequences of Chinese and Syrian hamster DNAs. A mixture of sheared, denatured ^{3}H-Syrian hamster DNA (147 µg/ml) and ^{14}C-Chinese hamster DNA (0.048 µg/ml) was incubated in 0.14M phosphate buffer at 50°C for 20.3 hours. 50.3% of the Syrian hamster DNA and 36.3% of the Chinese hamster DNA bound to hydroxyapatite and were eluted during a linear temperature gradient.

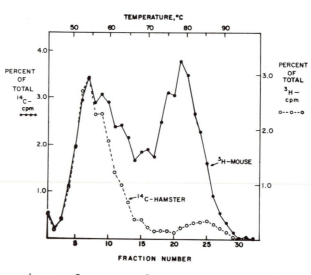

Figure 8. Comparison of repeated sequences of Chinese hamster and mouse DNAs. A mixture of sheared, denatured ^{3}H-mouse DNA (310 µg/ml) and ^{14}C-Chinese hamster DNA (0.13 µg/ml) was incubated at 50°C for 18 hours in 0.14M phosphate buffer. 49.6% of the mouse DNA and 25.9% of the hamster DNA bound to hydroxyapatite and were eluted during a linear temperature gradient.

bility and that these appear absent from Syrian hamster DNA.

Chinese hamster DNA shows even less homology to mouse
DNA than to Syrian hamster DNA, as shown in Fig. 8. Again
the elution of homologous and heterologous duplexes is very
similar at low temperatures. However, there are few heterologous duplexes of intermediate stability and there appear
to be none of high stability: all of the heterologous duplexes seem to be dissociated at temperatures below 70°C.
Together with the earlier mouse-rat experiment (Fig. 5),
these results also reflect the closer relationship between
DNAs of mouse and rat than between those of mouse and Chinese
hamster. The latter appear to share fewer sequences than the
former; and whereas roughly 20% of mouse DNA appears to be
without measurable counterpart in rat DNA, about 27% of
mouse DNA appears to be without counterpart in Chinese hamster DNA.

Comparison of rat DNA with DNAs from more or less closely
related species reveals these same patterns. Rat DNA shares
a considerable amount of low thermal stability material with
DNA of Mastomys coucha (the "multimammate mouse", a native
of sub-Saharan Africa), as shown in Fig. 9, but about 14% of
the total Mastomys DNA reassociates to high thermal stability
and consists of sequences apparently absent from rat DNA.
The line leading to the present-day Mastomys coucha, incidentally, appears to have diverged from the mouse-rat line at
about the same time as did those leading to Mus and Rattus,

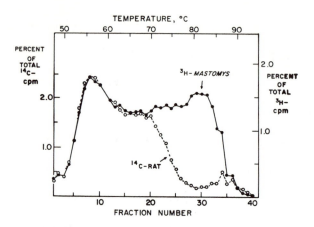

Figure 9. Comparison of repeated sequences of rat and Mastomys
DNAs. A mixture of sheared, denatured ^{3}H-Mastomys coucha DNA
(220 μg/ml) and ^{14}C-rat DNA (0.035 μg/ml) was incubated at 50°C
for 20.5 hours in 0.14M phosphate buffer. 43.3% of the Mastomys
DNA and 40.8% of the rat DNA bound to hydroxyapatite and were
eluted during a linear temperature gradient.

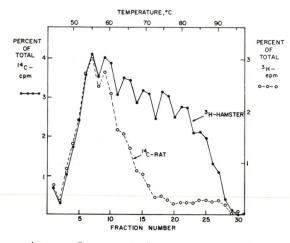

Figure 10. Comparison of repeated sequences of rat and Syrian
hamster DNAs. A mixture of sheared, denatured ^{3}H-Syrian hamster
DNA (144 μg/ml) and ^{14}C-rat DNA (0.035 μg/ml) was incubated in
0.14M phosphate buffer at 51°C for 23 hours. 53.1% of the total
hamster DNA and 36.9% of the rat DNA bound to hydroxyapatite and
were eluted during a linear temperature gradient.

so that the three are about equally distantly related.

There are fewer shared sequences between rat and Syrian hamster DNAs, as shown in Fig. 10, and still fewer between rat and guinea pig DNAs (Fig. 11). Yet both Syrian hamster and guinea pig DNAs contain considerable material able to reassociate to high thermal stability with homologous DNA.

Rate of Change of Repeated DNA

From each of these experiments one can draw a rough estimate of the amount of repeated DNA present in one member of a pair but apparently absent from the other. These estimates, of course, depend heavily on such factors as piece size of the DNA and degree of interspersion of various classes of repeated sequences. As Fig. 12 shows, the numbers are large indeed; 10-20% of the genomes of rodent species whose last common ancestor existed some 10-20 million years ago consists of repeated elements which reassociate to high thermal stability and which appear to be absent from other rodent DNAs.

Changes in repeated DNA among the _primates_ stand in striking contrast to this rapid rate of change among the rodents, as pointed out by Kohne.[40] Repeated DNA of human and gibbon, for example, whose lines are estimated to have diverged about 30 million years ago, are practically indistinguishable. Kohne has calculated from data of Hoyer and Roberts[41] that only about 2% of the human genome has apparently arisen since the human-gibbon divergence. Further,

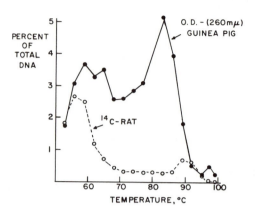

Figure 11. Comparison of repeated sequences of rat and guinea pig DNAs. A mixture of sheared, denatured guinea pig DNA (284 µg/ml) and ^{14}C-rat DNA (0.018 µg/ml) was incubated at 50°C in 0.14M phosphate buffer.   43.1% of the guinea pig DNA and 12.9% of the rat DNA bound to hydroxyapatite and were eluted in 3°C steps.

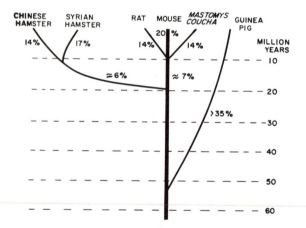

Figure 12. Apparent addition of repeated DNA during rodent evolution. Numbers are percent of total DNA which reassociates to high thermal stability and which is apparently not found in DNA of other species. Approximate divergence times are indicated.

there is less than about 5% of gibbon DNA, if any, which is
new since that time.[42]   However, changes do occur in repeated DNA in the primate line, as the 20% component of green
monkey[31] reminds us. Repeated DNAs of human and rhesus monkey
are also clearly distinguishable, as shown in Fig. 13. On
average, though it appears that the rate of addition of repeated DNA may be roughly ten times lower among the primates
than among the rodents.

   This difference in rates is strikingly correlated with
the difference in divergence rates of the single-copy DNAs
of the two lines. In an extensive series of experiments,
Kohne, et al.[43] used the thermal stability of hybrid duplexes
to determine the rate of change of single-copy DNA within the
primate line. It too is approximately ten times slower than
observed in the rodent line. The reassociated duplex formed
between single-copy DNAs of human and chimp, for example,
has a thermal stability only 1.7°C lower than that of the
homologous human DNA duplex; and lines leading to man and
chimp are estimated to have diverged about 15 million years
ago. In contrast, a hybrid formed between single-copy DNA
of mouse and rat is at least 15°C less stable than the homologous duplex, for an estimated divergence time of ten million
years. It appears, then, that these two measures of evolutionary change in DNA - rate of addition of repeated DNA to
the genome, and rate of sequence change in single-copy DNA -
both show an order-of-magnitude difference among the primates

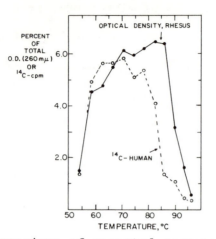

Figure 13. Comparison of repeated sequences of human and rhesus monkey DNAs. A mixture of sheared, denatured rhesus monkey DNA (757 µg/ml) and ^{14}C-HeLa cell DNA (0.05 µg/ml) was incubated at 50°C for 4.3 hours in 0.14M phosphate buffer. 53% of the rhesus DNA and 41% of the HeLa DNA bound to hydroxyapatite and were eluted in 4°C steps.

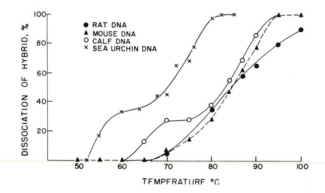

Figure 14. Thermal stability of the hybrid formed between rat ribosomal RNA and various DNAs. Filters bearing 70 µg of the indicated DNA were incubated at 67°C for 12 hours in 2 ml 5xSSC containing 0.9 µg rat ^{3}H-ribosomal RNA. After RNase treatment, thermal denaturation of the hybrid was carried out in 0.4xSSC.

as compared to the rodents. The basis for this difference is
not clear; Kohne, et al.[43] have pointed out, however, that
generation times are at least ten-fold shorter among the
rodents than the primates, and that the rates of change of
two DNAs may be more strongly correlated with number of
generations elapsed since their divergence than with absolute
time.

Addition of Repeated Sequences to the Genome

The results presented here are consistent with the model
proposed several years ago by Britten and Waring[44] and elaborated by Britten and Kohne[1], which states that the evolution
of eukaryotic organisms has involved a continuing process of
addition of repeated sequences to the genome. They envisioned that a new family of repeated sequences arises either by
the very large-scale amplification of an existing sequence
or sequences or through the outright introduction of new
sequences. Very early in its history, members of this new
family would be identical, but with time the accumulation of
mutations would result in a family of merely similar members;
very much longer time might result in the loss of any detectable relationship between them. It was proposed that this
process has occurred repeatedly in all evolutionary lines of
eukaryotic organisms, and that it is the origin of at least
the majority of repeated DNA sequences. The mechanism by
which a new family might arise is unknown, though several
possibilities can be imagined. Likewise, whether a supposedly

young family, e.g., mouse satellite, has any function, and
whether a family can be preserved at a relatively constant
level in the genome if it has no function are unknown.

Other interpretations of the data are certainly possible.
It can be argued, for instance, that a family consists of
very similar members not because it is quite new but because
of constraints imposed by its function. As with the ribosomal RNA genes, variation in sequence might not be tolerated by the cell; a family might thus be very old indeed and
yet be composed of very similar sequences. However, this
analogy with the rRNA genes does not hold. The rRNA genes
are not only extremely similar within a single species, but
also between species. As shown in Fig. 14, the thermal stability of the hybrid between rat rRNA and mouse DNA is practically as high as that of the homologous hybrid; indeed, the
hybrid between rat rRNA and calf DNA is reduced only slightly
in stability.[45] Thus, the ribosomal RNA genes constitute a
family of highly conserved repeated sequences present in
much the same form in very many lines. On the contrary, the
great majority of repeated DNA which reassociates to high
thermal stability appears to differ markedly from line to
line. Thus, one is still faced with the problem of the
origin of this material, and some sort of addition or amplification mechanism again appears attractive.

What is the direct evidence that a DNA component, e.g.,
mouse high thermal stability material, is not present in the

DNA of another species, say rat? Might there not be some
sequences in rat DNA similar enough to those of the mouse
satellite to permit reassociation between them? The results
of Flamm, et al.[24] demonstrate that such sequences, if in
rat DNA at all, are too distantly related to bind to mouse
satellite at 60°C in 0.12M phosphate buffer. In fact, it
appears that possible relationships between high thermal
stability components and heterologous DNA must be quite distant, if indeed they exist at all: extended incubation with
mouse DNA at 50°C in 0.14M phosphate buffer leaves most of
rat or Mastomys high thermal stability material unreacted.
Nevertheless, similarities may exist. The necessity for an
addition or amplification mechanism would not be obviated by
such a finding, for the origin of the repeated sequences must
still be accounted for. But it could greatly influence thinking about the frequency of addition events and about the subsequent course of the amplified material.

Since possible similarities between high thermal stability components and heterologous DNA might be more easily
detected in closely related organisms, I have chosen to
study DNAs from several species of the genus Mus. Individuals of Mus caroli and Mus cervicolor, both natives of South-
east Asia, have been collected by J. T. Marshall and very
kindly provided by Dr. Michael Potter of N. I. H., and I
have compared their DNAs with that of Mus musculus. Relatedness to Mus musculus can be roughly estimated by the extent

of divergence of the single-copy DNAs of these species from

that of M. musculus. It is found that ^{3}H-M. musculus single-

copy DNA will reassociate with M. caroli (or M. cervicolor)

DNA, but that the thermal stability of the hybrid is about

5°C lower than that of the homologous duplex. Under similar

conditions, the stability of the hybrid formed between ^{3}H-

M. musculus single-copy DNA and rat DNA is at least 15°C be-

low that of the homologous duplex. As a rough estimate,

therefore, these Mus species appear to have been separated

less than a third as long as mouse and rat - or for perhaps

a few million years.

Repeated DNA which reassociates to high thermal stability

is found in all three species, but there are clear differences

among these components. Mus caroli DNA, for example, does

contain sequences able to pair with high stability with Mus

musculus DNA, but as shown in Fig. 15 these are appreciably

fewer than found in M. musculus DNA itself. The reciprocal

experiments show that both M. caroli and M. cervicolor DNAs

also contain many more repeated sequences able to pair with

high thermal stability with homologous DNA than with M.

musculus DNA.   A result not seen previously, however, is the

relatively high proportion of low thermal stability heter-

ologous duplexes. Indeed, in several low temperature frac-

tions the percent of heterologous DNA is nearly twice that

of homologous DNA. Sutton[46] has already suggested that there

is some homology between M. musculus satellite and DNA of

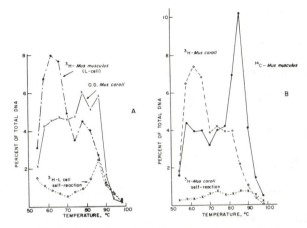

Figure 15. Comparison of repeated sequences of Mus musculus and Mus caroli DNAs. (A) A mixture of sheared, denatured M. caroli DNA (120 µg/ml) and 3H-L-cell DNA (0.0025 µg/ml) was incubated at 50°C in 0.14M phosphate buffer for 22 hours. 44% of the M. caroli DNA and 47.3% of the L-cell DNA bound to hydroxyapatite and were eluted in 4°C steps. Self-reaction of the 3H-L-cell was measured under the same conditions, except that M. caroli DNA was omitted from the incubation mixture. (B) A mixture of sheared, denatured 14C-M. musculus DNA (125 µg/ml) and 3H-M. caroli DNA (0.0035 µg/ml) was incubated at 50°C in 0.14M phosphate buffer for 23.3 hours. 48.5% of the M. musculus DNA and 42.1% of the M. caroli DNA bound to hydroxyapatite and were eluted in 4°C steps. Self-reaction of 3H-M. caroli DNA was measured under the same conditions, except that M. musculus DNA was omitted from the incubation mixture.

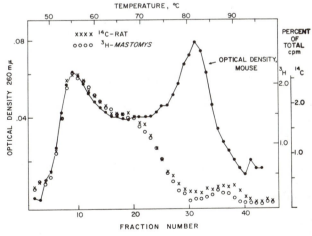

Figure 16. Comparison of repeated sequences of rat, Mastomys, and mouse DNAs. A mixture of sheared, denatured mouse DNA (225 µg/ml), 3H-Mastomys coucha DNA (0.042 µg/ml), and 14C-rat DNA (0.03 µg/ml) was incubated at 50°C for 21 hours in 0.14M phosphate buffer. 35.8% of the Mastomys DNA and 44.3% of the rat DNA bound to hydroxyapatite and were eluted during a linear temperature gradient.

M. caroli, and such a situation could account for the results

shown here. This is a fascinating avenue for future study,

for it may afford greater insights into both the mechanism

and the time of origin of the mouse satellite than has hither-

to been possible.

Repeated DNA of Low Thermal Stability after Reassociation

Repeated DNAs which reassociate to high thermal stabil-

ity, including satellite DNAs, are only part and often a

minor part of an organism's repeated DNA. Usually the major-

ity appears with a density indistinguishable from that of

total DNA and with a reassociated thermal stability much re-

duced from that of native DNA. It should be pointed out that

the total amount of repeated DNA observed is a function not

only of piece size, but is also usually a function of incu-

bation conditions.[17,47] Annealing mouse DNA at 60°C in

0.12M phosphate buffer to c_0t 50, for example, results in re-

association of about 30% of the DNA; lowering the incubation

temperature to 50°C allows increased reassociation of mater-

ial of quite low stability, to a total of about 50% of the

DNA. Similarly, we depend on hydroxyapatite to discriminate

between single- and double-stranded DNA; but DNA which is

double-stranded over only a small fraction of its length may

not, in fact, be retained by the column.

In contrast to the components which reassociate to high

thermal stability, this sort of repeated DNA does frequently

appear to be shared among organisms, as the results of the

rodent DNA experiments have shown. The less the evolutionary
separation of two species, the more of this material appears
to be shared. It is further observed that for those sequences
which are shared, the thermal stability is very similar indeed
in the heterologous and homologous duplexes. A particularly
striking example of this phenomenon is provided in Fig. 16,
where trace levels of ^{3}H-<u>Mastomys</u> DNA and ^{14}C-rat DNA have
been annealed with excess mouse DNA. The thermal fractions
were assayed for: optical density (260 mμ), which measures
dissociation of the mouse-mouse homologous duplex; ^{3}H, which
measures dissociation of the <u>Mastomys</u>-mouse heterologous
duplex; and ^{14}C, which measures dissociation of the rat-mouse
heterologous duplex. It is found that thermal stabilities of
both heterologous products are essentially identical, and
that in the region where they appear, they are identical to
the homologous product as well. Thus, it appears that the
reacting population in mouse DNA differs no more <u>on the</u>
<u>average</u> from the reacting population in rat or <u>Mastomys</u> DNA
than it does within itself.

What is this material? First, what accounts for its low
thermal stability upon reassociation? Is it duplexed over
most of its length, but lowered in stability by a high degree
of non-paired bases? or is it fairly precisely paired but of
low stability because the paired regions are very short? Or
is the answer intermediate between these two? This subject
is being actively pursued, and hopefully the next year will

provide some answers.

Second, what does it do? It is known that non-satellite repeated DNA appears generally distributed among the chromosomes,[21,22,30,32,48] that much of it appears to be transcribed, and that different families appear to be transcribed in different cell states.[49-56] Both Britten and Davidson[57] and Georgiev[58] have proposed that the repeated sequences are the genetic regulatory elements of eukaryotic cells.

Finally, what is the origin of the repeated sequences which reassociate to low thermal stability? Britten and Kohne[1] have proposed that _all_, or at least most, of the repeated sequences have a common origin; that they are added to the genome as families composed of identical or highly similar members; that these may or may not serve an immediate function within the cell; that with time and the accumulation of mutations, they increasingly diverge in base sequence from one another and appear to us as families of lower average thermal stability; and that in the meantime, with translocation of these sequences throughout the genome, they take on functions, regulatory functions possibly, of fundamental importance to the cell. Perhaps too, the physical arrangement in the genome of shared families _varies_ from species to species and helps to account for the differences we see _between_ species.

Acknowledgments

This work has been supported by Public Health Service Fellowship 5-F02-GM-24,603-02 and by National Science Foundation grant GB-27585.

References

1. R. J. Britten and D. E. Kohne, _Science_ 161, 529 (1968).

2. M. Waring and R. J. Britten, _Science_ 154, 791 (1966).

3. J. G. Wetmur and N. Davidson, _J. Mol. Biol._ 31, 349 (1968)

4. N. A. Straus, _Proc. Natl. Acad. Sci. U. S_. 68, 799 (1971).

5. E. H. Davidson and B. R. Hough, _J. Mol. Biol_. 56, 491 (1971).

6. R. J. Britten, This Symposium.

7. C. D. Laird, _Chromosoma_ 32, 378 (1971).

8. W. G. Flamm, P. M. B. Walker, and M. McCallum, _J. Mol. Biol_. 42, 441 (1969).

9. W. Hennig and P. M. B. Walker, _Nature_ 225, 915 (1970).

10. J. Marmur and P. Doty, _J. Mol. Biol_. 3, 585 (1961).

11. C. D. Laird, B. L. McConaughy, and B. J. McCarthy, _Nature_ 224, 149 (1969).

12. M. L. Birnstiel, M. Chipchase, and J. Spiers, in _Prog. Nucleic Acid Res. and Mol. Biol_. 11, 351. Eds., J. N. Davidson and W. E. Cohn (Academic Press, New York, 1970).

13. F. M. Ritossa, K. C. Atwood, D. L. Lindsley, and S. Spiegelman, in _International Symposium on The Nucleolus-- Its Structure and Function_. Eds., W. S. Vincent and O. L. Miller. _Natl. Cancer Inst. Monogr_. 23, 449 (1966).

14. D. D. Brown and C. S. Weber, _J. Mol. Biol_. 34, 661 (1968).

15. L. H. Kedes and M. L. Birnstiel, _Nature New Biology_ 230, 165 (1971).

16. B. H. Hoyer, B. J. McCarthy, and E. T. Bolton, _Science_ 144, 959 (1964).

17. M. A. Martin and B. H. Hoyer, _J. Mol. Biol_. 27, 113 (1967)

18. S. Kit, _J. Mol. Biol_. 3, 711 (1961).

19. J. J. Maio and C. L. Schildkraut, J. Mol. Biol. 40, 203 (1969).

20.   K. W. Jones, Nature 225, 912 (1970).

21.   K. W. Jones and F. W. Robertson, Chromosoma 31, 331 (1970).

22.   M. L. Pardue and J. G. Gall, Science 168, 1356 (1970).

23.   W. G. Flamm, M. McCallum, and P. M. B. Walker, Proc. Natl. Acad. Sci. U. S. 57, 1729 (1967).

24.   W. G. Flamm, P. M. B. Walker, and M. McCallum, J. Mol. Biol. 40, 423 (1969).

25.   F. E. Arrighi, M. Mandel, J. Bergendahl, and T. C. Hsu, Biochem. Genetics 4, 367 (1970).

26.   Y. Coudray, F. Quetier, and E. Guille, Biochim. Biophys. Acta 217, 259 (1970).

27.   P. M. B. Walker, W. G. Flamm, and A. McLaren, in Handbook of Molecular Cytology. Ed., A. Lima-de-Faria. (North Holland Publishing Co., Amsterdam, 1969), p. 52.

28.   G. Corneo, E. Ginelli, and E. Polli, J. Mol. Biol. 48, 319 (1970).

29.   G. Corneo, E. Ginelli, and E. Polli, Biochemistry 9, 1565 (1970).

30.   W. Hennig, I. Hennig, and H. Stein, Chromosoma 32, 31 (1970).

31.   J. J. Maio, J. Mol. Biol. 56, 579 (1971).

32.   R. A. Eckhardt and J. G. Gall, Chromosoma 32, 407 (1971).

33.   N. Sueoka, J. Mol. Biol. 3, 31 (1961).

34.   N. Sueoka and T. Cheng, Proc. Natl. Acad. Sci. U. S. 48, 1851 (1962).

35.   M. Smith, J. Mol. Biol. 9, 17 (1964).

36.   D. M. Skinner, W. G. Beattie, M. S. Kerr, and D. E. Graham, Nature 227, 837 (1970).

37.   Y. Miyazawa and C. A. Thomas, Jr., J. Mol. Biol. 11, 223 (1965).

38.   S. Kit, Nature 193, 274 (1962).

39. B. L. McConaughy and B. J. McCarthy, *Biochem. Genetics* 4, 425 (1970).

40. D. E. Kohne, *Carnegie Inst. Wash. Year Book* 69, 485 (1970).

41. B. H. Hoyer and R. B. Roberts, in *Molecular Genetics, Part II.* Ed., J. H. Taylor. (Academic Press, New York, 1967), p. 425.

42. N. R. Rice, Unpublished observations.

43. D. E. Kohne, J. A. Chiscon, and B. H. Hoyer, *Carnegie Inst. Wash. Year Book* 69, 488 (1970).

44. R. J. Britten and M. Waring, *Carnegie Inst. Wash. Year Book* 64, 316 (1965).

45. N. Reed, *Hybridization of Ribosomal RNA of the Rat*, Ph.D. Thesis, Harvard University, 1968.

46. W. Sutton, Personal communication to R. J. Britten.

47. P. M. B. Walker and A. McLaren, *J. Mol. Biol.* 12, 394 (1965).

48. F. E. Arrighi, T. C. Hsu, P. Saunders, and G. F. Saunders, *Chromosoma* 32, 224 (1970).

49. B. J. McCarthy and B. H. Hoyer, *Proc. Natl. Acad. Sci. U. S.* 52, 915 (1964).

50. H. Denis, *J. Mol. Biol.* 22, 285 (1966).

51. V. R. Glisin, M. V. Glisin, and P. Doty, *Proc. Natl. Acad. Sci. U. S.* 56, 285 (1966).

52. A. H. Whiteley, B. J. McCarthy, and H. R. Whiteley, *Proc. Natl. Acad. Sci. U. S.* 55, 519 (1966).

53. R. B. Church and B. J. McCarthy, *J. Mol. Biol.* 23, 459 (1967).

54. E. H. Davidson, M. Crippa, and A. E. Mirsky, *Proc. Natl. Acad. Sci. U. S.* 60, 152 (1968).

55. R. B. Church and B. J. McCarthy, *Biochem. Genetics* 2, 55 (1968).

56. R. W. Shearer and B. J. McCarthy, *Biochem. Genetics* 4, 395 (1970).

57. R. J. Britten and E. H. Davidson, Science 165, 349 (1969).

58. G. P. Georgiev, J. Theoretical Biol. 25, 473 (1969).

DISCUSSION

TAYLOR: How much is known about variations in base composition of the various fractions of fast reannealing DNA's? Specifically has anyone examined the base composition of fractions eluted from hydroxyapatite columns of the type you have shown?

RICE: I am not aware of any such measurements, but there are data from other types of experiments which apply to this problem. For instance, repeated sequences of high thermal stability after reassociation often result from DNA components of eccentric base composition, i.e., satellites; but frequently they do not. Similarly, satellites themselves sometimes exhibit a very asymmetric base composition in their separated strands; but sometimes they do not. Further, there have been at least two reports demonstrating that DNA fractions of varying buoyant density, derived from neutral CsCl gradients, contain repeated sequences (M. A. Martin and B. H. Hoyer, J. Mol. Biol. 27, 113 (1967)); B. L. McConaughy and B. J. McCarthy, Biochem. Genetics 4, 425 (1970)). At present, therefore, there is no evidence for a specific base composition requirement for repeated sequences.

MELTON: I wish to ask you a question about the unique fraction of DNA, since you are in effect studying it also,

by subtraction. Do you believe current data are compatible
with the hypothesis that there is a relatively constant
unique genome size in all mammals, and perhaps in all vertebrates as well if you care to extend yourself? By this I
mean that the well known large differences in genome size,
or haploid DNA content, between different vertebrates might
all be due to different contents of repetitious DNA sequences
only, and that when these are subtracted the remaining,
single-copy DNA contents will after all be fairly similar in
each case (namely, something less than one picogram of DNA
per unique genome).

MC CARTHY: A comment on Dr. Melton's questions. I
think that it is important to point out that the proposition
is difficult to discuss since the distinction between unique
and redundant is arbitrary and definable only operationally.
Therefore, it makes little sense to talk about per cent
unique since these estimates can vary from essentially zero
at low criteria to greater than 95% if the experiment is
carried out under highly specific conditions.

MELTON: I understand and appreciate Dr. McCarthy's
objection, but I believe my suggestion still has meaning
despite the imperfectness of reiteration. Since even at a
very low criterion Dr. Rice found no significant cross-
reaction between onion and sea urchin DNA's, then the
partial complementarity within repetitious DNA families of

a given species still betrays some relatedness. If the mismatching within each family arose secondarily after a salta-
tory event involving perfect reiteration, then what I am
calling the unique genome is all that DNA which does not
derive from such a history, and is not rapidly reannealing
under any criterion.

RICE: As the criterion is lowered from 60° C and
0.14 M phosphate buffer, to 50° C, to 45° C, one observes
increasing reassociation of rodent DNA at, say, $C_{o}t$ 100.
This trend continues as the criterion is lowered still
further, but since one now sees significant cross-reaction
between, say, guinea pig and $\underline{E}$. $\underline{coli}$ DNA's, the results become very difficult to interpret. We are thus stymied in
the attempt to find a DNA fraction which "is not rapidly
annealing under any criterion". Dr. Melton's suggestion is
a very interesting one, but regrettably it is difficult to
approach experimentally.

GALINSKY: Many species have diverged due to an evolutionary accumulation of translocations and inversions on
the chromosomes. These chromosomes will not pair during
meiosis, e.g. horse X donkey producing a sterile mule.
Your technique of matching small sections of DNA strands and
using this as an indicator of evolutionary divergence in
time could be very misleading in such cases since you might
obtain very good strand DNA matching even though the chromosomes themselves will not pair.

RICE: It is true that two DNA's which differ only in a rather large-scale rearrangement of sequences rather than in the sequences themselves could appear identical under the condition I've described. However, there are other experimental approaches which can help to elucidate the extent of sequence rearrangement; some of these will be described this afternoon by Dr. Britten.

MILKMAN: Have hybridization experiments of this type been performed with different strains of a given species?

RICE: I have compared the DNA's of two of the inbred strains of M. musculus musculus, C3H and DBA/2, and found no detectable differences in either their repeated sequences or their single-copy DNA. Similarly, any differences in the DNA's of the C3H mouse and that of the Japanese subspecies M. musculus molossinus (provided by Dr. Michael Potter of N.I.H.) are so small as to be undetectable by the methods I've described.

DNA SEQUENCE INTERSPERSION AND A
SPECULATION ABOUT EVOLUTION

R. J. Britten
Carnegie Institution of Washington
Department of Terrestrial Magnetism
Washington, D. C.

Introduction

I understand that I have the privilege of making some chairman's remarks. They will be brief and consist of a report of a new set of measurements and a speculation on the origin of novelty in evolution. The measurements are part of a study of the arrangement of repeated DNA sequences - their interspersion or scattering throughout the genome.

In the sea urchin Strongy locentrotus purpuratus, as previously has been shown in the calf, most DNA fragments of about 4000 nucleotide length appear to include repeated sequence regions. In fact, at least half of such fragments contain sequences belonging to a minority class (10% of the genome) of the repeated DNA. In the mouse DNA there appears to be a very fine scale intermixing between single copy DNA sequences and repeated sequences of low thermal stability.

Interspersion in the Urchin Genome. The sea urchin Strongylocentrotus purpuratus, a favorite of developmental biologists, appears suitable for measurements of transcription aimed at understanding the mechanism of gene regulation. In preparation for such studies we have

begun an exploration of the arrangement of its DNA sequences.

Measurements of the reassociation kinetics for total DNA are the first

step and the results of measurements made in a number of laboratories

are shown by the curve on the right in Figure 1. The degree of agree-

ment achievable with hydroxyapatite assay is reasonably good.

A beginning has been made on a program of fractionation and de-

tailed measurements of the individual repetitive components. The

results up to this time may be simply summarized by indicating the

major components that have thus far been identified. The highest fre-

quency component appears to be present in about 10,000 copies, based

on its kinetics of reassociation by hydroxyapatite assay. Then there

follows a range of components with frequencies around 1,000 copies

and finally some material present in about 50 copies. All of these com-

ponents are made up of DNA that melts, after reassociation, over a

very wide range of temperatures. Thus a wide range of degrees of

base pairing appears to be typical of this repeated DNA. The repetitive

DNA makes up about 40% of the total at this criterion (HAP, 60°C,

12M PB) and the remaining 60% appears to be single copy DNA.

The left curve on Figure 1 gives the results of an initial estimate

of the interspersion. For this measurement a relatively high concen-

tration of short (400-500 long) fragment DNA was mixed with a small

quantity of much longer labeled fragments (about 4,000 nucleotides

long). Controls showed that relatively little reassociation occurred

between long fragments. Therefore, the binding of the long fragments

to the hydroxyapatite was due to the reassociation with them of one or

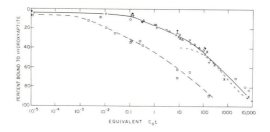

Figure 1. The reassociation of DNA from the sea urchin Strongylocentrotus purpuratus. Measurement was made by assaying the
fraction of the DNA that was bound to hydroxyapatite at 60°C
in .12 M PB after incubation for various times and concentrations in .12 M PB at 60°C except as noted. The dotted curve
represents the expected reassociation of the single copy DNA,
based on the known genome size of this urchin, assuming that
60% of the DNA is single copy. The right-hand curve shows measurements made with 50 K sheared DNA (approximately 450 nucleotides average fragment size). The left-hand curve (- □ -) shows
the binding to hydroxyapatite of an added small quantity of
longer labeled DNA fragments (about 4000 nucleotides). Measurements done in several laboratories with 50 K sheared DNA:
(+) DNA from gonads; (●) labeled embryo DNA present in the
same reassociation mixture; (Δ) D. Kohne .14 M PB rate corrected for salt conc.; (o) A. Aronson; (x) Eric Davidson.

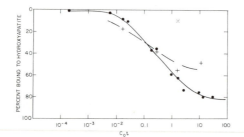

Figure 2. Measurement of the interspersion of a fraction of the
repeated DNA of the sea urchin Strongylocentrotus purpuratus.
The fraction of the urchin DNA corresponding to about 10% of
the genome and present in about 1000 copies was prepared from
sheared H3-labeled embryo DNA, as shown in Table 1. The solid
circles (●) show the reassociation of this fraction assayed by
passage over hydroxyapatite (60°C, .12 M PB). In addition to
the large quantity of the 450 nucleotide long fragments of this
selected fraction there was present a small quantity of long
(about 4000 nucleotides) fragments of total urchin DNA. The
crosses show the reassociation of the long fragments with the
short ones. A control incubation (x) of long fragments by themselves showed very little reassociation. It appears that representatives of this selected set of repeated sequences are present on a majority of 4000 nucleotide long fragments.

more short fragments. The high degree of binding by C_0t 100 of the

long fragment DNA shows that much of the single copy DNA has been

bound during the early part of the reaction. This indicates, as pre-

viously observed for calf DNA,[1] that a large fraction of the single copy

DNA sequences are present on long fragments that also contain repeti-

tive sequences.

This rather intimate interspersion of the different classes of

sequences was corroborated by measurements with an intermediate

frequency fraction of the repetitive DNA. For this purpose 400-500

nucleotide long DNA was prepared by successive hydroxyapatite frac-

tionation as indicated on Table 1. A mixture of H^3 labeled embryo DNA

(20 hrs. development) and unlabeled DNA from male gonads was frac-

tionated. As shown on the table no significant differences were observed

in the reassociation of the two kinds of DNA indicating that the popula-

tion of repetitive DNA is the same in early embryo as in sperm (domi-

nant source of DNA in the gonadal tissue).

Figure 2 shows the reassociation of this fraction as measured by

HAP assay at 60° in .12 M PB. Also present in the incubation mixture

was a small quantity of C^{14} labeled total urchin embryo DNA which had

been lightly sheared and selected on an alkaline sucrose gradient. This

DNA had an average fragment size of about 4,000 nucleotides. Clearly

a large fraction of the long fragments contain intermediate frequency

repetitive sequences.

The class of repetitive DNA utilized in this test corresponds to

little more than 10% of the total DNA and is present on the average in

about 1000 copies. Nevertheless about 50% of the long fragments

TABLE 1

FRACTIONATION OF URCHIN DNA

STEP	C_ot	MATERIAL	FRACTION BOUND EMBRYO (LABEL)	GONAD
1	126.	TOTAL DNA	45.7	49.4
2	3.6	Bound fraction from step 1	57.0	57.0
3	4.24	Bound fraction from step 2	73.0	72.6
4	.019	Bound fraction from step 3	22.9	27.1
5	.025	Unbound fraction from step 4	16.8	14.2

The unbound fraction of the last step is the intermediate fraction. It has a half reaction C_ot of about 0.3 and reacts to 80% at C_ot 30. It is somewhat heterogenous.

appear to have at least one representative of this set of repeated DNA

sequences somewhere in their length. How many different long frag-

ments are there? The genome size of S. purpuratus is about 10^9 nucleo-

tide pairs. Dividing by their length (4,000) we find 2.5×10^5 different

fragments. If about half of these contained a member of the intermed-

iate repeated sequence set it is clear that there must be hundreds of

separate short repeated sequence elements each one of which is present

in about 1,000 copies. Only in this way can we explain more than

100,000 repeated sequence elements present on 100,000 different types

of fragments.

These measurements show that the intermediate frequency re-

peated sequence families are made up of relatively short sequences

scattered widely throughout the DNA. For the moment "short" in this

context signifies that on the average the sequence elements must be

very much shorter (by a factor of more than a hundred) than the length

that can be calculated from the kinetic complexity of this class of re-

peated DNA. The kinetic complexity is simply an estimate of the total

sequence length based on the rate of reassociation of the intermediate

fraction. The $C_0 t$ for half reaction shown on Figure 2 is .3 or about

15 times faster than would be observed for E. coli DNA under these

conditions. Thus the kinetic complexity is about 300,000 nucleotide

pairs. At this stage of exploration we are left with rather broad limits

as to the actual lengths involved but there can be no question that this

particular class of repeated sequences is very widely scattered through-

out the genome. It appears quite likely from the curve shown on

Figure 1 that the other repeated sequence classes are also widely

scattered throughout the DNA.

Small Scale Interspersion in Mouse DNA.   Evidence from previous

measurements[2] has indicated that fragments of mouse DNA as short as

400-500 nucleotides contain on given fragments more than one class of

repeated sequence.   That work showed that both high and low thermal

stability repeated DNA regions were present on some fragments.   The

present measurement indicates that most fragments that contain regions

of repeated DNA that reassociate with low thermal stability also contain

regions that do not show repetition (single copy DNA) under the con-

ditions of these tests.

Intermediate frequency classes of radioactive mouse DNA

(sheared to 500 long fragments) were prepared in the following way.

First the satellite was removed by reassociating to $C_o t$ 3 x 10^{-3} and

passing over hydroxyapatite at 60°C in .12 M PB.   The DNA that did not

bind was reassociated to $C_o t$ 8.9 and again passed over HAP under the

same conditions.   The bound fraction was recovered, reincubated to

$C_o t$ 4.9 and again passed over HAP.   This final reassociated fraction

(26% of the total DNA) was eluted from the HAP in three fractions by

raising the temperature to 68°, 78° and 98° C.   The 68° and 98° fractions

used for the experiment were representative of the low and high thermal

stability repeated DNA of the mouse.

The experiment tested these fractions for the presence of single

copy DNA sequences by measuring their reassociation kinetics after

fragment size reduction.   The average fragment size was reduced by

minutes followed by 0.1 N NaOH for 5 minutes. [3] In order to achieve

the large C_0t required for single copy DNA the reassociation tests were

done in the presence of total unlabeled mouse DNA which had been

similarly sheared and chemically reduced in fragment size. The

results are shown in Figure 3. The high thermal stability fraction re-

associates with the total DNA in the way a preparation of intermediate

frequency repeated DNA does. It reaches 85% reassociation and is

better than half reacted by C_0t 0.1. Thus no appreciable number of

fragments which were purely single copy DNA were released from this

fraction.

The low thermal stability fraction, however, exhibited quite a

different behavior. A little more than half reassociates rapidly and

appears as repeated DNA, while the remainder reassociates as single

copy DNA. The extent of reassociation of the 68° fraction at C_0t 25,000

is just about the same as that of the total DNA.

The conclusion can be drawn that, on the average, about half of

the length of an original low melting 500 long fragment is made up of

single copy DNA. Each of these fragments, of course, also contains

stretches of repeated DNA. We can not say at this time how these two

classes of DNA sequences are arranged. It may be that each low melt-

ing repeated sequence is quite short and adjacent to much greater

length of single copy DNA. Such a result is suggested by the other

interspersion experiments done with 4,000 long fragments of calf and

sea urchin DNA. If indeed, as seems probable, the low melting

fraction is the principal widely interspersed repeated DNA then the

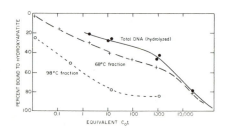

Figure 3. Measurement of the fine-scale interspersion of repeated and single copy sequences in the low thermal stability
fraction of reassociated repeated mouse DNA. (●) binding of
total mouse DNA which had been sheared at 50,000 psi and then
fragmented further by acid depurination; (o) added high thermal stability labeled fraction; (+) added low thermal stability
labeled fraction prepared as described in the text.

single copy regions must be fairly long and the repeated regions short.
There is not enough length of low melting repeats to go around, unless
they make up a small fraction of the length of the fragments.

A capsule view of the history and organization of the repeated DNA
can be now described but should be considered tentative and subject to
modification as new evidence comes along. Events of very excessive
replication of short sequences occur and quite rarely the product is
integrated into the genome as a cluster of tandemly organized precisely
repeated sequences. (Some of the satellites, as well as some of the
precisely and rapidly reassociating components that have main band
density, have characteristics that support such a view.) At later times,
events of rearrangement occur and some of these components become
clustered at the centromeres, if they did not originate there. More
events of rearrangement occur and the components become widely distributed throughout the chromosomes. (It can not be even guessed at
this time whether the centromeric location is a usual part of this
process). As this process of rearrangement (diffusion) continues the
observed thermal stability falls as base substitutions accumulate and
events of unequal crossover and translocation reduce the length of the
repeated sequences. During the rearrangement the now short lengths
of low thermal stability repeated DNA are inserted among pre-existing
single copy regions. (An unlikely alternative is that the tandem
clusters evolve entirely by base substitution. The observed interspersion would then be explained by the selective preservation of certain
short repeated regions while the intervening regions are substituted
beyond recognition). Whether the repeated DNA in its now widely

interspersed state is simply spacer between genes or carries out an

active role is the subject of intensive current investigation.

Speculation on the Appearance of New Patterns in the Genome.

Table 2 lists the possible roles for repeated DNA in higher organisms.

The categories are not mutually exclusive and are intended to be so

broad as to include most of the possibilities that have been suggested.

Four of these potential roles (4, 5, 6, and 7) use the sequence relation-

ship present in repeated DNA to establish control or maintain order

within the genome itself or in its expression. In these cases the re-

peated DNA is considered to form a means of communication within

the genome which leads to the establishment of temporal and spatial

organization at cellular and higher levels.

One of these suggested roles, gene regulation, has been formu-

lated as a model[4]. While a detailed proposal was attempted in that

paper, the mechanisms will not be examined here. It is sufficient to

assume that the repeated DNA sequences permit significant interactions

among the DNA and RNA sequences and thus affect the organization of

the phenotype.

In recent years it has become more certain that repeated DNA

has often been added to the genome of each species, as proposed a

number of years ago. [5,6] The rate of addition may in some cases

amount to several percent of the total DNA per million years. It

appears that, in general, the added DNA consists of very many copies

of relatively short sequences. The time rate of the events themselves

is not known but may be relatively sudden. It appears that the new

sets of sequences are tandemly arranged in clusters. It seems

TABLE 2

POTENTIAL ROLES OF REPETITIVE DNA

(Intended to be an overlapping and suggestive list)

1.   IN PRODUCTION   (Not yet functional)

2.   CARRIED ALONG   (Parasitic or Garbage)

3.   REPEATED GENES   (Ribosomal, etc.)

4.   STRUCTURAL   (Sequence specific linkages)

5.   HOUSEKEEPING   (e. g. Chromosomal Synapsis)

6.   PUNCTUATION   (e. g. Synthesis Initiation)

7.   GENE REGULATION   (Control Interconnections)

8.   EVOLUTIONARY I   (New genes from old parts)

9.   EVOLUTIONARY II   (New control connections and patterns)

probable that such clusters would be inactive for the roles suggested on

Table 2.   Rearrangement of the sequences might be necessary.

Individual fragments of the cluster, after translocation to many regions

in the genome, over a period of time could establish new interactions

leading to new organizational patterns at the level of the phenotype.

If this hypothesis is correct and such potent DNA sequence rela-

tionships are introduced into the genome then these events would form

a significant part of the molecular processes underlying evolution.

Events of multiplication of short DNA sequences would provide sources

of nucleic acid interaction and order that would be quite distinct from

the effects of adding random DNA sequences or even the duplication

and insertion of single extra copies of longer stretches of DNA.

The appearance of new repeated sequences may be considered

as a bubbling up, during evolution, of new classes of sequence relation-

ship and potential patterns of organization.   If such new patterns were

selectively advantageous they would be incorporated.   The result would

often be an increase in complexity of organization.   In biological evolu-

tion many new patterns of organization have been added, for example,

in the rise from a primitive metazoan to a mammal.   It can not be

said, of course, whether the process speculated about here would be

adequate for such a large scale and continuing increase in complexity

of organization.   However, it seems to have much more potentiality

than, say, gene duplication combined with base substitution.   It does

make use of the unexplained events of sequence multiplication that are

now thought to occur universally in higher organisms.   Further, a

testable prediction can be made that there will be observed a rearrangement of DNA sequences on quite a fine scale of length in short
evolutionary periods.

In each population, during the course of evolution there must be
a consistent set of solutions to molecular, developmental, structural
and behavioral problems. A continuous flow of new sequence relationships could be important to the formation of the control networks responsible for the integration of all of these processes.

Acknowledgement

Most of these concepts have been developed jointly with Eric
Davidson. A joint paper [7] examines the issue of the origin of evolutionary novelty from the point of view of a model of gene regulation and
changes in regulatory networks.

References

1. R. Britten and J. Smith, Carnegie Institution of Washington Year Book 68, 378 (1970).

2. N. Rice, Carnegie Institution of Washington Year Book 69, 479 (1971).

3. J. Ullman, thesis, University of Washington, Seattle (1971).

4. R. Britten and E. Davidson, Science 165, 349 (1969).

5. R. Britten and M. Waring, Carnegie Institution of Washington Year Book 64, 332 (1965).

6. R. Britten and D. Kohne, Carnegie Institution of Washington Year Book 65, 104 (1966).

7. R. Britten and E. H. Davidson, Quart. Rev. Biol. 46, 111 (1971).

BNL 16036

GENE FUSION

J. D. Yourno

Biology Department, Brookhaven National Laboratory, Upton, N. Y. 11973

GENE FUSION AS AN EVOLUTIONARY PROCESS

Evidence has accumulated in recent years for the existence of a class of long, chimaeric polypeptide chains in nature. Such polypeptides are usually responsible for multiple functions and fold into multiple, different globular regions. Often a unique function or set of functions of the multifunctional polypeptide is associated with each type of globule. For example, the multifunctional DNA polymerase polypeptide chain of the bacterium Escherichia coli has recently been found to be folded into two globular regions connected by an unfolded joining region which is susceptible to mild proteolysis[1,2] (Fig. 1). The two globules have no binding affinity and can be separated from one another in non-dissociating solvents following mild proteolysis of the enzyme. Each retains a different catalytic activity of this multifunctional enzyme.[1] The β-galactosidase protomer of E. coli is a single polypeptide chain. Yet it appears to fold into two or more dissimilar globular regions which can be separated by proteolysis of an unfolded joining region[3-5] (Fig. 1). These globular regions have a strong complexing tendency which is the basis for a novel form of complementation between

---

Research carried out at Brookhaven National Laboratory under the auspices of the U. S. Atomic Energy Commission.

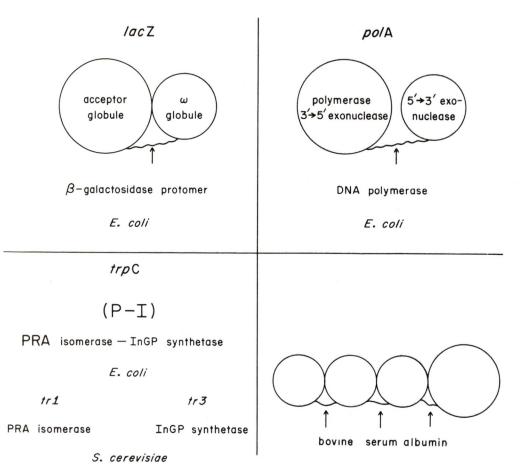

Figure 1. Diagrammatic structure of some chimaeric polypeptides. Where associated genes are known, they are designated above. Globular regions of a polypeptide chain are represented by circles which are drawn tangent where binding affinities exist. Postulated unfolded joining segments are represented by connecting lines. Arrows indicate points of proteolytic cleavage. Except for β-galactosidase and PRA isomerase-InGP synthetase the left to right order of globules is not meant to imply NH_2-terminal to COOH-terminal polarity. The number of smaller globules and the order of globules in bovine serum albumin is not known. The tertiary structure of the bifunctional enzyme PRA isomerase-InGP synthetase has not been investigated, but is presumably also multiglobular. See text for references.

polypeptide fragments of β-galactosidase.[3,4] In <u>Saccharomyces</u>

<u>cerevisiae</u> (yeast) the enzymes PRA isomerase and InGP synthetase,

mediating the third and fourth steps in tryptophan biosynthesis, are

demonstrably separate entities encoded by unlinked genes.[6] In <u>E. coli</u>,

however, both enzyme activities are carried on a single bifunctional

polypeptide chain[7,8] (Fig. 1). An example from higher organisms is

bovine serum albumin. This polypeptide chain folds into two or three

similar and one larger globular region, all of which have binding

affinities for one another after proteolysis of presumably unfolded

joining regions.[9]

These observations have engendered speculation on the mechanism

of evolution of such long and complex polypeptide chains.[5,10] An

equivalent question is the mechanism of evolution of the "long" genes

encoding these long polypeptide chains. It seems eminently plausible

that long genes have evolved from short genes. The most plausible

mechanism for gene elongation is fusion of gene segments.[5,10-21] Two

fundamentally different processes of gene fusion can account for the

observations.

The first process involves growth of a single gene. This occurs

through duplication of a gene segment entirely within the borders of

the single gene; viz., partial duplication or intragenic fusion.[11]

There is ample evidence for this process from the observation of seg-

ments in polypeptide chains which retain homologous amino acid sequences.

One group of human haptoglobin α chains, for example, shows a molecular

polymorphism of three types, α^{1F}, α^{1S}, and α^{2FS}. The α^2 chain is

almost twice the length of the α^1 chains and can be divided into two

segments of almost identical amino acid sequence which are similar to

those of the α^1 chains. The gene for the α^2 chain very likely arose

by partial duplication of an α^1 gene.[12,13] The constant and variable regions of the immunoglobulin light chains retain partial sequence

homologies and are considered by some investigators to have evolved by

partial duplication of an ancestral gene.[13,14] What is more remarkable

is that by the same criteria the heavy immunoglobulin chains show

evidence of having evolved by one or two rounds of partial duplication

of a light chain gene.[14-16] Similarly, internal sequence homologies

suggest that myoglobin and the hemoglobin chains,[17] and several

other polypeptide chains evolved by one or more rounds of partial gene

duplication[18-21] (Table 1). The potential of longer, partially

duplicated polypeptide chains for increasing diversity and efficiency

of function is presumably the prime selective factor in their evolution.

This mechanism provides a general explanation for the evolution of

polypeptide chains. Indeed, evidence for very short ancestral poly-

peptides of the order of five to seven residues length has been found

in the ferridoxin,[18] protamine[19] and hemoglobin[20] chains.

Partially duplicated polypeptides often have multiple, homologous

binding sites and may therefore fold into multiple, homologous globular

regions. The haptoglobin α^2 chain[13] and the transferrin chain[21] are

among the closest known approximations to a true double polypeptide

and are considered products of fairly recent duplication. Other

partially duplicated polypeptides which may fold around multiple

nucleation centers show evidence of extensive divergence of the

duplicated segments. The variable and constant regions of the

immunoglobulin chains, for instance, retain sequence homologies yet

carry out different functions.[14-16]

Thus multifunctional chimaeric polypeptides can be explained as

the products of partial gene duplication and a long history of internal

divergence. It is undoubtedly an oversimplification to draw infer-

ences about this process strict as to mechanism and general as to

Table 1. Some polypeptide chains which may have evolved
by partial gene duplication

Haptoglobin α^2 (13)	Cytochrome c (20)
Immunoglobulin λ, k (13,14)	Ferredoxin (18)
Immunoglobulin γ, α, μ (14,15,16)	Protamine (19)
Myoglobin (17,20)	Transferrin (21)
Hemoglobin chains (17,20)	Serum albumin* (9)

*Evolution of this polypeptide chain may have been by
fusion of both homologous and nonhomologous gene segments.

application from the examples of recently evolved, highly specialized

vertebrate binding proteins cited above. Nevertheless, it is conceivable

that even those chimaeric polypeptides concerned with basic cell functions

evolved by a process of more or less assymetric partial duplications at

an early date when these systems were being laid down.

The second process which could elongate a gene and its product

polypeptide is fusion of two or more genes; intergenic fusion. When

different genes are fused, in one step a fused chimaeric polypeptide

could result which is responsible for both functions originally asso-

ciated with distinct polypeptides.[5,10] Selection for intergenic fusion

may sometimes operate where complexing of dissimilar polypeptide chains

is obligatory for function. Fusion would expedite complex formation,

changing it from a second order to a first order process, as has been

hypothesized for the β-galactosidase protomer[5] and bovine serum

albumin.[9] Although the different globules of DNA polymerase do not

have binding affinity, an analogous argument can be made in this case.

Because of the unusual activity of this multifunctional enzyme the

different globules function interdependently.[1]

Mechanisms of gene fusion can be visualized at two basic levels of

gene function, repair-replication and recombination. Events such as

loop formation in repair-replication or unequal crossover in recombina-

tion can lead to duplication or deletion of gene segments. Events

which delete or otherwise remove the border between distinct genes

produce intergenic fusion, those which duplicate a gene segment

entirely within the borders of a single gene produce intragenic fusion

or partial duplication. There is strong evidence for such a fusion

from the haptoglobin α^1 chains, based on partially homologous genetic

tracts which could have paired strongly and permitted nonreciprocal

recombination.[13]

There are many plausible variations on these basic mechanisms for
fusion of gene segments. For instance intergenic fusion of duplicate
genes, tandem or otherwise, would be equivalent to intragenic fusion.
Hemoglobin Lepore contains a polypeptide chain which appears to be the
product of a crossover between the linked genes for the β and γ chains.[22]
In addition, other fusion processes differing more or less in detail
are conceivable, for example transposition or translocation.

The evidence for the evolutionary importance of intergenic fusion,
as embodied in the structure of modern chimaeric polypeptides, is
weaker than that for partial duplication. We have recently observed,
in collaboration with Dr. J. R. Roth, experimental fusion of adjacent
genes in a bacterium by mutational obliteration of the gene border.
These observations support the contention that fusion of different
genes is a prime force in the evolution of chimaeric polypeptides.[23,24]

EXPERIMENTAL GENE FUSION

The histidine (his) operon of Salmonella typhimurium contains nine
contiguous structural genes encoding ten enzymes for histidine biosynthesis.[25] The his operon is transcribed as a single polycistronic
mRNA.[26] Punctuation (stop,start) signals at the gene borders in
polycistronic mRNA ensure that each gene is translated as a distinct
polypeptide chain. Mutational erasure of the punctuation signals at
a border could lead to gene fusion and consequent fusion of the product
polypeptides. We have observed such a fusion of the hisD and hisC
genes, the second and third genes of the operon. These genes are
associated with the tenth and eighth steps, respectively, of histidine
biosynthesis. The single polypeptide produced by the fused genes is
responsible for both of the enzyme activities originally associated
with distinct protein molecules, products of the D or C gene (Fig. 2).
The hisD and hisC genes were fused by two sequential and compensating

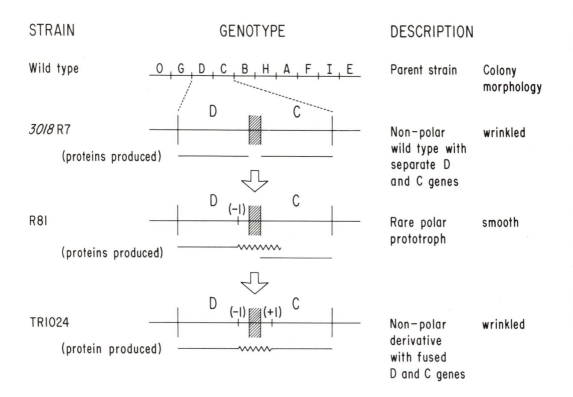

Figure 2. Genetic events leading to fusion of the hisD and hisC genes. Transcription and translation of the his operon proceeds from left to right. The inferred nature of polypeptide products is shown below each map. A straight line indicates in-phase reading and a pleated line out-of-phase reading.

Table 2. Enzyme levels in crude extracts

			Enzyme levels		
Strain no.	Description	Polarity effect	hisD (HDH)	hisC (AT)	hisB (HPP)
Wild type	Parent strain	No	100	100	100
R81	Rare polar prototroph	Yes	19	1	4
TR1024	Nonpolar derivative of R81	No	60*	80*	80

*These activities are due to a single bifunctional enzyme.

The genotypes of these strains are presented in Figure . All strains carry the operator regulatory mutation, his01242. Strains were grown on minimal medium containing 0.1 mM histidine. Enzymes were assayed as described by Martin et al.[40] Enzymes assayed are histidinol dehydrogenase (HDH), aminotransferase (AT) and histidinol phosphate phosphatase (HPP).

(- +) frameshifts bracketing the D-C border. The stop, start signals
separating hisD and hisC are consequently translated in the wrong
reading frame and are not recognized. The result is that the two
originally distinct genes are translated as one gene.

All Salmonella strains used carry the operator constitutive mutation, hisO1242. Nonpolar strains of this derivation form wrinkled
colonies on high glucose agar as a pleiotropic effect of derepression
of the his enzymes.[27] Polar his mutations (frameshift or nonsense)
cause the genes operator distal to the mutant gene to produce lowered
levels of enzyme. This counteracts the effects of derepression and
such strains form smooth colonies on high glucose agar.[28]

A strain carrying the first frameshift mutation, R81, arose
fortuitously and was isolated in our laboratory as a rare polar
prototroph. While R81 grows well without histidine supplements, strong
polarity in this strain was recognized by its smooth colony morphology.
That the polar mutation in R81 affected the hisD gene was indicated by
the altered electrophoretic mobility of the hisD product, histidinol
dehydrogenase (HDH) in crude extracts. As expected, the level of
enzymes produced by genes operator distal to hisD was severely
reduced by strong polarity. The polarity effect was much more drastic
on the immediately distal gene, hisC, than on succeeding genes, as
evidenced by the atypically low levels of the hisC product, aminotransferase (AT) (Table 2). Thus the polar mutation in the hisD gene
either extends into hisC or seriously disrupts translation initiation
of this gene. Genetic mapping of the polar mutation is consistent with
these findings. The polarity site in R81 maps at the region of the
D-C border in crosses with selected deletion recipients.

The altered HDH of R81 was purified by the standard procedure[29]
and examined for changes. Comparative mapping of tryptic peptides

revealed that the C-terminal peptide, Glu-Gln-Ala, was absent from its
normal position. Carboxypeptidase slowly released Ala and Gln from
normal HDH, but catalyzed the rapid release of Lys and Leu from R81
HDH. Thus the polar mutation at the end of the hisD gene alters the
C-terminus of R81 HDH.

That this polar mutation is a frameshift, probably in the (-) phase,
is indicated by its reversion response to selected mutagens. Since R81
is a prototroph and grows well without histidine supplements, a
technique allowing selective growth inhibition of polar mutants by the
histidine analogue, aminotriazole[30] was used by Dr. Roth to isolate
nonpolar revertants from R81. Nonpolar "wrinkled" revertants of R81
arose spontaneously and were readily induced with the mutagen ICR-191,
which causes both (+) and (-) frameshifts. The mutagen N-methyl-N'-
nitro-nitrosoguanidine, which can elicit (-1) frameshifts, was
ineffective.[31,32] This reversion response suggests that a compensating
frameshift, probably (+), is necessary to revert the R81 mutation to
a nonpolar state and therefore that the R81 mutation is itself a
frameshift, probably (-). All nonpolar revertants have greatly
increased levels of HDH and AT and normal levels of enzymes produced
by genes distal to C (Table 2). In these strains the D and C genes
were found to be fused, producing a fused chimaeric polypeptide
responsible for both HDH and AT activity.

The fusion of the D and C genes was discovered by isolation and
examination of the product, fused protein from a nonpolar revertant of
R81. Crude extract HDH in these strains all showed an unusual slow,
multiple banding pattern on polyacrylamide gel electrophoresis. Wild-
type HDH migrates as a single faster band. One spontaneous nonpolar
revertant, TR1024, was selected for further study. The HDH from
TR1024 showed several aberrant properties during its isolation by the

Table 3. Enzymatic activities of purified
proteins

Purified enzyme	Specific activity	
	HDH	AT
Wild type	2100	<0.014
Wild type AT	<0.018*	1750
TR1024 fused enzyme		
Peak I	2500	540
Peak II	2100	620
Peak III	1000	175

*This assay was performed with a sample
estimated to be 80% pure.

Specific activity is expressed as μmoles
substrate converted per minute at 37° per
μmole of enzyme. Calculations are based on
the following assumed molecular weights:
HDH, 90,000; AT, 70,000; HDH-AT, 160,000.
Data on AT is that of Martin and Goldberger[41]
expressed as defined above.

standard procedure.29 The multiple HDH species persisted throughout

the purification to near homogeneity and were divided into three

chromatographic fractions according to molecular weight. Yet all these

fractions yielded a single and identical band on SDS polyacrylamide

electrophoresis and gave identical typrtic peptide maps. Since the

multiple species contain only one type of polypeptide chain, these

must represent different aggregational states of the enzyme. The most

unusual feature of the purified HDH was its bright lemon-yellow color.

Normal purified HDH is colorless. Purified AT has a lemon-yellow color

due to bound pyridoxal phosphate, a common cofactor for transfer

enzymes. These observations, in connection with previous data suggest-

ing that the R81 mutation is near the D-C border, prompted us to

assay purified HDH of TR1024 for AT activity. Strong AT activity was

found in each of the purified HDH fractions (Table 3). Therefore,

HDH and AT are associated in the TR1024 enzyme.

That the HDH and AT polypeptide chains are fused in TR1024, i.e.

in covalent linkage, can be demonstrated by two types of experiment.

The molecular weight of the fused polypeptide is almost exactly

the sum of that of the individual HDH and AT polypeptide chains. By

SDS gel electrophoresis[33] approximate molecular weights of 40,000,

49,000, and 88,000 were obtained for the purified AT, HDH, and HDH-AT

polypeptides respectively (Fig. 3).

The amino acid sequences of the HDH-AT polypeptide are common to

HDH and AT. Tryptic peptide maps of HDH-AT show spots which are

essentially the sum of those of normal HDH and normal AT (Fig. 4).

Furthermore, several peptides of the fused enzyme were found to have

an amino acid composition identical to that of the homologous parent

peptide.

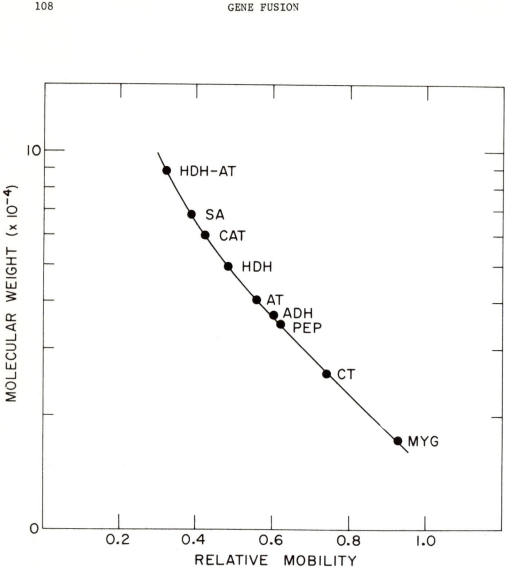

Figure 3. Subunit molecular weight estimation by SDS gel
electrophoresis. The log molecular weight is plotted versus relative
mobility. Standards with indicated molecular weights are: SA, serum
albumin (68,000); CAT, catalase (60,000); ADH, yeast alcohol dehydrogenase (37,000); PEP, pepsin (35,000); CT, chymotrypsin (26,000); MYG,
myoglobin (17,000). Arrows indicate experimental samples.

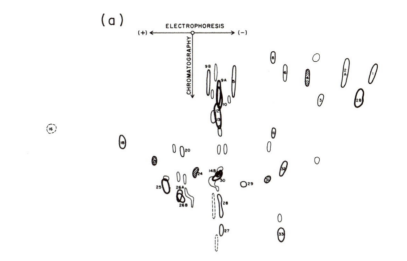

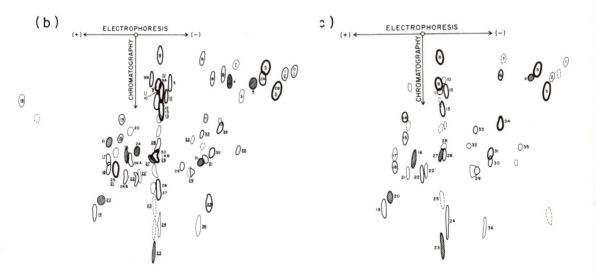

Figure 4.   Tryptic peptide maps of HDH (a), HDH-AT (b), and AT
(c). A tracing giving peptide designation is shown.   HDH-AT peptides
which correspond to those of AT are indicated by underlined numbers.

Hence, fusion of the adjacent $\underline{D}$ and $\underline{C}$ genes by mutational erasure of the gene border results in fusion of the product polypeptides. Remarkably, the fused polypeptide folds and associates so as to retain both catalytic activities of the originally distinct polypeptides. Normal HDH appears to be a dimer of identical subunits.[34,35] Previous studies,[36] in connection with this work, establish that AT is likewise a dimer of identical subunits. The multiple, electrophoretically slow peaks of HDH-AT represent either different unordered aggregational states of an active oligomer or various highly order multimeric states. The R_F of the multiple gel species is an inverse function of molecular weight. The fastest and smallest purified species elute from Sephadex G-150 columns at a diffuse molecular weight position of approximately 240,000. This suggests that the smallest species consist mainly of dimers or trimers and that the fused polypeptide, like each parent polypeptide, must associate to gain enzyme activity. The slowest and largest purified species elute at a molecular weight position in the order of 1,000,000.

As described above, mild proteolysis has proven to be a powerful probe into the tertiary structure of polypeptides. Multiglobular polypeptides can often be preferentially digested at unfolded connecting regions, thus releasing the well-folded fragments from covalent linkage. Purified HDH-AT was, therefore, mildly proteolyzed with trypsin in an attempt to release active HDH and AT globules from covalent linkage.[37] A highly active HDH fragment was released by this treatment, but AT activity was rapidly and completely destroyed. These relative stabilities to proteolysis are paralleled with normal HDH and AT. The results are consistent with multiglobular structure for HDH-AT. The active HDH fragment released from the fused enzyme has a molecular weight in the range of 125,000 smaller than that

calculated for the fused dimer (160,000) but significantly greater than
that of normal HDH dimer (about 90,000) (Fig. 5). On SDS gel electrophoresis, the active HDH fragment was found to consist of HDH subunits
of roughly normal size (45,000) and smaller polypeptides, none of
which corresponds to the AT subunit (Fig. 6). That is, cleavage of the
fused enzyme occurs at or very near the joining region, yet smaller
polypeptides remain   bound to the released, active HDH
globule. Normal HDH and AT have no demonstrated binding affinity.
While many possible explanations exist perhaps the most interesting
is that following fusion abnormal constraints are placed on the folding
or association of the HDH and AT polypeptide segments, leading to
disulfide or noncovalent interactions between the different globules.
Noncovalent interactions would mimic those of the β-galactosidase
and bovine serum albumin globules.

Because both the HDH and AT segments of the fused polypeptide
apparently must dimerize to gain activity, the possibility exists
that end to end association of subunits generates highly structured
multimers of increasing molecular weight. One model, based on a
flexible region connecting two folded noncovalently interacting regions
of the fused polypeptide, is shown in Figure 7. Further experiments
are necessary, however, to ascertain whether the various species of
the fused enzyme are structured multimers or merely unordered aggregates
of a smaller species.

Fusion of the D and C genes was accomplished by two compensating
frameshift mutations which bracket the gene border and thereby
eradicate the stop, start signals. The nucleotide sequence of an
intercistronic border remains to be defined. Recent evidence suggests
that the stop signal for D is UGA or UAA and that some intercistronic
space exists.[38] A plausible model for fusion is shown in Figure 2.

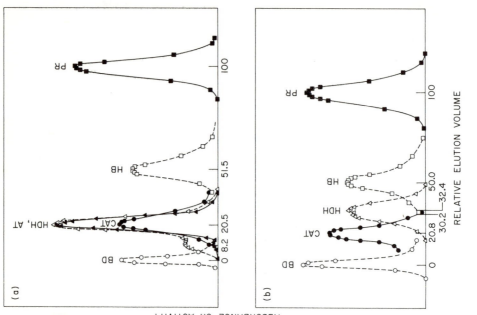

FIGURE 5 RIGHT

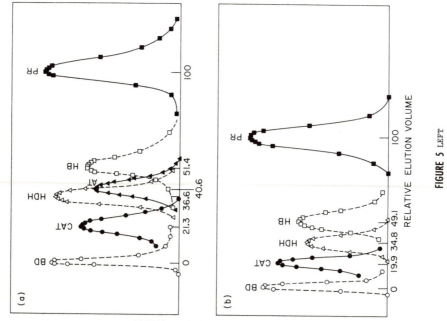

FIGURE 5 LEFT

Figure 5. Sephadex G-200 chromatography of purified untreated and proteolyzed HDH-AT, lowest molecular weight fraction. In each case a sample of 1 mg of native enzyme in buffer D^{29} was mildly proteolyzed 80 min with trypsin. The sample was then mixed with markers and passed through a Sephadex G-200 column in buffer $E.^{37}$ The markers used were blue dextran (BD), mol. wt. 2 x 10^6; catalase (CT), mol. wt. 240,000; hemoglobin (HB), mol. wt. 64,000, and phenol red (PR), mol. wt. 375. Peak positions are expressed as percentage of the internal volume of the column. Approximate molecular weights were calculated from the absolute elution volumes. Left, wild-type histidinol dehydrogenase (HDH) and wild-type aminotransferase (AT) controls. (a) Untreated wild-type histidinol dehydrogenase and aminotransferase control mixture, molecular weight about 90,000 and 75,000 respectively. (b) Treated sample, wild-type histidinol dehydrogenase control run only, molecular weight about 90,000. Right, HDH-AT. (a) Untreated sample, wild-type histidinol dehydrogenase about 240,000. (b) Treated sample, average molecular weight about 240,000. (b) Treated sample, molecular weight of histidinol dehydrogenase about 125,000.

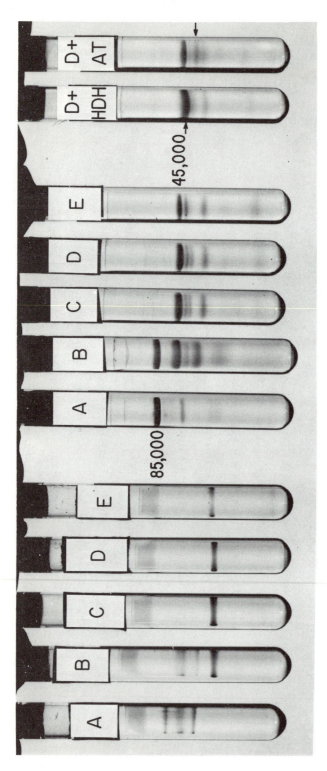

FIGURE 6

Figure 6. Kinetics of tryptic cleavage of the lowest molecular weight fraction of purified HDH-AT. Sample aliquots, digested for the indicated times, were subjected to polyacrylamide gel electrophoresis. Five standard gels stained for histidinol dehydrogenase are shown on the left. Samples of 50 μg of protein were run. Sodium dodecyl sulfate gels are shown on the right. Samples of 10 μg of protein were run. Note the multiple standard gel species in A which rapidly disappear during proteolysis with the concomitant appearance of a single, faster HDH fragment.   A, untreated; B, 10 min of digestion; C, 40 min; D, 80 min; E, 320 min. The approximate molecular weights of the fused subunit (85,000) and the released histidinol dehydrogenase subunit (45,000) are given next to the corresponding band. Arrows point to the position of the wild-type subunit of histidinol dehydrogenase or aminotransferase when run with the digest on sodium dodecyl sulfate gels.

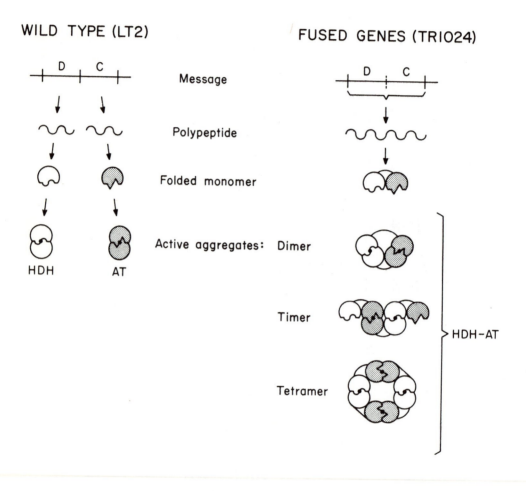

Figure 7. Possible molecular events accounting for the aggregation of bifunctional HDH-AT. Noncovalently interacting HDH and AT globules are drawn tangent. The joining region is unfolded and flexible.

In this model the (- +) frameshifts are assumed to be a deletion and addition of one DNA nucleotide pair. Since all nonpolar revertants of R81 examined produced fused enzyme it is possible that the original (-) frameshift of R81 is a longer deletion through the border and possibly even into hisC. In confirmation of these results, fusion of the D and C genes has now been repeated in similar experiments.[39] Experimental selection of the fused genes required a two step process. In nature such fusion events would more likely occur by a one step process, namely an in-phase deletion of the gene border.

It should be stressed that the fused D-C gene-protein system is only a convenient model for the evolution of complex polypeptides. There is no obvious selective advantage resulting from this experimental fusion. HDH and AT apparently do not complex normally and catalyze nonsequential reactions in histidine biosynthesis. We believe that the significance of this experimental gene fusion is rather in the demonstration that the process can indeed occur. These findings lend credence to suggestions that intergenic fusion is important in the evolution of multifunctional, chimaeric polypeptides.[5,10]

ACKNOWLEDGEMENTS

I thank Dr. Philip E. Hartman for comments on the manuscript.

REFERENCES

1. BRUTLAG, D., ATKINSON, M. R., SETLOW, P., AND KORNBERG, A., Biochem.
    Biophys. Res. Commun. 37, 982 (1969).

2. KLENOW, H. AND HENNINGSEN, I., Proc. Natl. Acad. Sci. U.S. 65,
    168 (1970).

3. ULLMAN, A., JACOB, F., AND MONOD, J., J. Mol. Biol. 24, 339 (1967).

4. ULLMAN, A., JACOB, F., AND MONOD, J., J. Mol. Biol. 32, 1 (1968).

5. GOLDBERG, M. E., J. Mol. Biol. 46, 441 (1969).

6. DeMOSS, J. A., Biochem. Biophys. Res. Commun. 18, 850 (1965).

7. CREIGHTON, T. E. AND YANOFSKY, C., J. Biol. Chem. 241, 4616 (1966).

8. CREIGHTON, T. E., Biochem. J. 120, 699 (1970).

9. WEBER, G. AND YOUNG, L. B., (a) J. Biol. Chem. 239, 1415, (b) ibid.
    1424 (1964).

10. BONNER, D. M., DeMOSS, J. A., AND MILLS, S. E., in Evolving Genes
    and Proteins, p. 305, H. J. Vogel and V. Bryson, Editors, Academic
    Press, New York, 1965.

11. OHNO, S., Evolution by Gene Duplication, Springer-Verlag, New
    York, 1970.

12. SMITHIES, O., CONNELL, G. E., AND DIXON, G. H., J. Mol. Biol. 21,
    213 (1966).

13. BLACK, J. A. AND DIXON, G. H., Nature 218, 736 (1968).

14. HILL, R. L., DELANEY, R., FELLOWS, R. E., AND LEBOWITZ, H. E.,
    Proc. Natl. Acad. Sci. U. S. 56, 1762 (1966).

15. WIKLER, M., KÖHLER, H., SHINODA, T., AND PUTNAM, F. W., Science
    163, 75 (1969).

16. DOOLITTLE, R. F., SINGER, S. J., AND METZGER, H., Science 154,
    1561 (1966).

17. FITCH, W. M., J. Mol. Biol. 16, 17 (1966).

18. ECK, R. V. AND DAYHOFF, M. O., Science 152, 363 (1966).

19. BLACK, J. A. AND DIXON, G. H., Nature 216, 152 (1967).

20. CANTOR, C. R. AND JUKES, T. R., Proc. Natl. Acad. Sci. U.S., 56, 177
    (1966).

21. MANN, K. G., FISH, W. W., COX, A. C., AND TANFORD, C., Biochemistry
    9, 1348 (1970).

22. BAGLIONI, C., Proc. Natl. Acad. Sci. U. S., 48, 1880 (1963).

23. KOHNO, T. AND YOURNO, J., Bacteriol. Proc. 59 (1970).

24. YOURNO, J., KOHNO, T., AND ROTH, J. R., Nature 228, 820 (1970).

25. AMES, B. N. AND HARTMAN, P. E., Cold Spring Harbor Symp. Quant. Biol. 28, 349 (1963).

26. MARTIN, R. G., Cold Spring Harbor Symp. Quant. Biol. 28, 357 (1963).

27. ROTH, J. R., ANTON, D. N., AND HARTMAN, P. E., J. Mol. Biol. 22, 305 (1966).

28. FINK, G. R., KLOPOTOWSKI, T., AND AMES, B. N., J. Mol. Biol. 30, 81 (1967).

29. YOURNO, J. AND INO, I., J. Biol. Chem. 243, 3273 (1968).

30. HILTON, J. L., KEARNEY, P. C., AND AMES, B. N., Arch. Biochem. Biophys. 112, 544 (1965).

31. OESCHGER, N. S. AND HARTMAN, P. E., J. Bacteriol. 101, 490 (1970).

32. YOURNO, J. AND HEATH, S., J. Bacteriol. 100, 460 (1969).

33. WEBER, K. AND OSBORN, M., J. Biol. Chem. 244, 4400 (1969).

34. LOPER, J. C., J. Biol. Chem. 243, 3264 (1968).

35. YOURNO, J., J. Biol. Chem. 243, 3277 (1968).

36. MARTIN, R. G., VOLL, M. J., AND APPELLA, E., J. Biol. Chem. 242, 1175 (1967).

37. KOHNO, T. AND YOURNO, J., J. Biol. Chem. 246, 2203 (1971).

38. RECHLER, M. AND MARTIN, R. G., Nature 226, 908 (1970).

39. RECHLER, M. AND BRUNI, C. B., J. Biol. Chem. 246, 1806 (1971).

40. MARTIN, R. G., BERBERICH, M. A., AMES, B. N., DAVIS, W. W., GOLDBERGER, R. F., AND YOURNO, J., in Methods in Enzymology, Vol. 23, H. Tabor and C. W. Tabor, Editors, Academic Press, New York (in press).

41. MARTIN, R. G. AND GOLDBERGER, R. F., J. Biol. Chem. 242, 1175 (1967).

DISCUSSION

BRITTEN: I would like to know if these two genes are evolution-arily related.

YOURNO: To my knowledge, not in the sense of the evolutionary relationships I have discussed. Certainly in the sense of being

members of a unit of regulation, a bacterial operon, they are evolutionarily related.

MAC INTYRE: (1) Does the growth rate of the strain with the gene
fusion differ from that of the wild type? (2) Have you measured the
relative growth rates of the mutant and wild type strains when in competition with each other?

YOURNO: (1) Not appreciably. With one or two exceptions, Dr. Roth
has found that the generation time of the non-polar revertants is normal, whether spontaneous or ICR-induced. This is not unexpected since
these genetically derepressed strains are over producing histidine
anyway. (2) No.

FITCH: You postulated two alternatives for the first mutational
step either in a frameshift or a deletion that extended into the aminotransferase gene. The latter alternative would alter the aminoterminal
sequence of the aminotransferase. Have you examined that sequence?

YOURNO: Dr. R. G. Martin and his collaborators are engaged in
this problem with a similar frameshift at the MsD-MsC border. The
technical difficulties are formidable, however, since very small amounts
of AT, as little as 1% of normal, are produced in these strains.
Martin's group hopes to obtain modest amounts of purified AT from this
frameshift mutant and to sequence the amino terminal section directly
on a sequenator.

A HETEROTIC MODEL FOR THE EVOLUTION OF DUPLICATIONS[*]

Janice B. Spofford
Department of Biology, University of Chicago,
Chicago, Illinois 60637

Abstract. Heterosis based on favorable hybrid dimers formed of monomeric polypeptides coded by alleles provides an adequate selective mechanism for the incorporation of a duplication, carrying one of these alleles, into a species genome. When the homodimers provide markedly different adaptive values, a stable polymorphism for chromosomes bearing versus lacking the duplication may result. Segregation for other alleles will persist at the original site and will be maintained at the new site if mutation introduces a new similar allele there. Conditions favorable to this process are more likely to be realized in random-breeding populations than with inbreeding.

INTRODUCTION

It seems a foregone conclusion that duplicated loci, of the sort that code for the hemoglobin chains or the immunoglobin chains, were acquired by evolving species as the result either of natural selection or of random processes. Since the rate of de novo production of duplications is likely to be extremely low, acquisition by drift seems far less likely than acquisition by some form of selection. What I wish to present is an elaboration of some implications of the speculation[1] that the benefit conferred by a duplication is permanent heterozygosity for a pair of heterotic alleles--in particular, alleles whose heterosis is due to polypeptide products that interact to form a hybrid multimeric protein more beneficial

[*]G. E. Time-Sharing System cost paid by Ford Foundation grant 67-375 for the support of population biology.

than either of the corresponding homomultimers.[2] Accordingly, some of the

divergence of the genetic information contained in such duplicate loci is

proposed to have preceded the physical twinning of the parent locus.

Whenever the most fortunate genotype is for any reason the heterozy-

gote, the species as a whole is less fortunate, since in a random-breeding

diploid population the highest fraction that can benefit is one-half, and

that only when both alleles are equally abundant. The remaining half are

homozygotes, and constitute a sort of "load" on the population to the

extent that they are less well adapted than the heterozygote.

When a species is confronted with this situation, it may succeed in

adopting one of the three main evolutionary stratagems for alleviating

this load, as diagrammed in Figure 1.

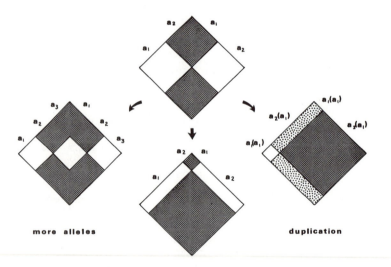

Figure 1. Maximum proportion of heterozygotes (shaded areas) in population

with (top) two alleles; (left) three alleles; (lower center) unequal gene

frequencies in uniting gametes; (right) homozygous duplication segregating

at original locus. Frequencies of alleles in sperm along upper left

border; in eggs, along upper right border.

One stratagem lies in diversification. The greater the number of alleles interacting in similarly heterotic fashion, the lower the total proportion of homozygotes. For example, with three equally abundant alleles, only one-third of the population is homozygous.

A second involves some form of asymmetry of allele frequencies in the uniting gametes, such as a meiotic asymmetry that yields different frequencies of the alleles in male than in female gametes, while preserving equality of allele frequencies in zygotes. A similar result would be achieved by negative assortative mating, the mating of unlikes. In either case, this stratagem forces the same solution on neighboring gene loci as well.

The third stratagem is to duplicate the locus, with a minimum of neighboring loci accompanying the heterotic locus into the new position unless of course these neighbors also form heterotic systems. This would .seem to be the ideal solution when available, giving as it does the possibility of a random-breeding (or even a self-fertilizing) population that is 100% "heterozygous."

ORIGIN OF DUPLICATIONS

It seems appropriate then to enumerate the known modes of generating duplications in the laboratory. Figure 2 presents the modes previously summarized by Watts and Watts.[3] All involve two or more chromosome breaks. Some also require crossing over and some may generate significant deficiencies on the duplication-bearing chromosome.

(a) Sister-strand crossing over within a ring followed by breakage and fusion, as well as (b) "unequal crossing over" of sister strands, generates a tandem duplication of identical genetic information.

(c) "Unequal crossing over" of homologs--more properly considered as inter-homolog reciprocal translocation--yields a tandem duplication that would contain different alleles if it occurred in a heterozygote.

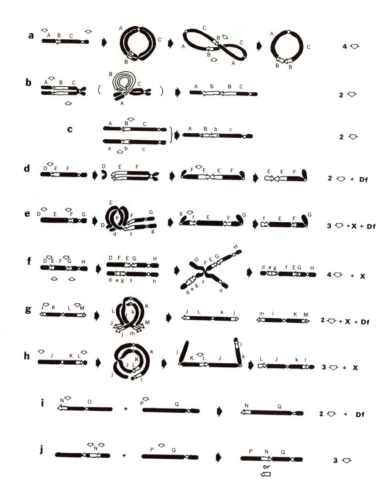

Figure 2.   Methods of generating a duplication for a short chromosome region.   Open segments represent the region eventually duplicated, with distal end arrow-tipped.   Open arrow heads point to positions of breakage. At the right, the total number of breaks (arrow head) and crossovers (X) required to produce the duplication-bearing chromosome from a structurally normal chromosome are indicated; the possible presence of a significant deficiency (Df) is also shown.

(d) An isochromatid rejoin, followed by breakage after centromere separation, generates a reverse repeat of identical genetic content at the end of the chromosome, after loss of its former tip. The lost tip may be small enough for the loss to be tolerable.

(e) A single crossover within a heterozygous paracentric inversion followed by breakage of the resulting dicentric may result in a duplication for part of the chromosome proximal to the inversion as well as in loss of the tip distal to the inversion. The duplicate regions are separated by the length of the inversion.

(f) In a heterozygote for two overlapping inversions, a single crossover within the common portion of the inversions can yield a duplication for both unshared parts of the two inversions. As in (e) the duplicate regions are separated by the length of the common part of the inversion.

(g) A single crossover within a heterozygous pericentric inversion yields a chromosome duplicated for one end outside the inversion and deficient for the other.

(h) A single crossover between a ring and a rod chromosome, followed by breakage of the resulting dicentric, also produces the duplicate regions in different arms of the resulting chromosome.

(i) A reciprocal translocation between non-homologs can yield a chromosome in which another tip replaces a negligible amount of the original tip. This (recipient) chromosome can later occur with normal chromosomes in genotypes thus hyperploid for the information in the translocated tip.

(j) An insertional translocation requires three chromosome breaks in the same nucleus. The piece may be inserted in either sequence at any point into a homolog or into a non-homolog, and, as in (i), eventually participate in genotypes hyperploid for the inserted region.

Of the ten modes listed, seven would combine different alleles if they occurred in heterozygotes. Four require the ability to heal ends formed by mechanical breakage. Six involve the loss of chromosome tips, which for the present purpose must not contain essential loci. The first four produce only adjacent duplications. All involve rare occurrences. Even "unequal crossing over," often invoked, is at least as infrequent as point mutation. Bender found new duplicates for the lozenge locus (combining alleles from both homologs) at a rate less than 10^{-4} after 4000 r of X-rays.

While unequal crossing over is an infrequent producer, it is a frequent dissociator of tandem duplications. Tandem duplications must be regarded as inherently unstable so long as the genetic length of the duplicated region permits crossing over at rates significantly above the per-generation mutation rates. Reverse repeats and duplications widely separated on the same chromosome or carried on non-homologs are not subject to such high rates of loss from asymmetric synapsis and crossing over.

The normal salivary chromosomes of Diptera provide numerous instances of reverse repeats. Bridges[4] identified eight reverse repeats of band sequences in his first major paper on the salivary chromosomes of <u>Drosophila melanogaster</u>, together with many doublets appearing to be reverse repeats of single bands. Lewis[5] found functionally similar loci to occupy doublet bands. Two different sequences of several bands were each repeated farther along the same chromosome, leading to frequent spiral synapsis. In <u>Sciara</u>, Metz reported abundant reverse repeats and found three cases of widely separated repeats in the same chromosome.[6] Four species have an X chromosome containing a region present in three widely separated positions,[7] though not all bands in all three locations synapse equally well. <u>Sciara</u> species also appear to be polymorphic for double <u>versus</u> single bands, where the double bands may well represent tandem duplications. Homologous chromosomes of different <u>Sciara</u> species differ by many segments one to four bands in length. Short homologous sequences of bands on non-homologous chromosomes would be more difficult to verify cytologically in wild-type

chromosomes, but have been seen following irradiation in the laboratory.

In view of the foregoing restrictions and considerations, appropriate

duplications will be treated as essentially non-recurrent events combining

different alleles of the parent locus into the same or different chromo-

somes capable of transmission in a single gamete, with any level of recom-

bination from 0 to 50%. Breakdown of the duplication by asymmetric cross-

ing over is not considered here.

THE FAVORED HETERODIMER MODEL

Let us make these assumptions:

(1) Each of the duplicate loci is translated into polypeptide at

the same rate within a cell.

(2) The monomeric polypeptides assort randomly into dimers.

(3) The total amount of polypeptide attributable to the locus is

regulated, so that genotypes differ in the proportions of possible

dimers but not in the total amount.

(4) The adaptive value W of the genotype is the weighted average

of the adaptive values of individuals, real or hypothetical, contain-

ing only one of the several dimers formed by the genotype. Thus the

hypothetical possessor of only hybrid dimer would have a value sur-

passing that of the heterozygote. This assumption shapes some of the

specific consequences detailed later.

(5) Both the duplicate and original site of the locus are auto-

somal, with $\underline{a}_1$ the allele in the duplicate site.

(6) The population is diploid and either (a) random-breeding

or (b) with a constant fraction of uniting gametes identical by de-

scent. It is large enough that stochastic processes are negligible

when the duplication is present in 0.1% of the gametes. Generations

are non-overlapping.

(7) The frequencies of alleles are at selective equilibrium before the duplication first appears. All alleles present form heterozygotes of roughly equivalent selective value.

To clarify these assumptions, let us translate them into the simplest of the specific models:

Complete linkage in a random-breeding population with two alleles, as for a reverse repeat. Figure 3 charts the types of dimers formed in each of the genotypes present during the course of selection. Note that heterozygotes for a "singlet" chromosome (carrying only the parent locus) and the duplication-bearing chromosome have a greater abundance of one of the homodimers than of the other. Adaptive values of duplication genotypes are calculated by assumption # 4, which assigns a value of $1 + (s_1 + s_2)/2$ to the hypothetical possessor of only hybrid dimers. Note that duplication heterozygotes have a lower value than singlet heterozygotes if s_1 is nearly equal to s_2, but that one of them has a higher value if one of the selection coefficients is more than double the other. Figure 4 details the initial genotypic composition of the population and the recurrence formulae for calculating the frequencies of the three gametic types in successive generations.

Two kinds of ultimate equilibrium are expected. (1) When the two singlet homozygotes have roughly equal adaptive value, the duplication-bearing chromosome will completely displace the singlets, and the average adaptive value of the population will rise to equal the adaptive value of the singlet heterozygotes earlier. (2) When one singlet forms homozygotes with an adaptive value more than twice as depressed as the other homozygote, an equilibrium will be reached in which the population is polymorphic for the duplication-bearing chromosome and the more adaptive singlet, with an average adaptive value surpassing that possible in any of the singlet genotypes.

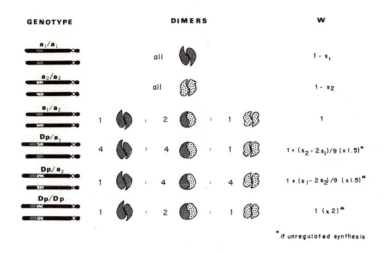

Figure 3.   Proportions of dimers in genotypes with $\underline{a}_1$ or $\underline{a}_2$ on singlets or both in duplication-bearing chromosome.

genotypes a_1/a_1 a_2/a_2 a_1/a_2 a_1/Dp a_2/Dp Dp/Dp

frequencies p_1^2 p_2^2 $2p_1p_2$ $2p_1p_d$ $2p_2p_d$ p_d^2

pre-duplication equilibrium frequencies:

$$\hat{p}_1{}^{\circ} = s_2/(s_1 + s_2) \qquad\qquad \hat{p}_2{}^{\circ} = s_1/(s_1 + s_2)$$

$$\bar{W} = 1 - s_1 s_2/(s_1 + s_2)$$

recurrence formulae during duplication incorporation:

$$p_1' = p_1(1 - s_1 p_1 - \frac{2s_1 - s_2}{9} p_d)/\bar{W}$$

$$p_2' = p_2(1 - s_2 p_2 - \frac{2s_2 - s_1}{9} p_d)/\bar{W}$$

$$p_d' = p_d(1 - \frac{2s_1 - s_2}{9} p_1 - \frac{2s_2 - s_1}{9} p_2)/\bar{W}$$

after duplication incorporation:

if $s_2/s_1 \leq 2$:   if $s_2/s_1 > 2$:

$\hat{p}_d = 1$ $\hat{p}_1 = 0$ $\hat{p}_2 = 0$ $\hat{p}_d = (7s_1 + s_2)/(5s_1+s_2)$ $\hat{p}_2 = 0$

$\bar{W} = 1$ $\hat{p}_1 = (s_2 - 2s_1)/(5s_1+s_2)$

$$\bar{W} > 1$$

Figure 4.   Specifications of two-allele complete linkage model.

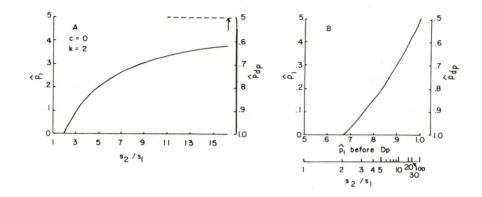

Figure 5. Equilibrium frequency of singlet chromosome with $\underline{a}_1$, when recombination frequency $\underline{c}$ = 0 and number of alleles $\underline{k}$ = 2. In A, $\hat{p}_1$ is plotted as a function of the ratio of the two selection coefficients. In B, it is plotted as a function of its equilibrium frequency before duplication. (From Spofford.[8])

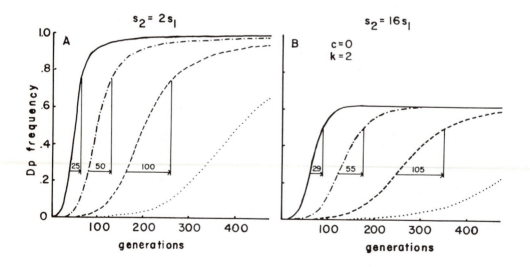

Figure 6. Duplication frequency kinetics starting with $\underline{p}_{dp}$ = .001 and $\underline{p}_{s1} : \underline{p}_{s2} = \underline{s}_2 : \underline{s}_1$ for $\underline{c}$ = 0, $\underline{k}$ = 2. $\underline{s}_1 + \underline{s}_2$ = .1 ($\cdots\cdots$), .2 (- - - -), .4 (— · —·) or .8 (———). (From Spofford.[8])

This last result had not been anticipated. It results from the superior fitness of the genotype with a higher proportion of the more adaptive of the homodimers. It can be mentioned parenthetically that other models for single-locus heterosis yield a similar result whenever the optimal proportion of genic products deviates widely from 1:1. In general, in a two-allele system, the higher the previous equilibrium frequency of the fitter allele, the higher its ultimate frequency in singlet form (see Figure 5).

The courses of selection with moderately symmetrical selection coefficients, and with highly asymmetrical, are plotted in Figure 6, as published earlier.[8] The number of generations required for the duplication frequency to rise from .25 to .75 is indicated in graph A, and from .25 to .50 in graph B, for each selection intensity. The speed of selection is related to the geometric mean of the s_i's.

The general model. Keeping in mind the formulation and emergent consequences of the simple model applicable to small reverse repeats just given, let us now progress to the more general formulation applicable to multi-allele systems, with recombination and partial inbreeding allowable.

Figure 7 charts the dimer phenotypes and adaptive values assigned to some of the genotypes that can result from additional alleles and from recombination, continuing the previous convention that a_1 is the allele in the new (duplicate) site and a_2 is the allele in the original, parental site in the gamete first carrying the duplication. Note in particular that heterozygotes combining three alleles are always superior to heterozygotes combining only two. If s_1 is comparatively large, three-allele duplication/singlet heterozygotes may be superior to three-allele duplication homozygotes, which again forecasts the possibility that at equilibrium the population will be polymorphic rather than monomorphic for the duplication.

Figure 8 details the initial genotypic composition and recurrence formulae. These reduce to the expressions for the first model when the recombination fraction $c = 0$, the number of alleles $k = 2$, and the coefficient of inbreeding level $F = 0$.

Taking the complicating modifications of the simple model one at a time, let us discover their consequences.

Recombination between the two sites in the duplication-bearing chromosome delays the phase of rapid increase in its frequency (Figure 9), but continues this rapid increase to much higher frequencies so that singlet chromosomes are excluded earlier from the population. A duplication inserted into a non-homolog has of course a 50% recombination frequency. Moderate levels of recombination, near .05, combine a fairly short initial delay with a rapid attainment of the final state. Higher levels of recombination continually regenerate less-favored from more-favored allele combinations with an effect the more pronounced the more asymmetric the selection coefficients. When recombination values exceed a minimum level indistinguishable from the locus-specific mutation rate, fixation of the duplication proceeds to completion so long as the a_1 homodimers are adaptive enough to render s_1 less than twice as great as s_2. When s_1 is less than half as great as s_2, segregation for both alleles persists at the original site.

Multiple alleles guarantee the persistence of segregation at the original site, though a_1 is retained there only with great asymmetry of the selection coefficients and only when linkage is complete. In the absence of recombination (Figure 10) the duplication-bearing chromosome becomes essentially one additional member of the multiple-allele system. It increases the average adaptive value of the population but at an equilibrium frequency that is lower with each additional pre-existing allele.

When the two sites can recombine, the presence of several alleles at

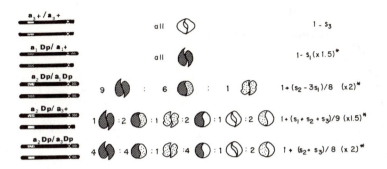

Figure 7. Additional genotypes with third allele and recombination.
Columns as in Figure 3.

gamete frequencies: p_{si} (carries a_i at original locus, singlet)

p_{di} (carries a_i at original locus; a_1 in Dp)

$$p_{d\cdot} = \sum_i p_{di}$$

zygote frequencies from $(1-F)(\sum_i p_{si} a_i + \sum_i p_{di} a_i a_1)^2 +$

$$F(\sum_i p_{si} a_i + \sum_i p_{di} a_i a_1)$$

preduplication equilibrium frequencies:

$$p_i{}^o = \frac{\tilde{s}}{k\, s_i} - \frac{F}{1-F}\left[1 - \frac{\tilde{s}}{s_i}\right]$$

giving $\bar{W} = 1 - \frac{\tilde{s}}{k}\left[1 + (k-1)F\right]$

recurrence formulae during duplication incorporation:

$$p_{si}' = \left\{p_{si}\left[F\,W_{si} + (1-F)\sum_j (W_{si\cdot sj}\,p_{sj} + W_{si\cdot dj}\,p_{dj})\right] - c\,(1-F)\sum_{j\neq i}D_{ij}\right\}/\bar{W}$$

$$p_{di}' = \left\{p_{di}\left[F\,W_{di} + (1-F)\sum_j (W_{di\cdot sj}\,p_{sj} + W_{di\cdot dj}\,p_{dj})\right] + c\,(1-F)\sum_{j\neq i}D_{ij}\right\}/\bar{W}$$

where $D_{ij} = W_{si\cdot dj}(p_{si}\cdot p_{dj} - p_{sj}\cdot p_{di})$

Figure 8. Specifications of general model. Symbols: $\tilde{s}$ is the harmonic
mean of s_i's; F, the inbreeding level (fraction of uniting gametes identical
by descent); for the frequencies p and adaptive values W, the subscript
"si" refers to allele a_i in singlet-bearing gamete and "di" to a_i at
original site in duplication-bearing gamete.

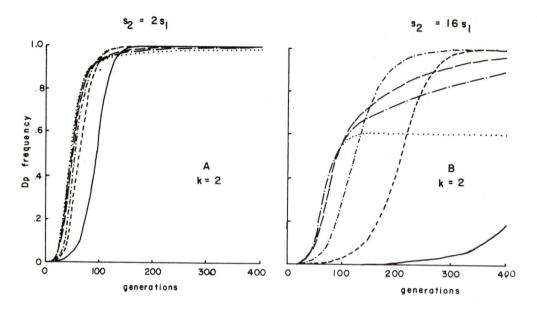

Figure 9. Duplication frequency kinetics starting with p_{dp} = .001 as in Figure 6, with $\underline{s}_1$ + $\underline{s}_2$ = .8, $\underline{k}$ = 2, and $\underline{c}$ = 0 (·····), .005 (—·—), .01 (——), .05 (—·—·—·), .10 (— — —), and .50 (———). (From Spofford.[8])

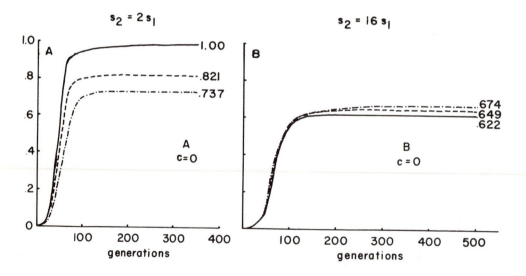

Figure 10. Same as Figure 9 except that $\underline{c}$ = 0 and $\underline{k}$ = 2 (———), 3 (— — —), or 4 (—·—·). (From Spofford.[8])

the original site permits heterotic interaction and thus continued segregation at that site, even in duplication homozygotes. From a series of
alleles, those with the greatest s_i's persist--usually the two greatest,
although when several are close in value they sometimes satisfy the general conditions for multiple allele equilibrium.[9] Since alleles with high
s_i's would be comparatively rare before the duplication became common,
their ultimate survival at the expense of alleles that were more common
at the outset may seem a paradox. The paradox disappears when it is realized that it is only because alleles "least fit" as homozygotes formed
extraordinarily favorable heterodimers that they were components of a
multi-allele system.

Partial inbreeding alters both the initial and final population com-
positions. In the absence of inbreeding, equal fitness of all heterozygotes guarantees that the individual allele frequencies are inversely proportional to their selection coefficients. Inbreeding exaggerates the
asymmetry of allele frequencies, leading to loss of alleles otherwise expected to be rare. Figure 11 displays the maximum asymmetry of selection
coefficients compatible with continuing polymorphism for the indicated
number of alleles under various levels of inbreeding. Figure 12 offers
another look at the type of restriction on heterotic systems imposed by
inbreeding. For a three-allele system in which one allele has a fitness
different from the other two, the frequency of that allele is seen to become 1 or 0 for higher levels of inbreeding, depending on the type and
degree of asymmetry of the s_i's. A delicate balance is maintained even
for F = 1 when s_i's are exactly equal. Thus, populations with significant
inbreeding appear less likely to possess heterotic allele systems favorable for the initial stages of duplication evolution.

On the other hand, where heterotic allele systems nevertheless exist,
inbreeding hastens the spread of a duplication throughout a population and

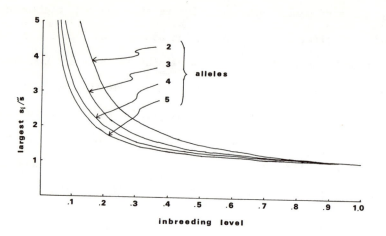

Figure 11.   Effect of inbreeding level (F) on the maximum asymmetry of selection coefficients (largest $s_1/\tilde{s}$) in a heterotic system with the indicated number of alleles.   With higher asymmetry there will be fewer alleles.

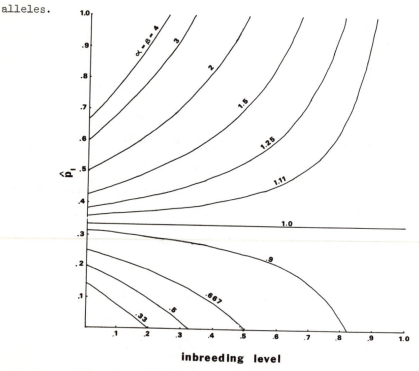

Figure 12.   Effect of inbreeding on equilibrium frequency of a_1 in three-allele heterotic system. $\alpha = s_2/s_1$; $\beta = s_3/s_1$.

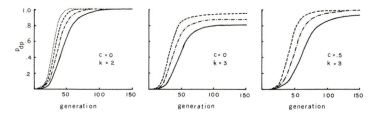

Figure 13.   Effect of inbreeding on duplication frequency kinetics, when $\underline{s}_1 = \underline{s}_2 = \underline{s}_3 = .4$ and $\underline{p}_{dp} = .001$ at start.   F = (———), .1 (—·—·),
.2 (————) or .3 (····).

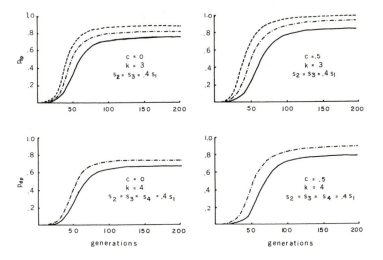

Figure 14.   Effect of inbreeding on duplication frequency kinetics, when $\underline{a}_1$ is "least fit" allele, and $\underline{s}_1 + \underline{s}_2 = .8$.   Symbols as in Figure 13.

lessens the likelihood that non-duplication-bearing homologs persist (Figures 13 and 14). The final stages of disappearance of these homologs can be exceedingly slow. With recombination in a four-allele system in which a_1 is the "least fit" allele, 10% inbreeding leads to ultimate mono-morphy for the duplication but at a rate scarcely detectable in the lower right graph in Figure 14. Nevertheless, segregation of the remaining alleles continues at the original locus even with the higher levels of inbreeding, as a result of the completely symmetrical selection on these alleles at the original site in duplication homozygotes (see Figures 3 and 7; all individuals homozygous for one allele at the original site and for another at the new site have W = 1).

Mutation at the new site is unlikely to occur until the duplication has become reasonably common. I have not investigated how an early muta-tion, to one of the pre-existing types or to a form not previously repre-sented but of similar heterotic properties, would alter the kinetics of duplication spread in the population. In either case, the outcome is clear from the adaptive values predicted on the favored-heterodimer model: the new allele will be selectively favored until its frequency equals that of a_1. An individual heterozygous at both new and old sites, combining four alleles in its genotype, has a higher adaptive value than one homo-zygous at either site.

VALIDITY OF THE MODEL

How widely applicable are the specifics of the favorable-heterodimer model to actual situations? How many of the predictions derived from the model are contingent on the specific assumptions embodied in it?

The proportion of the dimeric types of certain proteins in hybrids is in conformity with the first two assumptions and widely different in others. In the latter case, the array of types of expectations is un-changed, but the numerical values of selection coefficients are different

at which one outcome is to be expected rather than another.

If polypeptide chain synthesis is unregulated so that duplication genotypes have higher levels of the protein, and if these higher levels are more adaptive, then this becomes the main selection advantage of those duplications. Then even an originally monomorphic population would rapidly become monomorphic for the duplication.

The superior fitness of the heterozygote may derive from some other consequence of heterozygosity than the presence of hybrid dimers. In about half of the protein polymorphisms that Manwell and Baker[10] judge to be cases of positive heterosis, from genotype frequencies in populations, stable hybrid proteins are not formed. Presumably the simple presence of both proteins is advantageous. If one protein is advantageous in a greater fraction of the selective crises confronting individuals than is the allelic protein, a duplication heterozygote may have a more nearly optimal ratio of the proteins than a duplication homozygote. This leads to a spectrum of expectations roughly similar to that generated by the favorable-heterodimer model. Predictions of continued segregation at the original site, however, require further ad hoc assumptions as to how more benefits might be conferred by the presence of three proteins rather than by two.

The identification of polymorphic loci to which the theory elaborated here might apply is, unfortunately, obscured by the possibility that such a locus may be embedded in a chromosome region containing several other closely-linked heterotic loci. In this case the allele frequencies at one locus are strongly influenced by the adaptive values associated with the entire region held in severe linkage disequilibrium.[11] This would invalidate attempts to calculate adaptive values of the duplication genotypes from selection parameters estimated from population data. Inclusion of several heterotic loci in a single duplication would add further complication. Although this renders numerical prediction virtually impossible,

the simple model explored here remains useful in visualizing the process

of incorporation of a duplication into a species genome.

CONCLUSION

In brief, then, the contention that heterosis can provide an adequate

selective force for the spread of a newly-introduced duplication through-

out a population holds good for any mechanism causing single-locus heter-

osis, as was known already. But whether duplication fixation reduces the

incidence of polymorphism at a heterotic locus--or increases the number

of polymorphic loci--is very much a function of the molecular basis of the

heterosis.

REFERENCES

1. Fincham, J. R. S., Genetic Complementation, Benjamin Press, New York (1966)

2. Partridge, C. W. H., and N. H. Giles, Nature, 199, 304 (1963)

3. Watts, R. L., and D. C. Watts, J. Theor. Biol., 20, 227 (1968)

4. Bridges, C. B., J. Hered., 26, 60 (1935)

5. Lewis, E. B., Cold Spring Harbor Symp. Quant. Biol., 16, 159 (1951)

6. Metz, C. W., Carnegie Inst. of Wash. Pub. No. 501, 275 (1938)

7. Metz, C. W., Amer. Natur., 81, 81 (1947)

8. Spofford, J. B., Amer. Natur., 103, 407 (1969)

9. Mandel, S. P. H., Heredity, 13, 289 (1959)

10. Manwell, C., and C. M. A. Baker, Molecular Biology and the Origin of Species, University of Washington Press, Seattle, Wash. (1970)

11. Franklin, I., and R. C. Lewontin, Genetics, 65, 707 (1970)

DISCUSSION

O'BRIEN: An important assumption in the model you propose is the equivalence of selective coefficients of the original alleles to the alleles in the duplicated locus.   This assumption is somewhat disturbing in two aspects.   The first is that selection will act on the duplication not only as a function of the locus discussed but also as a function of the neighboring loci within the duplication which also respond to selective forces.   Secondarily there is a problem of position effects which also contribute to the fitness of a duplication.   In light of these points would you comment on the reality of this assumption of the model.

SPOFFORD: (1) This is really one of the features of the model that I am most disturbed about too and I can only say that, to start with, one tries the simplest assumptions.   The effect of neighboring loci in a duplication couldn't be trivial - since chromosome regions seem to be selected as wholes - and this is why I emphasized the abundance of one-band reverse repeats (the doublets) in Dipteran chromosomes where presumably only one major gene was present in the duplication.   (2) Not all rearrangements do involve position effects.   Many euchromatin to euchromatin ones do not.   Of course a position-effect of any sort would disturb the prediction of duplication fitnesses from the allelic fitnesses before the duplication.

MELTON: A tacit assumption you seemed to be making is that duplication is accompanied by dosage compensation - i.e., that doubling the gene dosage did not itself alter gene expression for example by doubling the amount of gene product.   Would you comment on whether you think dosage compensation is in general any more likely to occur than dosage control (strict or partial), and how the occurrence of dosage control of gene expression would affect your model.

SPOFFORD: I did as a matter of fact include, very rapidly, the possibility that gene dose is directly related to gene product, under the heading of unregulated protein production. In this case, any time a simple increase in product is beneficial - in a rate-limiting step for example - a duplication would be fixed selectively very rapidly without any need for allelic divergence or heterosis. Some loci show dosage compensation and some don't, in the case of X-chromosomes in *Drosophila*, for example. At this point I wouldn't like to guess what's the more common.

CAMPBELL: (1) To many of us, the most interesting question raised by your talk is whether heterosis has generally preceded duplication, as you suggest, or whether duplication has been followed by gene differentiation, as has frequently been supposed in the past. As I understand, in your theory the first alternative is a postulate rather than a proposition to be proved. (2) It seems reasonable that heterosis sets the stage for selection of duplication, but it is not obvious to me in which order duplication and mutation will occur, because duplication will allow subsequent selection of mutations for which the equilibrium level due to heterosis would be very small.

SPOFFORD: (1) What I tried to present is a selective drive so that a duplication, which occurs in the first place in only one individual, eventually gets transmitted to all members of the population - that is, this individual is an ancestor of the entire species later on so far as the duplication-bearing chromosome is concerned. Unless this selective drive comes from mere number of copies of the locus, in which case heterosis is not involved, heterosis seems a likely candidate to provide that drive and this does require that the early stage of gene differentiation come before the duplication. I don't have any evidence in hand otherwise yet, though there are some segregating duplication systems in <u>Drosophila</u> that should be investigated. (2) That certainly is true; one has to look for real polymorphisms with one allele held selectively in fairly low frequency to see how commonly they occur. Once a duplication is fixed, new heterotic alleles can be added fairly readily on the model just presented.

MULTIPLE ALLELES AND GENE DIVERGENCE IN NATURAL POPULATIONS

Ross J. MacIntyre
Section of Genetics, Development and Physiology
Cornell University
Ithaca, New York

Abstract. Some results from a new method for detecting intraspecific and interspecific variation in enzyme structure are presented in this paper. The method involves measuring the activities of homospecific and heterospecific enzymes which form after dissociation and reassociation of a dimeric enzyme, acid phosphatase-1, in Drosophila. Some consideration is given to the possible functional significance of this variation and a testable model is proposed to explain its presence in different Caribbean island populations of D. melanogaster and D. simulans.

INTRODUCTION

This paper will consider the variation in the products of structural genes, or more specifically, enzymatic or non-enzymatic proteins. I will try to show that there is no essential difference between intraspecific and interspecific variation, most of which has been detected in the last ten years. I will then raise the issue over whether this variation is adaptive and, proceeding from a prediction implicit in the so called neutral gene hypothesis, try to show that there is variation in at least one Drosophila gene enzyme‑system which should be selectively important. Because the experimental approach employed is novel it and some of its assumptions will be examined in some detail.

Intraspecific variation. Most of the gene-protein variation within and between populations of a single species can best be visualized by gel electrophoresis. Since the pioneering papers of Lewontin and Hubby in 1966[1,2] a rather extensive body of literature has accumulated. The

144

more detailed studies are listed in Table 1. All of these studies but
one, Lewontin's calculations on man's blood groups, involve gel electrophoresis. With regard to this table, note first of all that most of the
species examined in this sample of surveys are from the genus Drosophila.
The "number of loci sampled" refers to the number of different proteins
which could be visualized in gels following electrophoresis of extracts
from single individuals. The "number of populations sampled" is a rather
poorly defined datum. Usually what is meant is the number of localities
from which individuals were captured. Whether these represent separate
Mendelian populations, i.e. with no or reduced gene flow between them, is
generally not known. The "frequency of polymorphic loci" is the proportion of the number of genes which have more than one electrophoretic
variant in a population. This usually excludes systems in which the
rarer variant frequencies are less than 0.05. Finally, the "average
heterozygosity" is a figure calculated from all the gene frequency estimates from a single population. It is simply the proportion of loci, if
the number of loci sampled is assumed to be a representative sample of
the total genome, at which any individual from a particular population
will, on the average, be heterozygous.

These surveys, some restricted and some massive, come up with
remarkably similar estimates of the frequency of polymorphic loci, but
this can be somewhat misleading. When a fairly substantial number of
populations is included in a survey, different gene-enzyme systems show
different patterns of polymorphism. Table 2 contains some abstracted
data from a survey of Appalachian mountain populations of D. melanogaster[3].
In this table are four gene-enzyme systems each of which has a different
polymorphism pattern. The gene-enzyme systems are given in the first
column with alleles specifying the electrophoretic variants designated
as A and B (for "slow" and "fast", respectively). The localities

TABLE 1

Some Representative Population Surveys for Intraspecific Gene-Enzyme Variation

Species	Number of Loci Sampled	Number of Populations Sampled	Frequency of Polymorphic Loci	Average Heterozygosity	Reference
D. pseudoobscura	18	5	0.39	0.12	Lewontin & Hubby, 1966[2]
D. ananassae	20	2	0.4 - 0.5	--	Johnson et al, 1966[34]
Homo sapiens	10	1	0.3	--	Harris, 1966[35]
Homo sapiens	33	1	0.36	0.16	Lewontin, 1967[36]
D. melanogaster	10	8	0.54	0.23	O'Brien & MacIntyre, 1969[3]
D. nasuta	3	6	0.73	0.33	Stone et al, 1968[37]
D. ananassae	4	4	0.5	0.133	Stone et al, 1968[37]
D. pseudoobscura	24	3	0.40	0.12	Prakash et al, 1969[38]
mus musculus	40	1	0.3	0.11	Selander & Yang, 1969[39]
Limulus polyphemus	24	4	0.25	0.06	Selander et al, 1970[40]
D. persimilis	24	25	0.25	0.105	Prakash, 1969[9]
D. mimica	5	1	0.60	0.31	Rockwood, 1969[41]
D. willistoni	14	30+	.64	--	
D. paulistorum	13	30+	.62	--	
D. equinoxialis	13	30+	.69	--	Ayala et al, 1970[8]
D. tropicalis	12	30+	.75	--	

TABLE 2

Patterns of Polymorphism for Four D. melanogaster Gene-Enzyme Systems

Gene	Allele	LOCALITY							
		Ceres N. Y.	Painesville Ohio	Mt. Sterling Ohio	Mammoth Cave Ky.	Oxford N. C.	Manning S. C.	Red Top Mt. Ga.	Columbia Ga.
G-6-PDH	A	0.50	0	0.53	0.31	0	0.19	0.61	0.33
	B	0.50	1.0	0.47	0.69	1.0	0.81	0.39	0.67
LAP-D	A	0.75	0.69	0.81	0.88	0.65	0.64	0.79	0.75
	B	0.25	0.31	0.19	0.12	0.35	0.36	0.21	0.25
APH	A	1.0	1.0	0.75	0.42	0.65	0.68	0.50	0
	B	0	0	0.25	0.58	0.35	0.32	0.50	1.0
MDH	A	0	0	0.40	0	0	0	0	0
	B	1.0	1.0	0.60	1.0	1.0	1.0	1.0	1.0

sampled are arranged at the top from left to right in a rough North to
South transect. Glucose-6-phosphate dehydrogenase is polymorphic but
neither allele is prevalent, i.e. at a higher frequency, over the entire
region. On the other hand, Leucine aminopeptidase-D is also polymorphic
but A is always the more frequent allele. Furthermore, the gene frequencies are virtually identical over the areas sampled. Alkaline phosphatase
is gene-enzyme system which may show clinal geographic variation. The
A allele is more prevalent in the North but rare in the South. Finally,
Malic dehydrogenase is essentially monomorphic. A variant allele was
found in the sample from Mt. Sterling, Ohio, so here there is an apparent
local polymorphism. For other systems like Malic dehydrogenase only
intensive sampling will uncover rare alleles, and these will be a frequencies which could be maintained by a balance between the mutation
rates from common to rare alleles and selection against the carriers of
the new alleles. I want to emphasize that in most surveys, patterns
like those of Leucine aminopeptidase and Malic dehydrogenase are very
prevalent. The kind of pattern exhibited by Glucose-6-phosphate dehydrogenase is quite rare. Unfortunately, very few studies were designed to
detect the clinal variation seen in Alkaline phosphatase.

Interspecific Variation. An impressive amount of very well characterized interspecific variation in protein structure has been obtained in
recent years[4]. However, amino acid sequence comparisons and immunological
methods which provide the best information have generally been made
between species in different orders or classes. One might ask, does the
variation seen within species have its counterpart when closely related
species are compared? There is abundant information, but again mostly
from the genus Drosophila. Notable are the electrophoretic comparisons
of Xanthine dehydrogenase[5] and Amylase[6] and a survey of five enzymes in
several sibling species groups by Hubby & Throckmorton[7]. Another

example, Acid phosphatase-1, will be presented below in this report.

The variability seen here is simply an amplification of the intraspecific variation discussed above. Some species may have unique electrophoretic variants of an enzyme, while others may share electrophoretically
identical enzymes. Closely related species often have the same polymorphic patterns[8,9]. There is no evidence to support the idea that interspecific variation in protein structure is fundamentally different from
the variation seen at the intraspecific level.

Significance of this variation. The similarity between intraspecific
and interspecific variation is an important observation since the meaning
of the variation can be examined both by comparing the same enzymes from
different species and by comparing the electrophoretic variants from the
same species. It is the meaning of this variation that is very much at
issue at the present time. Several considerations have led some to the
conclusion that the variation just discussed is selectively neutral.

It has been pointed out above that at the population level, the
estimates of the frequencies of polymorphic loci are generally between
30 and 60 per cent. If the genome is adequately represented by the
sample of loci and if the variation were maintained by heterozygote
superiority, then simple genetic load theory predicts that the population
couldn't exist[2]. While several authors[10-12] have pointed how the segre-
gational component of the genetic load could be eased, others have tried
to devise ways by which selection could maintain the polymorphism without
entailing such a large cost. For example, Kojima has produced some
evidence that gene-frequency dependent selection can maintain polymorphisms[13]. Wills et. al.[14] and Franklin & Lewontin [15] have argued that
the unit of selection is really a block of linked genes and not the
single locus. Mukai has proposed an optimum level of heterozygosity
beyond which further polymorphism becomes selectively neutral[16]. The

problem with some of these models, e.g. gene frequency dependent selection
and optimum heterozygosity, is that it is difficult to envision how they
would work at the molecular level.   With regard to linkage disequilibria,
we have almost no experimental information about its extent in natural
populations. Nor are we likely to be able to obtain meaningful data on
this point in the near future. We simply don't have enough closely
linked genes identified. Then too, it is not very satisfying to have to
replace the single locus, whose gene product we can identify, with the
"correlated block" of genes whose boundaries may always be shifting and
whose allelic contents may be completely unknown. The danger here is
that the gap between theory and experimentation may become unbridgeable.

     With regard to interspecific variation in homologous proteins, two
observations are very difficult to reconcile with the contention that
natural selection has been responsible for the evolutionary differences.
First, the substitution of one allele for another entails a cost in terms
of genetic deaths[17]. Calculations on the rate of allele substitutions
from comparative amino acid sequence data during geologic time indicate
an intolerable genetic load would have been present if selection were
solely responsible for the substitutions[18]. Secondly, it has been noted
that sequenced proteins such as cytochrome-c and hemoglobin appear to
have evolved at constant rates[19]. Simple intuition makes this improbable
if selection were responsible for the observed differences.

     An attractive way out of this dilemma is to take the following
philosophical position: "Any product of a structural gene has a functional "core" whose integrity must be maintained for proper function.
(Certain invariant residues and sequences are, in fact, found in cytochrome-c and in hemoglobin). There are other amino acids whose functions
are not important, in fact, they may be trivial, like spacers, and their
positions could as effectively be occupied by several or perhaps many of

the other amino acids. These amino acids then could change under purely random processes producing high levels of polymorphisms within populations and becoming fixed at predictable rates during evolution. Then, the real stuff of meaningful evolutionary change, that is those genomic differences that make an elephant different from a mouse or D. simulans reproductively isolated from D. melanogaster may not be in the structural genes but in the controlling or regulatory genes. In short, during evolution, the blueprints change but the bricks remain essentially the same." The biologist who studies changes in structural gene products in an attempt to understand evolutionary processes may, if this interpretation is correct, be engaged in a considerable waste of time.

It is fortunate, however, that the neutral structural gene hypothesis, if I may call it that, makes a prediction about electrophoretic variants or different forms of an homologous protein. This prediction is that these proteins should be functionally equivalent, and if it were possible, they could be interchanged without any selective disadvantage to the two organisms. It is this prediction that I have chosen to examine using a single Drosophila gene-enzyme system.

Variation in Acid phosphatase-1. The gene-enzyme system has been called acid phosphatase-1[20]. The gene was mapped using two electrophoretic variants that are shown in Figure 1. In the figure are four zymograms representing four single flies. The individual on the left is a "slow" or AA homozygote. The "fast" or BB homozygote is on the right. The two heterozygotes (A/B) show a 3 banded pattern, characteristic of a dimeric enzyme. With electrophoretic variants, the gene has been mapped in both D. melanogaster and D. simulans and the homologous relationship has been established.

There is ample interspecific variation in the electrophoretic mobility of this enzyme. In Figure 2 are electropherograms of eight

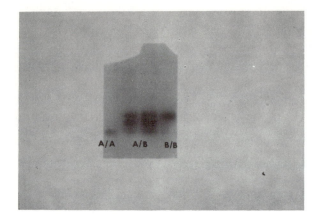

Figure 1:   A starch gel with acid phosphatase-1 patterns characteristic
            of "slow" of AA homozygotes, "slow"/"fast" or AB heterozygotes
            and "fast" or BB homozygotes.

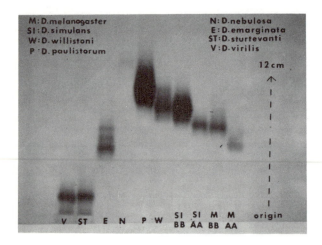

Figure 2:   Acrylamide gel showing positions of the adult acid phosphatases
            from eight Drosophila species. The AA and BB electrophoretic
            variants of D. melanogaster and D. simulans are also shown.

Drosophila species. Most are from the subgenus Sophophora, and several

are closely related, e.g., D. melanogaster and D. simulans, D. willistoni

and D. paulistorum, D. sturtevanti and D. emarginata. The electrophoretic

variants in D. melanogaster and D. simulans, are also shown in Figure 2.

Note that the electrophoretic variants are species specific. In addition,

the D. emarginata pattern has 3 bands. This pattern is found in all the

flies from the stock. We do not know what the molecular basis of this

subbanding is[21]. Finally, note that the enzymes from D. virilis and

D. sturtevanti have identical electrophoretic mobilities. These enzymes

will be compared in a different way below.

These prominent adult acid phosphatases are only assumed to be

homologous with the ones from D. melanogaster and D. simulans. Subunit

hybridization experiments, which will be described below, however, fully

support this assumption.

We have sampled natural populations of three species for intraspecific

variation in this enzyme. The three species are the cosmopolitan

D. melanogaster and D. simulans and the tropical species, D. nebulosa.

Electrophoretic variants have been found in each species.

The localities in the continental U.S. at which D. melanogaster and

D. simulans which have been recently sampled are listed in Table 3. Also,

Prof. Levins at the University of Chicago was kind enough to send some

collected lines of the three species from islands in the Eastern Caribbean.

The electrophoretic variants for each species are listed on the left in

Table 3. Each is also given an ratio value representing its mobility

over that of the BB enzyme from D. melanogaster. D. melanogaster has

been most extensively sampled. Note that this system is like the Malic

dehydrogenase shown earlier, i.e. mainly monomorphic with local polymor-

phisms. The A allele has been found only in the New York City area.

This pattern holds true in the Caribbean except for Mona Island where

TABLE 3

Recently Sampled Localities for D. melanogaster,

D. simulans and D. nebulosa

Species and Electrophoretic Variants	Continental U. S.		Caribbean Islands	
	Locality	Allele	Island	Allele
D. melanogaster				
A = 0.83[*]	Commack, N.Y.	A,B	Mona	A,B,C
B = 1.00	Garwood, N.J.	A,B	Palominitos	B
C = 1.07	Ceres, N.Y.	B	Hicacos	B
	Painesville, O.	B	Vieques	B
	Mt. Sterling, O.	B	Hassel	B
	Mammoth Cave, Ky.	B	Culebra	B
	Oxford, N.C.	B		
	Manning, S.C.	B		
	Red Top Mt., Ga.	B		
	Columbia, Ga.	B		
	Riverside, Cal.	B		
D. simulans	Ithaca, N.Y.	A,B	Ramos	B,C
A = 1.03	Garwood, N.J.	B,C	Guana	A,B
B = 1.17	Manning, S.C.	B	Maricao (Puerto Rico)	B
C = 1.31	Columbia, Ga.	B		
D. nebulosa			Hassel	A
A = 1.41			Cooper	B
B = 1.69			Guanica (Puerto Rico)	B

[*] Gel position relative to BB enzyme of D. melanogaster.

three alleles are segregating. D. simulans appears to be more polymorphic both on the mainland and in the Caribbean, and although we have now only three collections of D. nebulosa, already two alleles have been detected.

This represents only a superficial sampling at best. In most cases only a few flies were collected from the site. Nevertheless, the system provides variation at both the intra and interspecific levels. This is variation that we can use in order to determine the meaning of the evolutionary changes in the enzyme.

Rationale for the subunit hybridization experiments. How can we decide if these different forms of the acid phosphatase are, in fact, functionally equivalent? I'll describe the experimental approach first and then discuss why I suspect the differences in enzyme structure that it detects are physiologically important. In Figure 3, an outline of the rationale for the experiments is presented. An enzyme from two species (or two electrophoretic variants) which is assumed to be a dimer is dissociated into its constituent subunits. These are mixed and allowed to reassociate to form active enzymes, and the activities of the three enzymes which form, i.e. the two homospecific (or homodimeric) and the heterospecific (or heterodimeric) are measured. The observed activities can be expressed as a homospecific:heterospecific enzyme activity ratio which can be compared to one expected if subunit association is random. This is calculated from the observed results, i.e. $[X]$ and $[Y]$ and are determined from $\dfrac{XX + \frac{1}{2}XY}{Total}$ and $\dfrac{YY + \frac{1}{2}XY}{Total}$ and then $\dfrac{X^2 + Y^2}{2XY}$ equals the expected ratio. Finally, the data will be expressed as the difference between the two ratios. A real difference implies that either the subunits do not randomly associate or the enzymes have different activities, i.e. different subustrate turnover numbers.

It is my contention that two enzymes with different activities or different subunit affinities cannot be functionally equivalent. The

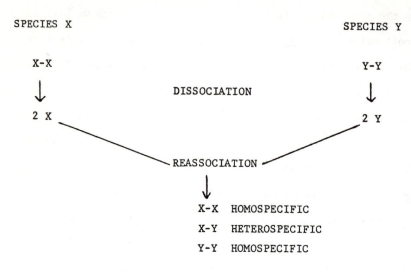

OBSERVED HOMOSPECIFIC:HETEROSPECIFIC RATIO=

$$\text{ACID PHOSPHATASE ACTIVITY OF } \frac{\text{X-X} + \text{Y-Y}}{\text{X-Y}}$$

$$\text{EXPECTED HOMOSPECIFIC:HETEROSPECIFIC RATIO=} \frac{p^2 + q^2}{2pq}$$

$$\text{WHERE} \quad p = \frac{\text{X-X} + \tfrac{1}{2}\text{ X-Y}}{\text{X-X} + \text{X-Y} + \text{Y-Y}}$$

$$q = \frac{\text{Y-Y} + \tfrac{1}{2}\text{ X-Y}}{\text{X-X} + \text{X-Y} + \text{Y-Y}}$$

Figure 3:   Outline of the experiments designed to detect interspecific
differences in enzyme subunit affinities.

apparently elaborate controlling systems which regulate enzyme activity attest to the importance of this property. Some preliminary evidence, however, which I'll discuss below indicates the differences we find are due to different subunit affinities. The question then is: Can random variation exist in amino acid sequences affecting subunit interactions? Unfortunately, almost no information is available on this point. In hemoglobin, however, a considerable number of amino acids hold the α and β subunits together[22]. Twenty-six of 143 α chain amino acids and 27 of 146 β chain amino acids are involved in non-covalent bonds between the subunits, making this a rather complex property of the molecule.

In addition, there is good evidence for evolutionary conservatism in these amino acids. With the use of the X-ray diffraction data of Perutz et. al. on horse hemoglobin[22] and the amino acid sequences listed in Dayhoff[23], I compared variability in the subunit contact amino acids both with those amino acids contacting the haem groups and the amino acids not involved in either of these functions. The data are given in Table 4. The conservatism in the amino acids forming intersubunit bonds is not significantly different from the degree of evolutionary change in the haem contacting residues. There is, however, a substantial difference between the variation in subunit contact amino acids and that in the amino acids at so called general positions.

Perutz and Lehman found 12 variant human hemoglobins involving amino acid substitutions in the contact region[24]. Of the 12 people from which the mutant hemoglobins were obtained (and these were obtained by screening blood samples from many populations), seven showed adverse clinical symptoms. It was stated that, in fact, that the stability of the molecule or the oxygen dissociation curve would be adversely affected in 10 of the 12 mutant forms.

Several studies in other systems, notably that of Cook and Koshland[25]

TABLE 4

Interspecific Variation in the Amino Acids of
Hemoglobin Considered According to Their Functions

α Chain (13 comparisons)	Number	Number that vary	Proportion of variable amino acids
Amino Acids Involved in β Chain Contact	26	3	0.12
Amino Acids Involved in Haem Contacts	19	1	0.05
Other Amino Acids	103	43	0.42
β-γ-δ Chains (24 comparisons)			
Amino Acids Involved in α Chain Contact	27	5	0.19
Amino Acids Involved in Haem Contact	21	2	0.10
Other Amino Acids	101	54	0.53

x^2 Tests for Homogeneity

Subunit Contact and Haem Contact Amino Acids

(3 degrees of freedom)

$x^2 = 2.18$, $P => 0.60$

Subunit Contact and Other Amino Acids

(3 degrees of freedom)

$x^2 = 65.41$, $P = < 0.001$

point to the extreme specificity involved in subunit recognition. But it is not a simple situation of the more and stronger the interchain bonds the better, for subunit interactions have been implicated in allosteric controlling mechanisms[26] and ligand binding[27]. In fact Cook & Koshland write:

> "Evolutionary selection therefore occurred for a precise alignment of amino acid residues at the binding region and this site, therefore, may be as highly desired and precisely oriented as the active site itself."[25]

The comparisons of variation in hemoglobin chains presented in Table 4 certainly support that statement.

Thus, it would seem, from admittedly too few bits of evidence, that very little, if any random variation in those amino acids involved in subunit contacts would be permitted.

MATERIALS AND METHODS

Details of both the assay system and the experimental procedures can be found elsewhere[28,29]. Only a brief resume' will be presented here.

Partially purified (5-10 fold) preparations of enzyme are used in the experiments. Enzyme activity is measured in both test tube and gel by coupling α-naphthol released from α-naphthyl phosphate to a diazonum salt. Dissociation of the enzyme is accomplished by raising the pH to between 10 and 11.5. The inactivation interval varies for each species enzyme but generally takes place over about 0.4 of a pH unit. Figure 4 shows the inactivation profiles for the enzyme from three species, D. melanogaster, D. simulans and D. virilis. Each species has its own characteristic inactivation curve. Care must be taken not to raise the pH too much above the point at which complete inactivation occurs. The amount of activity regained depends upon being very careful about this. Reactivation takes place upon dialysis against tris-maleate buffers at

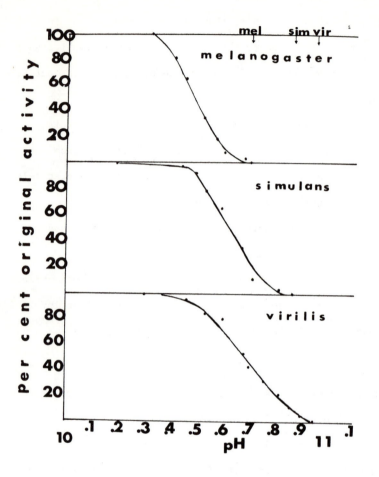

Figure 4:   Inactivation curves of Acid phosphatase-1 from <u>D</u>. <u>melanogaster</u>,

D. <u>simulans</u> and <u>D</u>. <u>virilis</u>.

0.1 N NaOH was added to 5.0 ml of a partially purified enzyme

preparation in 0.05 M Na Cl at 25°C. After mixing, the pH

was measured with a combination microelectrode. At the pH's

indicated on the graph, 0.2 ml aliquots were removed and

immediately assayed for acid phosphatase activity (see <u>Materials</u>

& <u>Methods</u>). Activities were corrected for the volume of Na OH

added.   The pH at which 100% inactivation occurs for each

species enzyme is indicated on the top line.

pH's near neutrality, although each species has a slightly different pH
optimum for reassociation. Usually between 20-85% of the initial activity
is regained during reactivation. It is easy to show that the inactivation
at high pH and the reactivation during dialysis involve the dissociation
and reassociation of subunits. When one mixes inactivated enzymes with
different electrophoretic mobilities and dialyzes this mixture, the
subsequent gel patterns invariably show in addition to the two original
enzymes a third heteromultimer with an intermediate electrophoretic
mobility[30].

It is important to know, for the experiments described below, if
the enzyme is really a dimer. That is, do we really dissociate the two
enzymes into only X and Y subunits? Several lines of evidence indicate
that with acid phosphatase, we are dealing with a dimeric system in which
the enzyme is composed of identical subunits. First, the molecular
weight, as determined by sucrose density gradient centrifugation and
electrophoretic retardation studies, is 100,000, a value compatible with
other known dimers[31]. Secondly, if the pH dissociated subunits are subjected to electrophoresis and the gel is incubated in a tris-maleate
buffer before staining, a zone of activity appears at a position different from the native enzyme. This indicates that the subunits have
reassociated in the gel to form active enzymes. Thus, presumably only
one kind of subunit is needed for reassociation making a heteromultimeric
structure unlikely. Furthermore, electrophoretic retardation measurements indicate the molecular weight of the subunit is 55,000, about one-
half the size of the native enzyme. Finally, complementation tests
between fifteen ethyl methane sulfonate induced zero or null activity
mutants for the enzyme indicate a single structural gene is involved
in its formation[32].

In a typical interspecific hybridization experiment, the following

regimen is used: Aliquots of partially purified enzyme from each species
are taken and adjusted to approximate quality in specific activities.
They are dissociated separately, taking the pH to just that point of complete inactivation. They are then assayed to confirm complete inactivation.
If inactivation is complete, equal amounts of each are then mixed and,
along with aliquots of each species' dissociated enzyme alone, are dialyzed
for 72 hours against a tris maleate buffer. The pH of that buffer is
usually intermediate between the pH optima for reassociation of the two
species' enzymes. At this point, all dialyzed preparations are assayed
to determine the percent reactivation, and finally, the dialyzed mixture
of subunits is subjected to electrophoresis and activities in the three
separated zones (i.e. the two homospecific and the heterospecific enzymes)
are determined by elution or densitometry. Usually a minimum of eight
determinations is made.

RESULTS

Preliminary tests. Several initial pairwise tests were made using
the enzymes from three species, D. melanogaster, D. simulans and
D. virilis, in order to test several assumptions. Gel patterns of the
reassociated mixtures of subunits are shown in Figure 5. On the left in
this figure are the gel positions of the three homospecific enzymes.
Then in the three slots on the right, are the patterns obtained after
subjecting the three reassociated mixtures of subunits to electrophoresis.
In order from left to right are D. virilis x D. simulans, D. melanogaster
x D. virilis and finally in slot 4, D. melanogaster x D. simulans. Note
in each, the presence of the heterospecific enzyme. This confirms that
the processes of dissociation and reassociation took place, and establishes
the homologous relationship.

In Table 5, are the detailed quantitative data for one of the repeats
from one of the tests, D. melanogaster x D. virilis. In each experiment

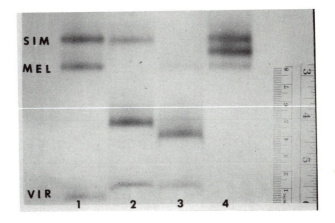

Figure 5:   Electropherogram of acid phosphatase-1 from D. virilis (vir),

D. simulans (sim) and D. melanogaster (mel) (slot 1), the

reassociated subunits from D. melanogaster and D. virilis

(slot 2), from D. simulans and D. virilis (slot 3) and D.

melanogaster and D. simulans (slot 4). In each mixture, the

heterospecific enzyme is in the middle of each pattern.

TABLE 5

Result of an Experiment Involving Reassociation of Acid Phosphatase-1 Subunits of
D. melanogaster and D. virilis

pH of reassociation	Subunits	Percent reactivation after dialysis	Percent of total acid Phosphatase in each zone*			Homospecific: heterospecific ratio*	Expected ratio**	Difference
			melanogaster homospecific	heterospecific	virilis homospecific			
	melanogaster	31						
5.9	virilis	34	9	70	21	0.43	1.00	-0.57
	mixture	52	(7-11)	(67-76)	(17-22)	(0.32-0.49)	(1.00-1.04)	(0.51--0.68)

* Ranges in parentheses: eight determinations.

** The expected ratio is calculated as follows: p (mel subunits) = mean proportion of melanogaster homospecific enzyme + ½ mean proportion of heterospecific enzyme, q (virilis subunits) = 1-p. Expected ratio = $\dfrac{p^2 + q^2}{2pq}$,

For example, p (melanogaster subunits) = $\dfrac{9 + \frac{1}{2}(70)}{100}$ = 0.44

q (virilis subunits) = 1.00-0.44 = 0.56

expected ratio = $\dfrac{(0.44)^2 + (0.56)^2}{2(0.44)(0.56)}$ = 1.00

the subunits were reassociated at each species' pH optimum for reassociation. The second column in the table indicates how much activity was
regained, both by subunits from one species alone and when mixed.   Then,
the relative activities in the three zones are given and the actual or
observed homospecific:heterospecific enzyme activity ratio is calculated.
The calculations showing how the expected ratio is determined are at the
bottom. Finally, then, the difference between the observed and expected
ratios is given in the last column. Table 6 lists the summarized results
for the three tests: Note the repeatability of the experiments and the
independence of the results from the pH optimum of reassociation. The
data are expressed ultimately as a difference between observed and
expected ratios. The sign of the difference is important; a minus indicates more than expected heterospecific enzyme activity is present. A
plus sign indicates more than expected homospecific enzyme activity has
been measured.

     With D. melanogaster and D. simulans another assumption implicit in
the methodology can be directly tested. This assumption is that the
methods used in purification, dissociation and reassociation do not
impair the functional integrity of the subunits. Specifically, with
these two species it is possible to make several in vitro vs in vivo
comparisons by examining the differences between observed and expected
homospecific:heterospecific ratios in three tests and comparing them
with those obtained from the three corresponding interspecific hybrids
or intraspecific heterozygotes. The data are summarized in Table 7.
In the first case, the electropherograms of D. melanogaster heterozygotes
for the alleles specifying the AA and BB electrophoretic variants were
compared with the pattern obtained when partially purified, dissociated
and reassociated AA and BB forms of the enzyme were used as the sources
of the mixed subunits. In the second case, D. simulans A/B heterozygote

TABLE 6

Summary of Results from Acid Phosphatase-1 Subunit Reassociation
Experiments Involving D. melanogaster, D. simulans and D. virilis

Interspecific test	Reassociated at pH	Observed homospecific: heterospecific* enzyme ratio	Difference from expected ratio
melanogaster x	5.9	(1) .43 (2) .42	− .57 − .58
virilis	6.5	(1) .50 (2) .41	− .50 − .59
simulans x	6.3	(1) .60 (2) .79	− .40 − .21
virilis	6.5	(1) .79 (2) .75	− .21 − .25
melanogaster x	5.9	(1) 1.44 (2) 1.38	+ .44 + .38
simulans	6.3	(1) 1.50 (2) 1.33	+ .42 + .33

* Results of 2 experiments for each test

TABLE 7

Comparison of _in vitro_ and _in vivo_ Results with
Regard to Reassociation of Acph-1 Subunits

Test	Difference between observed and expected homospecific:hetero-specific ratio, ± s.d.	
A.		
D. melanogaster		
AA x BB (_in vitro_)	+ 0.31 ± 0.20	t = 0.48
Acph-1^A/Acph-1^B heterozygotes (single flies)	+ 0.22 ± 0.19	P = > 0.5
B.		
D. simulans		
AA x BB (_in vitro_)	+ 0.23 ± 0.06	t = 1.96
Acph-1^A/Acph-1^B heterozygotes (mass homogenates)	+ 0.32 ± 0.12	P = < 0.10 > 0.05
C.		
Interspecific hybrids		
BB (_melanogaster_) x BB (_simulans_) (_in vitro_)	+ 0.32 ± 0.03	t = 0.22
D. melanogaster ♀♀ x D. simulans ♂♂ (single flies)	+ 0.33 ± 0.12	P = > 0.5
D. simulans ♀♀ x D. melanogaster ♂♂ (single flies)	+ 0.28 ± 0.04	t = 2.22 P = < 0.05 > 0.01

patterns were compared in a similar way with the corresponding dissociated
and reassociated AA and BB enzymes from this species. Finally, reciprocal
interspecific hybrids were compared with the in vitro results from the
D. melanogaster x D. simulans test. Note that the results agree well in
every case except one, but this is close to the acceptable level. The
in vitro results appear, then, to faithfully refect in vivo enzyme
subunit associations.

     Interspecific tests. It was evident that in the tests with
D. melanogaster, D. simulans and D. virilis, there were rather striking
differences between the observed and expected ratios. To date, we have
performed all possible tests with eleven selected Drosophila species.
All these results   will be presented elsewhere, but a selected group
will be shown below to demonstrate the sensitivity of the technique and
the fact that it detects variation not related to differences in net
charge. The data are those derived from the tests of D. virilis with six
species from the subgenus Sophophora and D. Sturtevanti with those same
six species.

     Figure 6 shows the reassociated patterns of D. virilis and the
other species. The three zones are evident in most cases except when
sub-banding complicates the picture, e.g. D. virilis x D. emarginata or
D. virilis x D. willistoni. The patterns, however, can always be separated
well enough to clearly define and measure the activities in the zones.

     The quantitative data from the twelve tests are summarized in
Table 8. The differences between observed and expected homospecific:
heterospecific ratios and the standard errors are given. Note, first of
all, that with either D. virilis or D. Sturtevanti subunits, reassociation
tests show an obvious difference between enzymes from the sibling species,
D. melanogaster and D. simulans (.74 vs .46 and .54 vs .01). On the
other hand, no difference can be found between the D. willistoni and

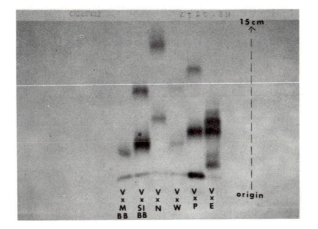

Figure 6:   Acrylamide gel showing the six patterns of homospecific and
            heterospecific enzymes which formed during reassociation of
            subunits from the D. virilis acid phosphatase with subunits
            from the enzymes of six other Drosophila species.   See
            Figure 2 for species designations.

TABLE 8

Results of Tests Involving D. virilis and D. sturtevanti
Difference of Observed Homospecific:Heterospecific Ratio
from that Expected if Subunit Association were Random
$\pm$ Standard Error

	D. virilis	D. sturtevanti
D. melanogaster	-0.74 $\pm$ .02	-0.54 $\pm$ .02
D. simulans	-0.46 $\pm$ .03	-0.01 $\pm$ .05*
D. willistoni	-0.12 $\pm$ .02	-0.25 $\pm$ .06
D. paulistorum	-0.11 $\pm$ .03	-0.29 $\pm$ .04
D. nebulosa	-0.22 $\pm$ .02	+0.12 $\pm$ .03
D. emarginata	-0.23 $\pm$ .03	+0.42 $\pm$ .08

* not significantly different from zero

D. paulistorum acid phosphatases, even though the enzymes differ in their
electrophoretic mobilities. D. nebulosa, a close relative of these two
species, has an enzyme quite different from either as shown by both tests.
Finally, as mentioned above, D. virilis and D. sturtevanti have enzymes
with the same electrophoretic properties. The subunit reassociation tests,
however, in every case save that with D. willistoni indicates they must
be rather different.

Intraspecific tests. The results from the interspecific tests show
that the sensitivity of the method is quite good. Hopefully, then, it
can be used as a tool to detect new multiple alleles in populations, on
the one hand, and demonstrate the presence or absence of what should be
selectively important differences between polymorphic electrophoretic
variants on the other. We have made a beginning on these projects using
the mainland and Caribbean populations listed in Table 3. We decided to
look for this variation in the following way: D. virilis subunits are
hybridized with dissociated acid phosphatase-1 extracted from different
geographic populations of another species. If repeatable differences in
the results from different populations could be demonstrated, it would
indicate the populations contained different acid phosphatase-1 alleles,
even though the enzymes might have the same electrophoretic properties.

Since the differences between multiple alleles measured in this way
might be rather small, we wanted to make sure the repeatability of the
test was good enough to give us a reasonable chance of detecting the
variation. Enzyme was purified from five different groups of flies
collected from the inbred Riverside, California strain. The five preparations were separately dissociated and each was mixed with an aliquot of
dissociated enzyme from D. virilis. Quantitative data from the reassociated
pattern of each were collected. They are presented in the first five
rows of Table 9. There were no significant differences between the

TABLE 9

Results of Subunit Hybridization Experiments: $\underline{D}$. melanogaster
from Riverside, California with $\underline{D}$. virilis (Five Repeats) and
$\underline{D}$. melanogaster, $\underline{D}$. simulans and $\underline{D}$. nebulosa
(electrophoretic variants) with $\underline{D}$. virilis

Species	Locale	Electrophoretic Variant	$D \pm$ s.e.[*]
$\underline{D}$. melanogaster	Riverside, Cal.	B	$-0.73 \pm .05$
"	"	"	$-0.77 \pm .02$
"	"	"	$-0.78 \pm .01$
"	"	"	$-0.78 \pm .03$
"	"	"	$-0.78 \pm .04$
$\underline{D}$. melanogaster	Commack, N.Y.	A	$-0.77 \pm .03$
"	"	B	$-0.78 \pm .03$
$\underline{D}$. simulans	Ithaca, N.Y.	A	$-0.32 \pm .03$ }**
"	"	B	$-0.46 \pm .03$
$\underline{D}$. nebulosa	Hassel Island	A	$-0.27 \pm .04$
"	Cooper Island	B	$-0.40 \pm .09$

[*] Difference between observed and expected
  homospecific:heterospecific enzyme activity
  ratios $\pm$ standard error.

** Difference is significant at the .01 level.

observed and expected homospecific:heterospecific enzyme ratios of the
five replicates. The D values are different from those in Tables 5 and 6
because quantitation was carried out by elution in the one case and densitometry in the other. See reference 28 for details. Since the repeatability
of this test appeared to be adequate, we then looked for differences
between electrophoretic variants in the three species. The data from
these tests are also in Table 9. The AA and BB variants from D. melanogaster
were not significantly different from each other nor from the enzyme made
by the B allele from Riverside, California. On the other hand, the
electrophoretic variants from D. simulans gave significantly different
differences of expected from observed ratios. This indicates the subunits
made by the two alleles may have different affinities for the D. virilis
subunit. Finally, the variants of D. nebulosa have not been thoroughly
analyzed. Only four determinations were made on each so the standard
errors are quite large. These tests must be repeated to see if the
difference is real. With regard to the question therefore, "do electrophoretic variants differ in those properties detected by the subunit
hybridization technique?" the answer is: some do and some don't. This
again demonstrates the independence of the two kinds of variation in
enzyme structure.

If this is true, the method ought to detect variation in electrophoretically identical enzymes from different or indeed from within the
same population. To date, we have looked at nine mainland populations of
D. melanogaster and four Caribbean island populations, all of which are
monomorphic for the Fast or BB form of the enzyme. Two island populations
of D. simulans have also been examined. Table 10 contains the data from
the first tests. It should be emphasized that these are only preliminary
results. In the mainland D. melanogaster populations, no evidence for
variation was found. However, the island populations are more interesting.

TABLE 10

Results of Subunit Hybridization Experiments: D. melanogaster
and D. simulans (Geographic Populations) with D. virilis

Species	Locale	Electrophoretic Variant	D± s.e.*
D. melanogaster	Ceres, N.Y.	B	-0.70 ±.06
"	Painesville, Ohio	B	-0.70 ±.03
"	Mt. Sterling, Ohio	B	-0.73 ±.02
"	Mammoth Cave, Ky.	B	-0.71 ±.01
"	Oxford, N.C.	B	-0.78 ±.05
"	Manning, S.C.	B	-0.79 ±.05
"	Red Top Mt., Ga.	B	-0.73 ±.03
"	Columbia, Ga.	B	-0.75 ±.02
"	Garwood, N.J.	B	-0.71 ±.02
D. melanogaster	Palominitos Island	B	-0.73 ±.01
"	" "	B	-0.67 ±.03
"	Hassel Island	B	-0.67 ±.01
"	Hicacos	B	-0.67 ±.02 ⟩**
"	Culebra	B	-0.85 ±.01
D. simulans	Maricao, Puerto Rico	B	-0.28 ±.01 ⟩***
"	Ramos Island	B	-0.37 ±.01

* See Table 9.

** Significantly different at .05 level from all
except Oxford and Manning.

*** Difference significant at .01 level.

The differences obtained from one of the Palominitos lines and from both Hassel and Hicacos islands are lower, but not significantly so from the mainland values. The enzyme extracted from the Culebra Island population is clearly different from all examined so far except those from Manning, S.C. and Oxford, N.C. We will have to repeat these three tests but it looks as if the Culebra population contains a different acid phosphatase-1[B] allele. The two populations of D. simulans also appear to have different forms of the BB enzyme. More data are needed here, however as only four determinations have been made on each.

There is much more material from the island populations to be analyzed. We are in the process of making A, B and C alleles from the polymorphic populations homozygous so we can extend these tests.

It should be obvious that the technique tells us only that a different allele is present if a difference is found. The population may be fixed for that allele or it may be polymorphic. Similarly, some populations which appear to be the same may contain different alleles at low frequencies. In this respect the procedure of subunit hybridization is not as precise as simple gel electrophoresis. If a difference is found or suspected in a D. melanogaster population, however, isogenic lines can be readily extracted with balancer chromosomes and analyzed individually to get an approximate idea of the number of different alleles and their frequencies. In fact we plan to extract 20 such lines from the Culebra population to see if one or more different acid phosphatase-1 alleles are responsible for the difference of -0.85. Also, if this difference is real, it allows us to unambiguously determine if the difference segregates with the acid phosphatase-1 locus. To date, the only evidence we have that the differences are due to differences in the structure of the acid phosphatase-1 molecule and not to some other factor or factors in the extracts, is that the same results are obtained after further purification of the enzyme[23].

DISCUSSION

There are some obvious unanswered questions about the technique.
One of the most troublesome is that we cannot be sure that differences in
subunit affinities are being measured. The amounts of protein in the
zones of activity in the gel are too small to measure directly. Thus, the
data are not in terms of specific activities. This means that, for
example, in the D. virilis x D. melanogaster test, that instead of 3 to
4 times as much of the heterospecific enzyme forming, amounts expected on
the basis of random subunit association might be present, but the enzyme
might have hydrolyzed 3-4 times the number of substrate molecules as did
the homospecific forms.

This is a difficult problem to attack experimentally, unless absolutely
pure proteins can be obtained. In lieu of this, we have some indirect
evidence from some experiments on the D. melanogaster x D. virilis test
that subunit affinities are involved in the differences. First, we
found that we could clearly separate the three enzymes formed after
reassociation on phosphocellulose columns. When we measured the K_m's of
the three enzymes, they turned out to be identical, 1×10^{-4}M. This
evidence is only suggestive, however, since K_m is only indirectly related
to the turnover number of an enzyme.

A second indirect approach involved determining the initial rates of
reactivation after pH induced dissociation. We found that we could
measure these rates by placing the dissociated subunits on one side of
the membrane in a rotating dialyzing chamber. Tris-maleate buffer at a
constant temperature was pumped through the other side. Aliquots could
be withdrawn from the sampling port and immediately mixed with an equal
volume of 2M tris. This raised the pH to about 10 and effectively
stopped any further reassociation. As expected, the initial rates
varied directly with subunit concentration and temperature. The rationale

then, behind the experiments was as follows: If a mixture of D. virilis
subunits reassociates at a particular initial rate, and if a mixture
D. melanogaster subunits regains activity at another initial rate, then
in a one to one mixture from the same two preparations of subunits, the
initial rate should be intermediate if the heterospecific enzyme does not
form faster but is simply more active than either homospecific enzyme.
If, however, the heterospecific enzyme forms at a faster rate because of
the greater affinities of the two unlike subunits, then the initial rate
of reactivation should be more rapid. Said another way, the concentrations
of free subunits necessary for the initial rapid recovery of activity will
decrease below that level sooner if the heterospecific enzyme is forming
more rapidly.

In Figure 7 are the results of the three tests, i.e. subunits from
each species alone and in a 1:1 mixture. In this experiment, the
D. melanogaster preparation had the slowest initial rate, followed by
the D. virilis subunits. The most rapid rate is in the mixture.

Figure 8 contains the data plotted in such a way that they are
independent of the total enzyme activity of the preparations. That point
at which the initial linear rate slows is set at 100%. Here, it can be
seen that the mixture reaches that point in just over nine minutes. If
the heterospecific enzyme were simply more active but formed at the same
rate as the D. melanogaster or D. virilis enzymes, it should reach the
"tailing off" point at about twenty minutes since the concentration of
subunits in the mixture is equal to the mean of the D. melanogaster and
D. virilis subunit concentrations alone. This evidence is still not
conclusive since this measures only the forward reaction or dimer formation whereas, when the activities of the reassociated enzymes in the
gel are measured, they probably represent an equilibrium between the
formation and dissociation of the enzymes. The reverse reaction may

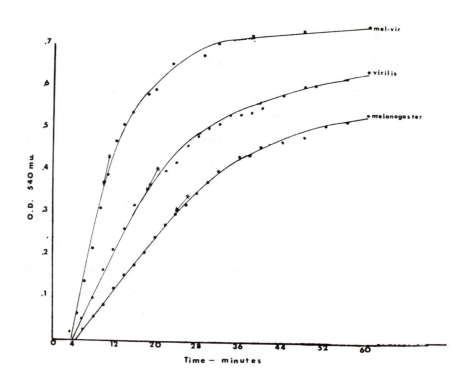

Figure 7:   Increase with time of acid phosphatase activity of reasso-

ciating subunits of D. melanogaster, D. virilis and a 1:1

mixture of the two (mel-vir). See text for a description of

the experiments. The arrow indicates a decrease in the

initial, linear rate of reactivation in each test.

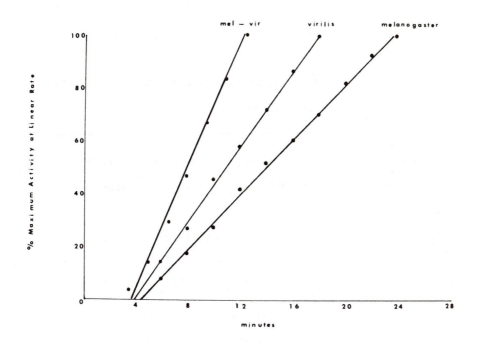

Figure 8: The initial, linear increases in acid phosphatase activity plotted against time. In each test the acid phosphatase activity attained when the rate began to decrease was set at 100%. See arrows on Figure 7.

also be important in determining the final ratio. We hope to either
obtain pure proteins or to estimate their amounts immunologically in the
near future. Nevertheless, we feel the available evidence favors the
interpretation that the test measures differences in subunit affinities
and not specific activities.

This method of measuring variation, like all other except amino
acid analysis, does not reveal how many and what kind of amino acid substitutions are responsible for the observed differences. It is obvious
that the hemoglobins offer a unique opportunity to correlate specific
amino acid substitutions with their affect on subunit affinities in heterospecific dimer or tetramer formation.

Nevertheless, the differences seen in this study of acid phospha-
tase-1 can be rationalized on a model, which itself can be experimentally
tested. The differences may mean that each species has a selectively
determined, or physiologically important subunit-dimer equilibrium.
Thus, under certain conditions, e.g. high phosphate concentrations,
the equilibrium may shift toward the enzymatically inactive subunits.
There is precedent for such a regulation mechanism[33]. Each species'
enzyme may have evolved in such a way that delicate adjustments in this
equilibrium might have been accomplished by the substitution of amino
acids which affected subunit affinities. If the kind of intersubunit
bonds rather than some absolute number of them is critical in this
equilibrium, then the frequent examples of apparent excessive heterospecific enzyme formation are perhaps not so surprising. Consider the
D. melanogaster x D. virilis test. As diagrammed in Figure 9, if the
subunits of each species were justapositioned so that a balance between
attracting and repelling forces exists, and,if in each species oppositely
charged residues represented the repelling forces, then the stronger
ionic bonds forming between the heterospecific subunit combination would

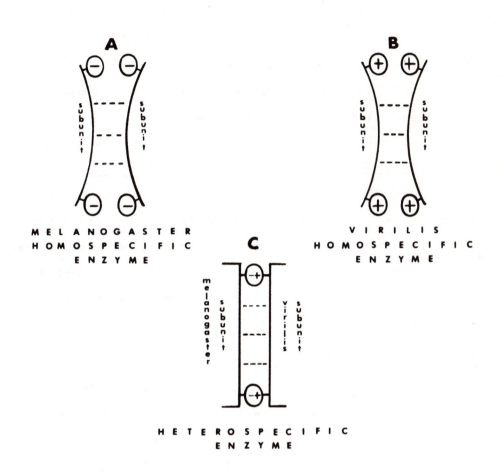

Figure 9:   Hypothetical scheme of intersubunit bonds between the
            homospecific enzymes (A = D. melanogaster, B = D. virilis)
            and the heterospecific enzyme (C) formed in the reassocia-
            tion of D. melanogaster & D. virilis subunits. (+) and (-)
            are positively and negatively charged side chains of amino
            acids. ---- indicates weaker hydrogen or hydrophobic bonds
            between subunits.

explain the results of that test.

Most importantly, I think we have a reasonable chance of detecting a changeable subunit-dimer equilibrium in the acid phosphatase-1 system either during development or under different media conditions. I have already mentioned that free subunits can reassociate and form active enzymes in gels after electrophoresis. It may be even more effective to "rescue" subunits of D. melanogaster by incubating extracts with those "sticky" dissociated subunits from a preparation of D. virilis enzyme.

If an equilibrium can be detected and measured and experimentally altered, then we can see if the frequencies of certain alleles which make subunits that alter this equilibrium can be correlated with environmental conditions that would make such an alteration selectively advantageous. While admittedly fanciful, the model does at least lead us away from pure description.

CONCLUSION

A method has been described which detects differences between homologous enzymes of closely related species and even between the same enzyme from different geographic populations within a single species. These differences, because they will affect either the subunit-multimer equilibrium or the substrate turnover number of the enzyme, should not be selectively neutral.

In addition, these findings imply that evolution involves meaningful, i.e. functionally important, changes in the product of a structural gene, and not just in the controlling or regulatory genes of an organism.

ACKNOWLEDGMENTS

I want to thank Mrs. Margaret Dean for expert technical assistance, my colleagues Bruce Wallace and Richard Hallberg for many helpful discussions, and finally Dr. Clement Markert, whose comment at the 1970 Isozyme Meetings redirected my thinking toward the question of the evolutionary significance of structural gene variation.

REFERENCES

1. Hubby, J. and Lewontin, R. Genetics 54, 577 (1966).

2. Lewontin, R. and Hubby, J. Genetics 54, 595 (1966).

3. O'Brien, S. and MacIntyre, R. American Naturalist 103, 97 (1969).

4. Bryson, V. and Vogel, H. J. Evolving Genes and Proteins, Academic Press (1965).

5. Duke, E. and Glassman, E. Genetics 58, 101 (1968).

6. Doane, W. Problems in Biology:RNA in Development, E.W. Hanly Ed. p. 73, (1969).

7. Hubby, J. and Throckmorton, L. American Naturalist 102, 193 (1968).

8. Ayala, F. J., Mourao, C. A., Richmond, R. and Dobzhansky, Th. Proc. Nat'l. Acad. Sci. 67, 225 (1970).

9. Prakash, S. Proc. Nat'l. Acad. Sci. 62, 778 (1969).

10. Milkman, R. Genetics 55, 493 (1967).

11. Sved, J. A., Reed, T. E. and Bodmer, W. F. Genetics 55, 469 (1967).

12. King, J. Genetics 55, 483 (1967).

13. Kojima, K. and Yarborough, K. Proc. Nat'l. Acad. Sci. 57, 645 (1967).

14. Wills, C., Crenshaw, J. and Vitale, J Genetics 64, 107 (1970).

15. Franklin, I. and Lewontin, R. Genetics 65, 707 (1970).

16. Mukai, T. Genetics 61, 479 (1969).

17. Haldane, J. B. S. Jour. Genetics 55, 511 (1957).

18. Kimura, M. Nature 217, 624 (1968).

19. King, J. L. and Jukes, T. H. Science 164, 788 (1969).

20. MacIntyre, R. Genetics 53, 461 (1966).

21. Markert, C. L. Annals N.Y. Acad. Sci. 151, 14 (1968).

22. Perutz, M. F., Murhead, H., Cox, J. M. and Goaman, L. C. G.
     Nature 219, 131 (1968).

23. Dayhoff, M. Atlas of Protein Sequence and Structure, Vol. 4
     National Biomedical Research Foundation (1969).

24. Perutz, M. F. and Lehmann, H. Nature 219, 902 (1968).

25. Cook, R. A. and Koshland, D. E. Proc. Nat'l Acad. Sci 64, 247 (1969).

26. Noble, R. W. Jour. Molecular Biology 39, 479 (1969).

27. Haber, J. E. and Koshland, D. Proc. Nat'l. Acad. Sci. 58, 2087
     (1967).

28. MacIntyre, R. Biochemical Genetics 5, 45 (1971).

29. MacIntyre, R. Genetics, August 1971. In Press.

30. MacIntyre, R. & Dean, M. Nature 214, 274 (1967).

31. Klotz, I. and Darnall, D. W. Science 166, 126 (1969).

32. Bell, J. and MacIntyre, R. Isozyme Bull. 4, 19 (1971).

33. Iwatzuki, N. and Okazaki, R. J. Moleuclar Biol. 29, 139 (1967).

34. Johnson, F. M., Kanapi, C. G., Richardson, R. H., Wheeler, M. R. and
     Stone, W. S. Proc. Nat'l. Acad. Sci. 56, 119 (1966).

35. Harris, H. Proc. Roy. Soc., B, 164, 298 (1966).

36. Lewontin, R. C. Amer. J. Hum. Genetics 19, 681 (1967).

37. Stone, W. S., M. W. Wheeler, F. M. Johnson and Kojima, K. Proc.
     Nat'l. Acad. Sci. 59, 102 (1968).

38. Prakash, S., Lewontin, R. C. and Hubby J. Genetics 61, 841 (1969).

39. Selander, R. and Yang, S. Y. Genetics 63, 653 (1969).

40. Selander, R., Lewontin, R. and Johnson, W. E. Evolution 24, 402 (1970).

41. Rockwood, E. Univ. Texas Publ. Studies in Genetics 5, 111 (1969).

DISCUSSION

SCANDALIOS: How meaningful is the "lack of differences" in the Km's when you really don't know the endogenous substrate for the enzyme?

MAC INTYRE: It is true that we don't know what the physiological substrates of this enzyme are at the present time. However, in all the experiments I have described, a single substrate, α-naphthyl phosphate was used. The issue, then, about whether the differences we find are due to different subunit affinities or different specific activities must be resolved with the use of α-naphthyl phosphate. Thus, the fact that the Km's of the three enzymes which form during the reassociation of D. melanogaster and D. virilis subunits are the same when α-naphthyl phosphate is the substrate represents rather strong indirect evidence that we are, in fact, detecting differences in subunit affinities.

SCANDALIOS: Have you checked the heat denaturation of the enzyme from different species?

MAC INTYRE: No, we have not compared the enzymes from the different species with regard to heat denaturation.

DOES THE FIXATION OF NEUTRAL MUTATIONS FORM A SIGNIFICANT PART OF OBSERVED EVOLUTION IN PROTEINS?

Walter M. Fitch, Department of Physiological Chemistry
University of Wisconsin Medical School
Madison, Wisconsin, 53706

INTRODUCTION

Kimura (1) noted a discrepancy between the apparent rate at which amino acids have been substituted in the evolution of proteins and the rate allowable according to the cost of such natural selection when calculated by the procedure of Haldane (2). Kimura noted that the discrepancy would disappear if there were essentially no advantage conferred by the new allele during the period when it was going to fixation. Such as allele is said to be selectively neutral and the mutation it carries is said to be a neutral mutation. This immediately raises the issue, apart from the discrepancy it was designed to resolve, whether neutral mutations do in fact exist. The issue has been vigorously joined. The most vigorous proponents have been King and Jukes (3), Jukes and King (4), King (5) as well as Corbin and Uzzel (6), while the major opponents have been Maynard-Smith (7), O'Donald (8), Clarke (9, 10), and Richmond (11). Twenty minutes will not do justice to any of these people so I shall take a different tack.

A major, if theological, objection to the idea of the fixation of neutral mutations is that it removes the moral necessity of a biologist to explain what brought about the observed evolutionary change, for in the same way that Heisenberg's uncertainty principle permitted free will to survive in the face of strict determinism, so here, an act of God,

186

however stochastic, intrudes upon the strict selectionist. I have been asked to present something of the arguments of the proponents of neutralism, but in view of their apostate status, I shall reverse the preceding metaphor and regard myself as a devil's advocate and let Dr. Milkman be the avenging angel with his terrible swift sword. I cannot resist, however, pointing out that sword is merely an anagram for words.

To deny the existence of any neutral mutations is to argue that a nucleotide change that does not alter the amino acid encoded must, *in every conceivable case*, be subject to selective processes. This extremity is, in my opinion, untenable. But if some mutations are neutral, it is also untenable to deny that chance will determine their fate.

Granting, however, that neutral mutations exist, it does not necessarily follow that such Non-Darwinian events are a significant part of the evolutionary changes we observe, that is, they may be real, but irrelevant. The question thus phrased is quantitative, not qualitative. If they are observable anywhere in data presently available, it is more likely that they would be seen in significant numbers in the structural changes in proteins rather than in gross morphological changes. I shall therefore restrict my remarks to protein evolution and present a few data that suggest that neutral mutations may indeed form a significant fraction of the evolutionary changes observed in proteins.

THE IMPERMANCE OF CODON VARIABILITY

Three years ago, after my talk to this symposium (12), Wyckoff (13) presented new information obtained by examining the sequence of rat ribonuclease (14) in the context of the three-dimensional structure of bovine ribonuclease which he, Richards, and their collaborators had recently determined (15). The impressive observations were that no alteration of the position of the backbone of the peptide chain was necessary to accommodate the rat sequence and that differences in their sequence frequently occurred as spatially related pairs of substitutions. Three of these are shown in figure 1. Dr. Wyckoff challenged me to explain the significance of such paired replacements. Three years later I have returned, not to answer the question, but rather to restate Dr. Wyckoff's remarks in a different context.

  Figure 2 shows the context of the bottom diagram of figure 1, namely that the methyl group of methionine$_{79}$ of the cow occupies the same part of a thoroughly filled space as that of isoleucine$_{57}$ of the rat. If we assume that these sequences are orthologous, that is, are derived from a common ancestral RNAse gene in the common ancestor of rats and cows, then presumably there was, at some point in time, a transition state form of the enzyme when one of the paired replacements had occurred but not the other. The two possibilities are shown in the middle portion of the lower diagram in figure 1. But if we also assume that a disturbance of the position of the peptide backbone might be deleterious then the simultaneous occurrence of methionine$_{79}$ and isoleucine$_{57}$ would not be allowable. This conclusion is indicated by the X's blocking the pathways to the forbidden intermediate.

  The middle diagram of figure 1 is detailed in figure 3 where the lower molecule represents rat RNAse with the critical lysine$_{41}$ in its proper position in the active site. It is assumed, as depicted in the upper right of figure 3, that if rat glycine$_{38}$ were to mutate to aspartate, the negative charge would, by creating a charge-charge bond, be able to interact with the lysine to pull it out of the active site and thereby render the molecule inactive. Such an aspartate might be rendered harmless by screening the lysine from the deleterious aspartate by interposing another positive charge such as the arginine$_{39}$ shown in the upper left of figure 3. In fact, cow RNAse has the sequence depicted

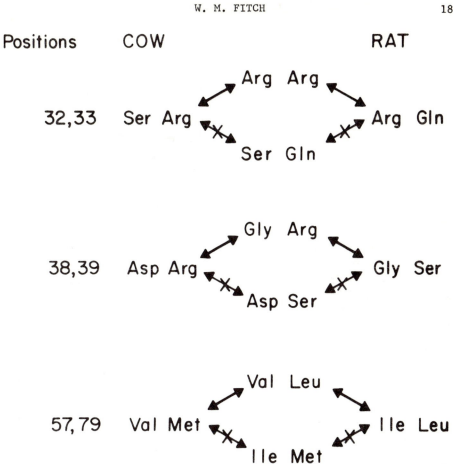

Figure 1. Paired Replacements Between Bovine and Rat Ribonuclease. Pairs of amino acids, spatially related in the three-dimensional structure of pancreatic ribonuclease, are shown for the cow and rat. Each pair differs such that they must differ by two nucleotide replacements in their codons. The change of only one nucleotide at a time would lead to two possible transition states as shown in the middle. The upper transition state was postulated in each case on the basis of arguments given in the text. The amino acid residue positions are shown on the left. Figure from ref. (17).

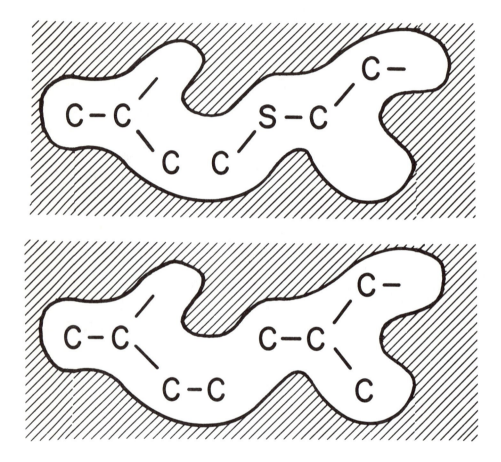

Figure 2.   Spatial Confinement of Residues 57 and 79 of Bovine and
Rat Ribonuclease.   The hatched area depicts space filled by other residues.
The open space is filled by valine and methionine in the bovine RNAse
(upper) and by isoleucine and leucine in the rat RNAse.

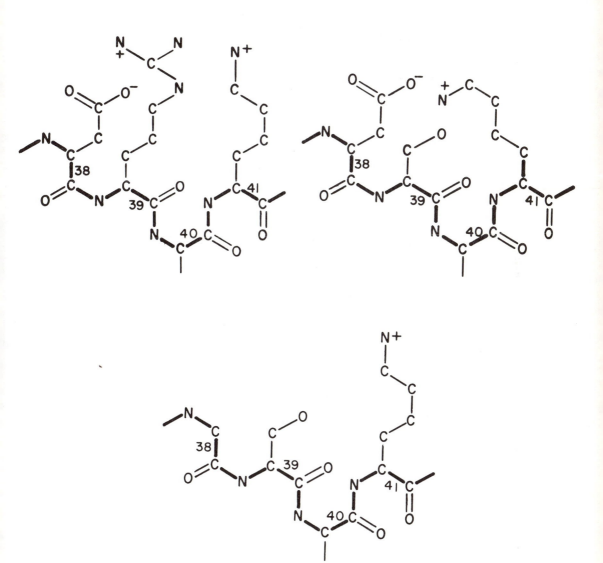

Figure 3.   Factors Affecting Active-Site Lysine$_{41}$ of Ribonuclease.
Molecular representations are only diagrammatic. All hydrogen atoms
are omitted for clarity. The upper left diagram shows bovine RNAse
residues 38-41. The lower diagram is for rat RNAse. The upper right
diagram shows a theoretical intermediate form in which a negatively
charged aspartate pulls the positively charged lysine out of its active
site position. See text for discussion.

in the upper left of figure 3. If our rationalizations have merit, then
the upper right sequence is a forbidden transition state and we must
assume, as shown in the middle portion of figure 1, that $glycyl_{38}$ -
$arginine_{39}$ was the real transition state.

The amino-terminal peptide (S-peptide) is necessary for activity.
It is held in place at least partially by a negatively charged aspar-
$tate_{14}$ bonding to the positively charged $arginine_{33}$ in the cow. As
shown in the upper part of figure 1, position 33 is not arginine in rat
RNAse, but glutamine. Nevertheless, Wyckoff notes that it is structurally possible for the rat arginine in position 32 to locate its
positive charge in the same position that the charge of $arginine_{33}$ is
located. In terms of holding the amino terminal peptide in place, it
may be irrelevant whether the positive charge arises from position 32
or 33 so long as at least one of them is present. But if this is so,
we must conclude that the double arginine may be an acceptable transition state but that the complete absence of a positive charge is not.

Following the formulation of these hypotheses regarding the transition states (16, 17), the sequence of pig RNAse appeared (18) which
afforded a potential opportunity to see whether any evidence regarding
these hypotheses had become available. In positions 57 and 79, the
sequence was identical to that of the cow (19) and therefore provided
no evidence. The sequences of pig RNAse in positions 32-33 and 38-39
are argininyl-arginine and glycyl-arginine respectively, precisely as
predicted for the transition states at these positions.

Although these observations only lend credence to our general
approach, they should, perhaps, sharpen our appreciation of two other
biologically reasonable conclusions. The first, as has since been
shown by a totally independent method (20), is that variable sites are
not always variable but depend upon their structural context. For ex-

ample, so long as aspartate$_{38}$ is present, arginine$_{39}$ can not mutate to lose its charge whereas aspartate can. Once aspartate has mutated to glycine, the arginine with its charge is no longer necessary and becomes free to change to serine. But once the arginine becomes serine, glycine could not mutate to aspartate without deleterious consequences. Thus fixing a mutation at one site may permit previously invariable positions to become variable. Conversely, positions capable of fixing a mutation may become invariable as a result of fixations elsewhere. The second biologically reasonable conclusion is that selection is everywhere circumscribing the possible. Nevertheless, it may be noticed that our explanations were directed toward avoiding the deleterious and at no point was the nature of any advantage revealed to us.

HOW MANY VARIABLE POSITIONS ARE THERE?

Since we have already seen that positions may be variable for only part of their evolutionary existence, the question of their number needs to be defined more clearly. Providing proper care is taken to recognize the biological complexities, it has already been shown that the addition of poisson distributions can provide an excellent fit to the mutation data and reveal the number of invariable residue positions in a gene examined over a range of species (17). By difference the remainder are variable. It should be clear that the variable and the invariable are not the same as the varied and the unvaried for a position may by variable but unvaried. Of course the counting of the variable positions over a range of species lumps together those that are variable in each species at any time during their evolution since their common ancestor. To know how many codons are variable at any one point in time in any one species requires us to narrow the range of species down to one, a process accomplished by extrapolation. This has previously been done

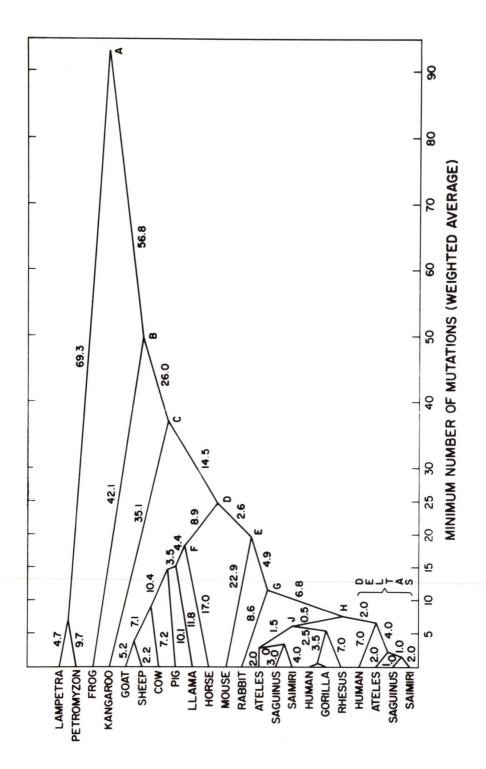

Figure 4.   Phylogeny of Genes for Twenty-two Beta and Delta Hemoglobins.   All sequences represented are beta hemoglobins except the
four delta hemoglobins at the bottom and the two lamprey hemoglobins
at the top.   The lamprey sequences are included since they are orthologous to the beta hemoglobins because this line diverged prior to
the gene duplication that lead to the alpha-beta hemoglobins.   The
numbers on each line give the expected number of mutations in the indicated interval.   They are not integers because there is more than one
most parsimonious way of accounting for the present-day sequences from a
common ancestor and each most parsimonious solution has been weighted
according to the procedure of Fitch (26).   The total mutations in each
most parsimonious solution is 436.   The hemoglobin sequences are as
published by the following authors (listed from top to bottom of figure):
Lampetra fluviatilis (28); *Petromyzon marinus* (29); *Rana* (30); *Macropus*
Giganteus (31);   Goat-A (32);   Sheep-B (33); *Bos*-B (34);   Llama (35);
Equus (36);   Mouse (37); *Oryctolagus* (38); *Ateles* (39); *Saguinus* (39);
Saimiri (39); *Homo sapiens* (23); *Gorilla* (40); *Macaca* (41);   *Homo*-δ (42);
Ateles-δ (39);   *Saguinus*-δ (39); and *Saimiri*-δ (39).   Where only compo-
sitions are known the order proposed by Dayhoff (24) or Boyer *et al.*
(39) was used except as expressly reformulated in table 1.   Each branch
point is located on the abscissa at the weighted average of all mutations
in the lines descending therefrom.

for cytochrome c (17), fibrinopeptide A(17), and alpha hemoglobin (21).
Today I shall present similar data for the beta-delta group of hemoglobins.

In figure 4 is shown the phylogeny of the species whose hemoglobins
were used. The sequences were not used to obtain the phylogeny except
that the order in which the pig and llama join the artiodactyls and the
order in which rabbit and mouse join to the primates was determined so
as to minimize the total number of mutations required. It is most
inappropriate to utilize these sequences to determine the phylogeny
because the only completely sequenced beta hemoglobins are the human,
rhesus, rabbit, kangaroo, frog and the two lamprey's. The horse, sheep
and cow are each more than two-thirds sequenced. The remainder are
largely known only by composition and their sequence is postulated by
maximizing their congruence to the known sequences, a procedure that
frequently requires assumptions as to which of several known sequences
a third compositional sequence is most closely related. Thus, to use
the sequences to determine the phylogeny is to become enmeshed in circular reasoning.

The sequences used are as given in the references in the legend to
figure 4 except for those rearrangements shown in table 1. The changes
permitted residues 10, 111 and 114 to be identical in all species and
reduced the range of variability in a number of other positions. The
numbering is as for mammaliam beta hemoglobins. To maximize the homology with the lamprey and alpha hemoglobins, a two residue gap must be
inserted between residues 18 and 19. It should now be clear, as suggested
earlier (22), that the placement of the gap between residues 19 and 20
as originally proposed by Braunitzer et $al.$ (23) and perpetuated by
Dayhoff (24, alignment 11) is incorrect. The utilization of the lamprey
sequences is not intended to imply that they are really beta hemoglobins.
Their inclusion is justifiable because they are orthologous to the beta

TABLE 1

CHANGES IN BETA HEMOGLOBIN SEQUENCES

	2	7	9	10	12	13	16	110	111	113	114	121	123	125	129
Saguinus	(His	(Glu)	(Ser	Ala	Thr	Thr	Gly)	(Leu)	Val	Val	Leu)	(Glu)	Thr	Gln	Ala
Mouse	(His	Ala)	(Ala	Ala	Ser	Gly	Gly)	(Ile)	Val	Leu	Leu)	(Asx	Thr	Ala	Ala
Horse	Gln	Glx	Ala	Ala	Leu	Ala	Asx	(Leu	Val	Ala	Leu	Asx	Thr	Glx	Ser
Pig	(His	Glx)	(Gly	Ala	Leu	Gly	Glx)	(Ile	Val	Val	Leu)	(Asx	Ala	Val	Asx)
Llama	(Asx	(Glx)	(Ser	Ala	His	Gly	Asx)	(Leu	Val	Val	Leu)	(Glx)	Thr	Ala	Asx)
Known	His_4	Glu_6	Ala_3	Ala_8	Thr_6	Ala_5	Gly_6	Leu_3	Val_4	Val_3	Leu_4	Glu_5	Thr_6	Gln_2	Ala_4
	Gln	Glx	Ala_3		Ser	Thr	Asx	Ile		Cys		Asx		Val	Ser
			Asn_2		Asn	Gly	Ser							Glx	
			Thr			Ser								Asp	
														Pro	

The parentheses denote the limits of individual tryptic peptides. Numbers are residue positions. Underlined amino acids are those that do not follow the alignments of Dayhoff (24) or Boyer *et al.* (39). The bottom portion shows the frequency of occurence of amino acids known from sequenced hemoglobins to occupy that position.

hemoglobins, that is, the ancestry of these genes has a one-to-one
correspondence to the ancestry of the species in which they occur. This
may be contrasted to the beta and the delta hemoglobins which are
orthologous in themselves but paralogous to each other as their homology
derives from a gene duplication and thus these two genes have evolved
together in parallel in the same individuals and would be expected to
give congruent phylogenies for their separate genes. Indeed, a most
parsimonious solution (*i.e.,* a phylogeny requiring the fewest mutations)
for the combined beta and delta hemoglobins of man, *Ateles, Saguinus* and
Saimiri gives precisely the same ancestral relationships among these four
 genera for both gene products. These relationships are preserved in
figure 4. Overall, there is an excellent agreement between this tree and
that of figures 1 and 2 in Barnabas *et al.* (25) except that their four
delta sequences do not cluster as shown here. The result is that Barnabas
et al. postulate that the delta hemoglobins of the ceboid and catarhine
lines arose independently on two separate occasions, whereas my figure 4
requires only one gene duplication. The same number of mutations
(nucleotide replacements) are required in the primate portion of the
tree regardless of whether the Barnabas *et al.* or my tree is accepted.
Overall, my tree requires 436 mutations. The Barnabas *et al.* tree,
restricted to the sequences I used, requires 438 mutations. This minor
savings arises solely from the pig-llama reversal.

Given the amino acid sequences, and the phylogeny shown in figure 4,
there are procedures for determining those ancestral sequences at each
ancestral node that will account for the evolution of extant sequences
from their common ancestor in a minimum number of mutations (26). In
the process, one obtains the number of mutations necessarily fixed in
each codon during the descent from any given ancestor. These are the
data required to utilize the method of Fitch and Markowitz (17) to

compute the number of variable and invariable codons for any range of
species. The ranges are denoted by the peak height of the common ancestor and were computed for those ancestors in figure 4 labelled with
capital letters. When the number of invariable codons is expressed as a
fraction of the total number of codons in the gene and plotted against the
range of species, the result shown in figure 5 is obtained.

Extrapolation of the line to the ordinate indicates that, at any one
point in time and in any one mammalian species approximately 75% of the
beta hemoglobin gene is invariable. Since the complete matching of all
these sequences requires a total of 157 codons, there are approximately
39 codons, on the average, that in a given gene are capable of fixing
mutations at any one time. Such codons are called *concomitantly
variable codons* or covarions for short. To say that mammalian beta
hemoglobin has about 39 covarions is to indicate that mutations at most
positions will be malefic, *i.e.*, either lethal or sufficiently deleterious
as to assure extinction of such alleles except by founder effects. The
mutations that may survive can occur only in the covarions.

                          RATES OF EVOLUTION

Comparing rates of evolution among several proteins can be somewhat
hazardous for several reasons, one of which is the uncertainty in the
paleontological dating of times of divergence. One way of avoiding this
difficulty is to make the comparisons in such a way that the common
ancestor is the same for every gene so that the relative rates are
correct even though the absolute rates are imprecise. For example, the
sequences of cytochrome *c*, fibrinopeptide A, and the alpha and beta
hemoglobins are known for both pig and horse and by our procedures we
can determine the number of mutations that must have been fixed in each
gene since the time of their common ancestor. In table 2 is given, for

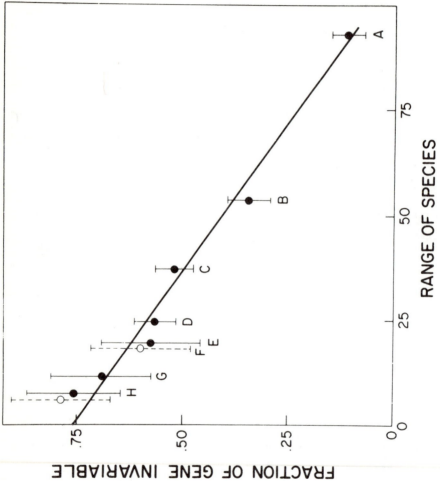

Figure 5.   Fraction of Gene that is Invariable as a Function of the Range of Species. Each point is determined by the method of Fitch and Markowitz (17) using the data from that part of the phylogeny in figure 4 descending from the ancestor with the same capital letter as the point in this figure. Range is defined by that point's abscissa value in figure 4. All points require the assumption of two poisson-distributed classes of variability except for points F and J (open circles) for which there was insufficient data to fit more than one poisson-distributed class of variability. For these two points, the two poisson-distributed result was estimated using the relationship between the two models shown in figure 6. Vertical bars show the standard deviation. For the open-circle points, the standard deviation is arbitrarily taken to be the largest of the known standard deviations among the closed circles. The line is a weighted least-squares fit with each point's weight inversely proportional to its variance. The equation for the line is y = -0.00700x + 0.755. The standard error of the intercept is 0.022.

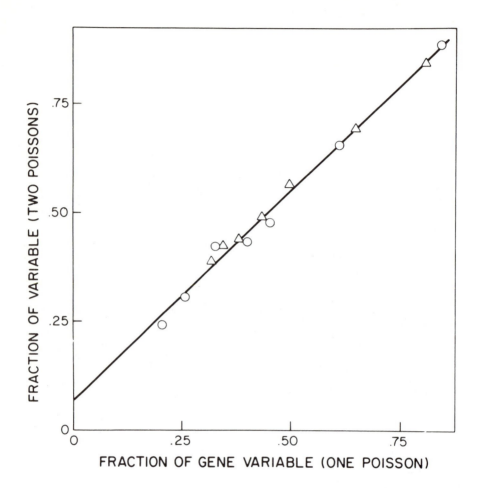

Figure 6.   Relationship Between the Fraction of Gene That is Variable
as a Function of Assuming One and Two Poisson-Distributed Classes of
Variability. Points are data from (∆) alpha hemoglobins (21) and (o)
beta-delta hemoglobins for those cases where the data were extensive
enough to fit both models. Line is an unweighted least-squares fit
with slope 0.970 and intercept 0.069.

TABLE 11

COMPARATIVE EVOLUTIONARY RATES

Protein	Fixations	Codons	Rate$_1$	Covarions	Rate$_2$
Cytochrome c	5	104	0.048	10	.50
Alpha Hemoglobin	22	141	0.156	50	.44
Beta Hemoglobin	31	146	0.212	39	.80
Fibrinopeptide A	13	19	0.684	18	.72

"Fixations" are the number of nucleotide replacements occuring in the indicated gene in both lines of descent since the common ancestor of the horse and the pig. "Codons" is simply the length of the sequence. "Rate$_1$" is the rate of fixation/codon since the divergence of horse and pig. "Covarions" are the concomitantly variable codons. "Rate$_2$" is the rate of fixation/covarion.

each of these four genes, the number of mutations fixed and the number of codons in the gene to give rate$_1$. It is clear that some genes are evolving faster than others. However, we have seen in the preceding sections several kinds of evidence that say that some positions are invariable. It is not logical to compute evolutionary rates on the basis of codons that can not evolve. Rates of fixation should be on the basis of the codons where mutations can be fixed. But the codons capable of fixing mutations are the covarions. The number of covarions in each of these genes is shown in table 2 along with the evolutionary rate on a per-covarion basis. Considering the limitations on the accuracy of determining the number of covarions, it may be concluded that these four gene products are evolving at approximately equal rates. One should also recognize that in the case of beta hemoglobin, the very process of ordering the unsequenced amino acids to optimize homology introduces a bias that maximizes the invariable and minimizes the variable positions so that 39 covarions for beta hemoglobin may be a little low and hence its evolutionary rate a little high.

Barnabas *et al.* (25) have suggested that evolution has proceeded more slowly in the primate than in the other mammalian lines so far as alpha and beta hemoglobins are concerned. No evidence for this differential rate was observed when alpha chains were analyzed in the same fashion as the beta chain sequences of this paper (21). However the data in figure 4 do support the contention of Barnabas *et al.* with respect to the alpha chain. The rate of accumulation of mutations in the ungulates is about 50% greater than among the primates since their common placental ancestor at D. If there was sufficient data to separate out the two lines of descent in the plot shown in figure 5, we might find that the number of covarions among the ungulates was considerably greater than among the primates toward which these computations are biased. It may be noted that point F falls below the best fitting line whereas the primate points H and J fall above the line. Since table 2 computes its evolutionary rate per covarion on the basis of two ungulates, the rate for the beta gene might well be brought more into line with the alpha hemoglobin and cytochrome *c* rates. Also clear from figure 4, in agreement with the suggestion of Barnabas *et al.* (25), is the more rapid evolution of the delta gene than that of the beta from which it arose.

NEUTRAL MUTATIONS AND RATES OF EVOLUTION

Up to this point, all the data presented are completely independent of any concepts of neutral mutations and can be accomodated quite nicely in a selectionist view. There is however a neutralist view that can be put to a test by these data. Kimura (1) has shown that the rate at which neutral mutations (if they exist) will be fixed is proportional to

the rate at which they occur. We can not easily know this rate but if
we could restrict our consideration to those codons at which neutral mutations could occur, the evolutionary rate per such "neutral" codon
should be the same. Furthermore, Kimura (1) has asserted that the over-
whelming majority of the amino acid replacements observed in protein
evolution were selectively neutral. But if that is true, then Kimura
must accept my determination of the number of covarions as a valid
estimate of the number of codons at which neutral mutations may occur.
Moreover, he must then accept that evolutionary rates on a per-covarion
basis should be the same in different genes. The data in table 2 fail
to disprove Kimura as they would have if they had been significantly
different. Indeed, they tend to support him for while the evolutionary
rates of two genes might by chance be similar, the likelihood that many
genes would have essentially the same evolutionary rate is low. Certainly
four genes should be sufficient to convince anyone that, at the very
least, the proposition that neutral mutations may account for a significant part of the protein evolution we observe deserves serious
consideration.

DOUBLE MUTATIONS

    Double mutations are defined as two observable nucleotide replace-
ments fixed in a single codon in the interval between two successive
nodes of a phylogenetic tree. Three years ago in this symposium I presented a stochastic model of evolution that mimicked, in every detail
measured but one, the evolution of cytochrome c (12). That one exception was that double mutations occurred far less frequently in the model
than in the actual evolution of cytochrome c. The reason for the discrepancy lay in the model's assumption that more than 75% of the cytochrome c codons were variable. We have seen that it is more like 10%.

Obviously, the probability that the next mutation will be fixed in the
same codon as fixed the last one is inversely proportional to the number
of covarions. Thus with cytochrome c having only 10 covarions, we
readily understand why the model evolution with 76 variable codons grossly
underestimated the number of double mutations actually observed.

We have also just seen that covarions do not remain forever variable,
so that a second factor in determining the rate at which double mutations
will occur is length of time, measured in fixations, that a covarion
will remain variable. We may define persistence of variability as the
probability that a mutation fixed elsewhere will not remove a given
codon from its covarion status. A formula for determining the expected
number of double mutations, given the number of covarions (c), the persistence of variability (v), and the number of mutations (m) fixed has
been developed (27).

Figure 7 shows the phylogeny of 20 cytochromes c shown at this symposium three years ago except that after the number of fixations shown
on each internodal leg is a slash followed by a number representing the
number of double mutations in that interval (12). For example the line
descending to the tuna shows 17 fixations and 1 double mutation. The
line descending to the snake shows 17 fixations and 4 double mutations.
Since no other line has 17 fixations, our observation is that on the
average there are 2.5 double mutations, given that 17 mutations were
fixed and this point and all others derivable from the figure are
plotted as X's in figure 8. It is our task, then, to discover values
of c and v that will best approximate the observed data. Plotted in
figure 8 are lines showing the values one would expect if the number of
covarions (c) were five for various values of the persistence of variability (v). For v = 1, the same five codons are always variable so
that the maximum number of codons that can get double mutations is five

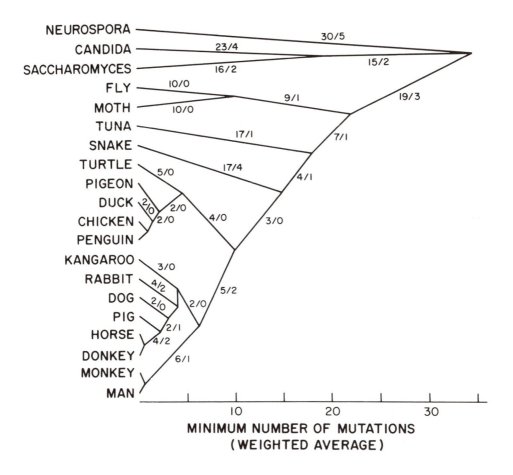

Figure 7.   Phylogeny for Twenty Cytochromes c.   The phylogeny is as previously given (12).   The number preceding the slash is the number of mutations fixed in the interval.   The number following the slash is the number of codons that had two of their nucleotides mutated in that interval.   Figure from ref. (27).

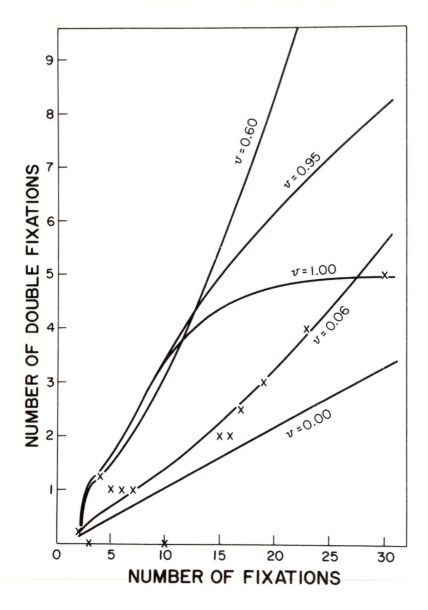

Figure 8.   Double Fixations as a Function of Number of Mutations Fixed.
Data are from figure 7, the X's being the observed number of double fixations averaged over all lines having a given number of mutations fixed
in their interval. The lines show the expected distribution of points
for five covarions and different values for the persistence of variability. See text for explanation. Figure from ref. (27).

and the curve for v = 1 therefore approaches the value 5 asymtotically.

For v = 0, a covarion loses its variability as soon as a fixation occurs

elsewhere so that double mutations only occur as two immediately succes-

sive fixations in the same covarion.

To determine which curve best fits the data, I have used the chi-

square statistic which involves the square of the vertical distance

between an observed point and the theoretical line. The lower the chi-

square value, the better the fit. Figure 8 shows only 5 curves for one

value of c and each curve will provide a chi-square value. Not only are

other values of v allowable, but other values of c are also possible as

well. In figure 9 are shown the chi-square values for about 70 such

curves. Other values were also determined but the ones shown clearly

define the region of the minimum chi-square for the various numbers of

covarions that might be postulated. When the minima are plotted as a

function of the number of covarions, the result is as seen in figure 10.

The minimum appears about c = 4.5 and a special computation at c = 4.5,

v = 0.04 is shown by the star in figure 9. Since chi-square is least

for these values, our best estimate for cytochrome c by the double mu-

tation data is that there are, on the average, about 4.5 covarions whose

persistence of variability is only 0.04. The probability of a chance

fit as good as that obtained is less than 1%. The accuracy of these

values is not great however since the minimum chi-square values are all

quite low in the range of c = 3→10. Thus our previous estimate that

there are 10 covarions in cytochrome c is not contradicted by the pre-

sent estimate of 4.5. The difficulty arises because changes in c can

be compensated by changes in v. If we knew either c or v accurately

from alternative data, the other variable could be determined fairly

accurately.

If we accept from the previous data the number of covarions as 10,

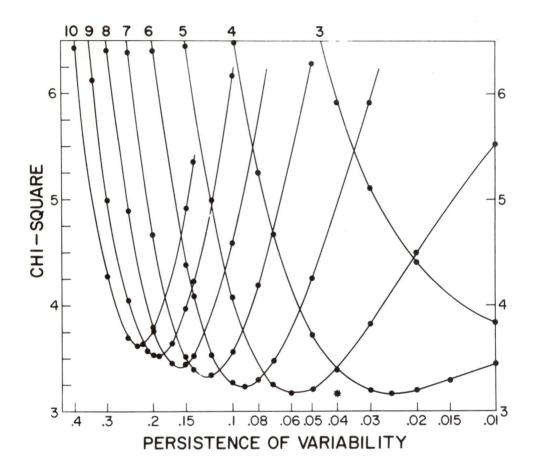

Figure 9.   Goodness of Fit of Double Fixation Data as a Function of
the Number of Covarions and Their Persistence of Variability.   Each line
connects points that assume the same number of covarions (top).   Chi-square
is the measure of goodness of fit (lower is better).   Values less than
3.57 (12 degrees of freedom) imply that fits this good would occur by
chance less than one percent of the time.   The * denotes the best fit
found (c = 4.5, v = 0.04).   Figure from ref. (27).

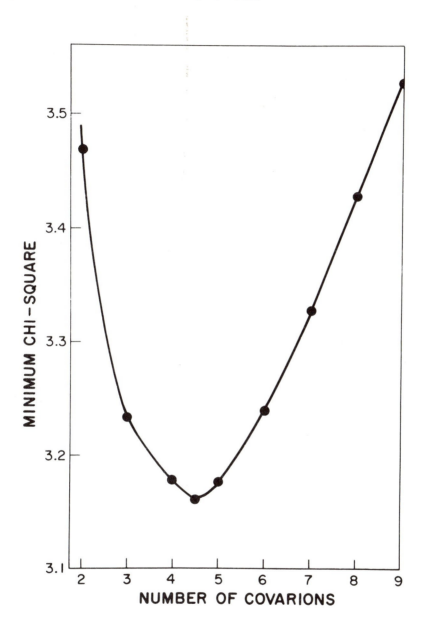

Figure 10.   Least Chi-Square as a Function of the Number of Covarions.
Each point is the lowest chi-square obtained for a given number of
covarions.   Curve shows that the minimum of the minima occurs at 4.5
covarions.   Figure from ref. (27).

then the persistence of variability is 0.25. Values of c which better
fit the data all lead to lower values of v so that we may conclude that
cytochrome c covarions are not likely to retain their variability very
long which leads in turn to the conclusion that most double mutations
are immediately successive fixations.

The problem of explaining two immediately successive fixations in
the terms of a strict selectionist theory in which every mutation fixed
confers an advantage is interesting in terms of these results. Consider
the replacement of proline (CCX) by valine (GUX) which occurs at position
44 of the rabbit. The intermediate possibilities are alanine (GCX) or
leucine (CUX). The strict selectionist asserts that either alanine or
leucine must have been superior to proline and replaced it and that it
in turn was replaced by a superior valine. This leaves us with an
apparent contradiction. On the one hand, the genetic code seems to
be fashioned so that single nucleotide replacements minimize deleterious
changes and maximize the possibilities of advantageous changes. On the
other hand, the optimum amino acid is very frequently two nucleotide
replacements away and one of the two intervening amino acids has an
intermediate fitness. In the present data there are 223 mutations of
which 64 (32 doubles) are involved in this particular form of change.
Thus 29% of the mutations fixed were to get to a more fit amino acid
two nucleotide replacements away from that originally encoded. Does
this really square with the idea that cytochrome c is highly evolved
and tolerates little change? And what we see can only be those potential improvements for which an intervening amino acid has intermediate
fitness. Surely there must be others for which the intervening amino
acids are deleterious in which case "you just can't get there from here".
But if these latter cases that we can't observe number anywhere near as
many as those that we have been able to observe in the former, then the

fraction of amino acid substitutions that confer an advantage but require two nucleotide replacements becomes unreasonable.

In conclusion, the relative equality of the rate of evolution per covarion in 4 different genes, the difficulty in explaining the high frequencies of double mutations in strict selectionist terms plus some of the earlier arguments (1, 3-6) we have not had time to consider, drives me to conclude that non-Darwinian evolution could well prove to be a major contributor to the evolution observable among proteins.

ACKNOWLEDGEMENTS

This project received support from National Science Foundation grant BG-7486. The University of Wisconsin Computing Center, whose facilities were used, also receives support from NSF and other government agencies.

REFERENCES

1.   Kimura, M., Nature 217, 624 (1968).

2.   Haldane, J. B. S., J. Genet. 55, 11 (1957).

3.   King, J. L. and Jukes, T. H., Science 164, 788 (1969).

4.   Jukes, T. H. and King, J. L., Nature 231, 144 (1971).

5.   King, J. L., Sixth Berkely Symposium on Mathematical Statistics and Probability -- Conference on Evolution, April 1971, in press.

6.   Corbin, K. W. and Uzzel, T., Amer. Nat 104, 37 (1970).

7.   Maynard-Smith, J., Nature 219, 1114 (1968).

8.   O'Donald, P., Nature 223, 900 (1969).

9.   Clarke, B., Science 168, 1009 (1970).

10.   Clarke, B., Nature 228, 159 (1970).

11.   Richmond, R. C., Nature 225, 1025 (1970).

12.   Fitch, W. M. and Margoliash, E., Brookhaven Symp. Biol. 21, 217 (1968).

13.   Wyckoff, H. W., Brookhaven Symp. Biol. <u>21</u>, 252 (1968).

14.   Beintema, J. J. and Gruber, M., Biochem, Biophys. Acta <u>147</u>, 612, (1967).

15.   Wyckoff, H. W., Hardman, K. D., Allewell, N. M., Inagami, T., Johnson, L. N. and Richards, F. M., J. Biol. Chem. <u>242</u>, 3984 (1967).

16.   Fitch, W. M. and Margoliash, E., Evolutionary Biol. <u>4</u>, 67 (1970).

17.   Fitch, W. M. and Markowitz, E., Biochem. Gen. <u>4</u>, 579 (1970).

18.   Jackson, R. L. and Hirs, C. H. W., J. Biol. Chem. <u>245</u>, 637 (1970).

19.   Smyth, D. G., Stein, W. H. and Moore, S., J. Biol. Chem. <u>238</u>, 227 (1963).

20.   Fitch, W. M., Biochem. Gen. <u>5</u>, 231 (1971).

21.   Fitch, W. M., Haematologie and Bluttransfusion, J. F. Lehmanns Verlag, Munich, in press.

22.   Fitch, W. M., J. Mol. Biol. <u>16</u>, 17 (1966).

23.   Braunitzer, G., Gehring-Muller, R., Hilschmann, N., Hilse, K., Hobom, G., Rudloff, V., and Wittmann-Liebold, B., Z. physiol. Chem. <u>325</u>, 283 (1961).

24.   Dayhoff, M. O., Atlas of Protein Sequence and Structure, National Biomedical Research Foundation (1969).

25.   Barnabas, J., Goodman, M., and Moore, G. W., J. Comp. Biochem. and Physiol. in press.

26.   Fitch, W. M., Systematic Zoology, in press.

27.   Fitch, W. M., J. Mol. Evol., in press.

28.   Rudloff, V., Zelenick, M. and Braunitzer, G., Z. physiol. Chem. <u>344</u>, 284 (1966).

29.   Li, S. L. and Riggs, A., J. Biol. Chem. <u>245</u>, 6149 (1970).

30.   Chauvet, J. P. and Acher, R., FEBS Letters <u>10</u>, 136 (1970).

31.   Air, G. M. and Thompson, E. O. P., Nature <u>229</u>, 391 (1971).

32.   Huisman, T. H. J., Adams, H. R., Dimmock, M. O., Edwards, W. E. and Wilson, J. B., J. Biol. Chem. <u>242</u>, 2534 (1967).

33.   Boyer, S. H., Hathaway, P., Pascasio, F., Bordley, J., Orton, C. and Naughton, M. O., J. Biol. Chem. <u>242</u>, 2211 (1967).

34.   Schroeder, W. A., Shelton, J. R., Shelton, J. B., Robberson, B. and Babin, D. R., Arch. Biochem. Biophys. <u>120</u>, 124 (1967).

35.   Braunitzer, G., Hilse, K., Rudloff, V. and Hilschmann, N., Adv. in Protein Chemistry, <u>19</u>, 34 (1964).

36.   Smith, D. B., Canad. J. Biochem., 46, 825 (1968).

37.   Rifkin, D. B., Rifkin, M. R. and Konigsberg, W., Proc. Nat. Acad.
       Sci. 55, 586 (1966).

38.   Braunitzer, G., Best, J. S., Flamm, U., and Schrank, B.,
       Z. physiol. Chem. 347, 207 (1966).

39.   Boyer, S. H., Crosby, E. F., Thurman, T. F., Noyers, A. N. Fuller,
       G. F., Leslie, S. E., Shepard, M. K. and Herndon, C. N., Science
       166 1428 (1969).

40.   Zuckerkandl, E. and Schroeder, W. A., Nature 192, 984 (1961).

41.   Matsuda, G., Maita, T., Takei, H., Ota, H., Yamaguchi, M.,
       Miyauchi, T. and Migita, M., J. Biochem 64, 279 (1968).

42.   Ingram, V. M. and Stretton, A. D. W., Biochim. Biophys. Acta. 63,
       20 (1962).

DISCUSSION

O'BRIEN: Would you be willing to give a minimum estimate of percentage of amino acid substitutions which are
selectively neutral?

FITCH: No.

O'BRIEN: What about with the four proteins which you
have discussed?

FITCH: I know of no way to compute the fraction of
neutral replacements in these data. I can only conclude that
my data on evolutionary rates are sufficiently close to each
other, in accordance with the expectations of neutralism, to
make it necessary to take seriously the views of the neutralists as an explanation for a part of the changes observed
among proteins.

WILLS: Don't you think it is somewhat naive to assume
that cytochrome c is such a perfectly adjusted molecule? A

selectionist will say that over three billion years of evolution conditions must have changed many times, so that double substitutions may easily represent one change as a response to a particular environmental intention followed by another change as a response to a new set of conditions.

FITCH: I would not wish to leave the impression that I thought cytochrome c is perfectly adapted. On the other hand, experiments utilizing heterologous elements as divergent as fungi and mammals show that substantial oxidative phosphorylation can nevertheless be obtained. This suggests that cytochrome c was probably highly evolved as early as the common ancestor of fungi and vertebrates. More pertinent, however, to your question is the nature of successiveness of the two nucleotide replacements making up double the mutation. If one accepts the premises in the computations I presented, the conclusion that follows is that regardless of the number of covarions assumed, the persistence of covarion variability is so low as to require the vast majority of the double mutations to have been fixed successively without other mutations being fixed elsewhere in the gene between the first and second fixations of the double mutation. It is this "immediacy" that leads me to assume that the intermediate amino acid had an intermediate fitness but that is indeed an assumption, not a logical necessity. The immediacy of the next fixation may well span millions of generations providing ample opportunity for environmental changes you suggest.

HOW MUCH ROOM IS LEFT FOR NON-DARWINIAN EVOLUTION?

Roger Milkman
Department of Zoology, The University of Iowa
Iowa City, Iowa 52240

There is a good deal of common ground in modern evolutionary theory. We all understand that stochastic events are inherent in populations--animate and inanimate--and that natural selection is inevitable in natural animate populations, as we know them. We are all certain that natural selection is a central feature of evolution, but we recognize the likelihood that not all changes over time are brought about by natural selection.

Evolution is a series of changes. Darwinian evolution is a series of adaptive changes that can be related to one another and to an objectively describable function. The adaptive value is often not permanent, but at the time when the properties of an organism are changing in Darwinian evolution, the changes have an adaptive significance. The purely stochastic changes in evolution we call "Non-Darwinian". They stem from random genetic drift (16). We might call them "evolutionary noise" in the same sense that Waddington (14) has used the term "developmental noise" for the stochastic developmental variation in phenotype, which, incidentally, was also first characterized by Sewall Wright (17). In large populations as in small ones, random processes can account for the complete replacement of one allele by another. On the way, two or more different alleles can coexist, even for long periods of time, so that genic polymorphism is predicted, as well as outright replacement. The process is random from the moment of origin to the new allele. Most are lost, but a very few drift to high frequencies.

Before going on, I should like to distinguish two levels of organization, the linear level and a more complex level (or set of levels). The linear level is that of sequences: bases or amino acids. The Central Dogma applies to this level, where DNA, RNA, and primary structure of proteins are considered. Causality is relatively simple here: there is a large degree of independence among codons as they determine their respective amino acids. At the complex phenotypic level, things are not so simple. Interactions rapidly obscure the experimenter's view. Yet it is at this level that the amino acid sequences have their effects. The approaches to variation observed at these two levels must differ.

Exponents of Non-Darwinian evolution claim that most amino acid substitution is stochastic, and that most protein polymorphisms are stochastic. Their evidence begins with an estimate of mutation rate per codon. This rate is reduced to the estimated neutral mutation rate per codon on the basis of educated conjecture that a certain set of amino acid substitutions will have no significant consequence to the properties of the protein in which they occur. (Here we come to a branch point: it can be claimed, in addition, that certain substitutions will cause changes that can be measured by an enzymologist or even by an experimental population geneticist, but which will not be important in natural selection. We shall consider this later.)

The neutral mutation rate, taken together with estimates of effective population size, leads to estimates of stochastic substitution rates, and also to estimates of the number of alleles expected at a given locus at any one time. King and Jukes (5) have clearly set forth

the fundamental considerations. Kimura has provided important estimates

(1, 3, 4). The analysis of amino acid substitutions over time in terms

of this approach has been refined by Fitch and others into covarion

theory, about which you have just heard. Attacks on the predecessors

of this approach should be considered obsolete. It should also be

mentioned that the vast overestimates of the number of gene loci per

organism and of the total cost of selection at many loci, which led

to the renewed consideration of stochastic evolution, were noted by

King and Jukes as unnecessary to the argument and probably incorrect.

The covarion approach, however, is not inconsistent with selection.

Covarions are the only codons at which mutations can be tolerated:

this includes favorable ones as well as neutral ones. Since selection

operates much faster than drift, it is possible that the favorable

mutations will always outrace the neutral mutations. Selection at some

covarions would thus lead to the unexpectedly high proportion of double

substitutions which Dr. Fitch has observed. Indeed, this is an important

potential source of evidence for selection.

The Non-Darwinian calculations predict amino acid substitutions

and they predict polymorphisms, the two observable indications of

change--evolution--at the linear level. They can account for what

we have seen.

Darwinian theory also predicts substitutions; and only the

narrowest and naivest form of Darwinian theory would stop there. Any

biologically realistic form of the theory would certainly predict

widespread polymorphism. But admittedly the major impetus to apply

the theory of natural selection more realistically has come from the

recognition of widespread polymorphism.

At this point one might ask why there is anything worth arguing

about? We can restrict our consideration of selection to changes

at the complex level and refuse to dispute the proportionate roles of
selection and chance in linear variation.

Actually, the connections between the linear and complex levels
are too strong. We want to study selection at the simplest possible
level; and we want to follow the consequences of stochastic processes
to their final fruition.

We are addicted to the adaptive view: long and detailed experience
has led biologists to think "If it's there, it has an adaptive basis".
But is this true at the linear level? This is an important biological
question. In terms of amino acid substitutions, it's important to
know how many and what sorts of changes are necessary to make an ape
into a man, or a worm into an insect. Here we are at the basic
mechanistic level of evolution. Can anyone care so little as to
turn away?

In terms of polymorphism, we must decide whether to look for
selection in each case of electrophoretic mobility variants (allozymes);
whether these are indeed the building blocks of evolution, or whether
most of them are of no consequence at higher levels. How loud _is_
evolutionary noise?

The extreme viewpoint of Non-Darwinian evolution is this: We
can explain essentially all variation at the linear level in terms
of chance. Why seek additional explanations? Exceptions can be made
for the trivial proportion of changes with substantial consequences
at higher levels, the rare changes chosen to participate in Darwinian
evolution.

Students of genic polymorphism, however, can search for evidence
that each polymorphism is maintained by one form of selection or
another: heterosis, frequency-dependent selection, and so on. To
find a large number of cases maintained by selection could not refute

the Non-Darwinian hypothesis in general, but it would save the day

for electrophoresis. Investigators of allozymic variation make no

claims about what they cannot see, but they feel that genic polymorphisms

are maintained by selection and thus figure directly in Darwinian

evolution. Evidence in this direction (see, for example 12, 13) must

be viewed as pertaining to a rather limited area of genic variation,

but it suggests that when it is possible to subject a broader array

of proteins to scrutiny, they, too, will behave as if selection

operates on most amino acid differences.

Naturally, I can't spend much time on the details of research,

but here are some cases. Petra and coworkers (11) have shown that

bovine carboxypeptidase A comes in two forms differing by three rather

widely spaced amino acids. Very few sets of allozymes have been

sequenced: the assumption too often is that they differ by only one

amino acid. But in the present case if no intermediate forms are

found, the carboxypeptidase A allozymes will fit more comfortably

with the selection hypothesis than with the stochastic one. And here

is a somewhat similar case. _Mytilus edulis_ has three prominent

leucine aminopeptidase allozymes. A mussel in a neighboring genus,

Modiolus demissus, has three also, but they all run far faster than

the fastest _Mytilus_ allozyme (10). Doesn't this suggest a more

orderly change than chance would predict?

Kojima and coworkers (7) have found strong evidence for frequency-

dependent selection in several cases of enzyme polymorphism. Franklin

and Lewontin (2) have pointed out, however, that it is extremely

difficult to distinguish effects on a locus from effects on the many

loci closely linked to it. Wills and Nichols (15) appear to have

done just this by demonstrating, for _Drosophila pseudoobscura_

octanol dehydrogenase, heterosis conditional upon the presence of

octanol in the medium. Non-Darwinians, however, would doubtless take
a skeptical view of conditional selection until it could be demonstrated
in nature. Koehn (6) has distinguished allozymes on the basis of
temperature-dependent properties and showed a nice relationship between
their distribution and environmental temperature.

Various investigators, including myself, have noted heterosis
in cage populations of Drosophila. In natural populations, too,
heterosis appears to be operating (8): here the linkage problem is
particularly difficult, due to the fact that in a population composed
of immigrants from various genotypically different populations, the
entire genome is in effect linked.

Finally, I am reasonably certain that at least 20 loci are
polymorphic in nature for alleles that in certain combinations make
the posterior crossveins defective in D. melanogaster (9). There is
evidence that these alleles do not markedly affect other characters.
If they are representative of a much larger class, then a substantial
portion of existing polymorphs have effects at higher phenotypic
levels--but the reply can be made that though the individual alleles
are common enough, the combinations that make considerable changes
are rare in nature. And it might be added further that initially
neutral variants could be combined to acquire adaptive significance.

Thus we come to the two opposing arguments, presently irrefutable
and therefore not worth making much of:

1. "There's an important difference there; you haven't studied
it sufficiently; cells are more sensitive than analytical chemists."

2. "Agreed, you can amplify the distinction until you are able
to demonstrate selection, but what makes you think that would happen
in nature? And if it did, it would still only be one case."

My own feeling is that there is one area where the Non-Darwinian

hypothesis can be tested explicitly. We require an organism that 1) is
very numerous, 2) has been very numerous for a long time, 3) is generally
asexual (to cut down the incidence of inbreeding), 4) is sometimes
sexual (to show it's all one species, or gene pool), 5) is generally
haploid (to eliminate heterosis), 6) has a small genome, 7) has a
genome of known size, and 8) can provide a large sample of homogeneous
material.

This organism is of course E. coli --- or some other suitable
bacterium or virus. Non-Darwinian evolution must entail a great deal of
polymorphism here: many alleles at each locus, perhaps hundreds. The
general calculations on which I base this statement are tabulated and
discussed in a paper by Kimura (3). Furthermore, this case of multiple
allelism should be general for many loci, and the argument would be
strengthened if the variations were discordant; that is, if all sorts of
combinations of alleles were found. The theory of natural selection would
predict far less polymorphism, and it would be more concordant: some sets
of alleles (one or perhaps two alleles per locus), all adapted to
particular conditions, would occur more often than other combinations.

Note that although each higher species is thought to contain few
covarions per gene at any one time, E. coli's variation should reflect a
diverging tree of covarions even larger than that estimated for all
metazoans. Thus the number of positions at which amino acids would not
be uniform in E. coli proteins should be very large if substitutions are
stochastic. If there are not scores of allozymes determined at each of
many loci in E. coli, then Non-Darwinian evolution cannot be very
important in higher organisms, because we shall have to conclude that very
few differences are neutral.

Thus, with all the arguments and counter-arguments, the qualifications
and disqualifications, there remains one area in which the Non-Darwinian

view has immediate impact; it is an area accessible to experiment,
and it is at the heart of evolutionary mechanisms. In the study of
allozymes, we should devote considerable effort to seeing how often
we can reasonably conclude that a polymorphism is maintained by
selection. After all, the major virtue of Non-Darwinian evolution
is as a backstop. What natural selection can't account for can be
accomodated by random events. I find it challenging to realize that
Darwinian evolution has not been proven quantitatively.

References

1. Crow, J. F., and Kimura, M. An Introduction to Population Genetic
    Theory, Harper and Row, New York (1970).

2. Franklin, I. and Lewontin, R. C. Genetics 65, 707 (1970).

3. Kimura, M. Genet. Res. 11, 247 (1968).

4. Kimura, M., and Ohta, T. Genetics 61, 763 (1969).

5. King, J. L., and Jukes, T. H. Science 164, 788 (1969).

6. Koehn, R. K. Science 163, 943 (1969).

7. Kojima, K., and Tobari, Y. N. Genetics 61, 201 (1969) and 63,
    639 (1969).

8. Marshall, D. R., and Allard, R. W. Genetics 66, 393 (1970).

9. Milkman, R. D. Advan. Genet. 15, 55 (1970) and Genetics 65,
    289 (1970).

10. Milkman, R. D. Biol. Bull. 139, 430 (1970).

11. Petra, P. H., Bradshaw, R. A., Walsh, K. A., and Neurath, H.
    Biochem. 8, 2762 (1969).

12. Selander, R. K., Smith, M. H., Yang, S. Y., Johnson, W. E., and
    Gentry, J. B. Univ. Texas Publ. 7103, 49 (1971).

13. Tashian, R. E., Goodman, M., Headings, V., DeSimone, J., and
    Ward, R. H. Biochem. Genet. 5, 183 (1971).

14. Waddington, C. H. The Strategy of the Genes. George Allen and Unwin, London (1957).

15. Wills, C., and Nichols, L. Nature (in press).

16. Wright, S. Evolution and the Genetics of Populations. Univ. Chicago Press (1969).

17. Wright, S. Proc. Nat. Acad. Sci. U. S. 6, 320 (1920).

DISCUSSION

KOEHN: I wonder whether the critical issue in this discussion has not been the relative importance of selective versus stochastic processes in protein evolution, but rather the relative information to be obtained from studying protein polymorphism in contemporary populations versus amino acid sequences of selected proteins from evolutionarily diverse organisms. The fact that "selective neutralists" talk convincingly in terms of linear sequences and nearly all observations on contemporary populations are inconsistent with neutrality may indicate that these approaches are giving us different kinds of information. Would either of the speakers care to comment on this point?

MILKMAN: Of course the differences revealed by electrophoresis are more likely to be important to a protein than amino acid substitutions as a whole, but I don't think that's the critical factor. Both theories predict substitution and polymorphism. Observations of both phenomena can be explained by both theories. The evidence is largely circumstantial, and plausibility is not proof. I don't find that "nearly all observations on contemporary populations are inconsistent with neutrality". We have to show (1) for a large number of cases of polymorphism (2) that most of them (3) are due to selection (4) affecting the locus in question. The "selective neutralists" haven't convinced me, either. I think the answer has to come from experiments such as yours.

FITCH: I would make two responses. One is that the rate of evolution determined from blood group polymorphisms in South American Indians (1.2 x 10^{-7} gene substitutions/locus/year) was approximately the same as that observed in molecular evolution (0.9 x 10^{-7} nucleotide replacements/locus/year). The "locus" in this case was fibrinopeptide A and there are many caveats including the awareness of the effect of the size of the locus and of the many approximate values assumed in making these calculations (see Fitch and Margoliash, Evolutionary Biology, Volume 4, page 100 for details and further discussion). I believe the result indicates that observed polymorphisms may frequently be merely a different view of the same evolutionary process that protein evolution perceives. The apparent conflict may stem from nothing more than an improper perspective,which brings me to my second response. Let us for the moment assume that every observed polymorphism is selectively maintained, much as are the Catostomid serum esterases of your own elegant work (Science 163, 943 (1969)). If one accepts neutral mutations at all, there is nothing to prevent neutral mutations from accumulating in each of the two heterotic alleles. And if, however rarely, one of those mutations should create an allele whose contribution to the fitness of its possessors was identical to the fitness contributed to the population by the previous balanced polymorphism, that new allele could eventually go to fixation and end the polymorphism. The result would be a long continued balanced polymorphism with numerous fixations of no necessary selective benefit (except of course the original one that produced the heterotic advantage). The case is hypothetical, but I believe this illustrates that apparent heterosis, among many isozymes examined, is not necessarily in conflict with the idea that many of the fixations observed in the evolution of proteins were neutral. This is one reason I believe that Richmond's argument (Nature 225, 1025 (1970)) is irrelevant. His conclusion is modest enough, that "Demonstration of a cline in gene frequencies provides indirect evidence for the action of natural selection at the locus in question". That's beautiful and I accept it completely but it fails as a "critique" of non-Darwinian evolution because there is no necessary contradiction.

WILLS: With reference to Dr. Milkman's comments concerning the ideal organism in which to search for neutral variants, even if a pattern of variation apparently indicative of selection were to be discovered in E. coli, Kimura would of course have an answer. Suppose the same frequencies of isoalleles were to be found in widely scattered E. coli populations. Kimura would simply respond (as he has in a similar connection) by postulating a high mutation rate to specific isoalleles and a high rate of migration. The most unequivocal demonstration of isoallelic selection coefficients from natural populations is the data of Prakash, Lewontin and Hubby on the non-random association of isoallelic variants with different inversion phylads in D. pseudoobscura. The association is not perfect, so exchange is possible between phylads. It is very difficult to see how this situation could have arisen by stochastic processes.

MILKMAN: I'd like to think that as the experiments on this question accumulate, the arguments will decrease. No experiments on polymorphism to date here had a scope comparable to the stochastic theory; they can be dismissed at present as anecdotal insofar as they apply to the evolution question, even though you and I interpret them to be representative of a far larger class of more conclusive experiments to come. Any polymorphisms in linkage disequilibrium with (sets of) chromosomal arrangements require cautious interpretation. Anyway, exponents of the non-Darwinian view would be happy to grant you an adaptive polymorphism or two if it came in a package of hundreds of undescribed loci.

FITCH: It seems to me that a major difficulty with non-Darwinian evolution is that it can explain nothing and that a major difficulty with Darwinian evolution is that it can explain everything. I wonder, Roger, if that doesn't pose a problem for your experiment to decide unequivocally whether non-Darwinian evolution occurs to any significant extent. If upon examining hundreds of strains of E. coli you find this vast polymorphism, won't you, as a good selectionist, suggest that these differences are a function of the differences in the habitats from which they were obtained?

MILKMAN: No, I won't.

FITCH: Yes, but the rest of you will. I can't speak for everyone, but I'm willing to commit myself, and I think this is a position that others might find reasonable. I really don't expect dozens of polymorphic loci to be found in E. coli, each with scores of alleles, each allele having a substantial frequency in some populations. Furthermore, I certainly don't expect the combinations of alleles at the various loci to be random. Now if it is found that E. coli does contain scores of alleles at many loci combined randomly, I'll conclude that the evolution of amino acid sequences is mainly non-Darwinian. To be sure, if it is then proven that these alleles are adaptive, I'll reverse myself once more and marvel at the relatively vast variety of circumstances to which a bacterial species can accommodate by means of genetic versatility. But I'd bet that E. coli covers its range with few common alleles, helped by not having the burden of differentiation.

BOYER: I'm uncertain whether the problem of non-adaptive vs. adaptive evolution at the structural level is best answered by a 1-unit choice. Both explanations seem to be needed to account for certain kinds of variation among the primate hemoglobins A ($\alpha_2 \beta_2$) and A_2 ($\alpha_2 \delta_2$).

Four of the six high allele frequency hemoglobin polymorphisms (two in β

and two in δ) encountered in a survey of seven primate species occur at

either β^6 or δ^6 (Boyer et al., Biochem. Genetics, in press, 1971). This

position is also the one most frequently involved in differences between

species. Among eleven species of primates whose β and δ amino acid

sequences have been delineated, $\sim$ 20% of changes accumulating since gene

duplication occur at codon 6. Four of these (<u>Saimiri</u> β^6, <u>Saimiri</u>-<u>Aotus</u>-

<u>Saguinus</u>, <u>Mystar</u> δ^6, <u>Saguinus nigricollis</u> δ^6 and <u>Callicehus moloch</u> β^6)

represent forward and/or back mutations between glutamic and aspartic

acids. These, together with the polymorphic changes, form an outlandish

recurrence of one homologous codon even in the co-varion setting sug-

gested by Fitch. Interpretation of such changes hinges on the finding

that hemoglobin A_2 forms only 1/16 to < 1/200 - depending on species -

of all hemoglobin. Because of the pancellular poverty of A_2, it has

been presumed that the A_2-δ chain is functionally less important and

therefore less visible to natural selection through A-β (Boyer et al.,

Science, 1969). Accordingly, quite apart from Kimura's arguments, the

substantial number of δ changes found at structural codon 6 and else-

where, can be attributed to non-adaptive factors. No one, however,

doubts the role of selection in furthering the frequency of 6 Glu $\longrightarrow$

Val mutation characteristic of human sickle hemoglobin. Thus, there is

evidence at the same homologous codon for concurrence of both non-

adaptive noise and an adaptively clear tone. In this and other respects

the β-δ model, and additional "major-minor" gene duplicant systems like

it, offer attractive means for dissecting the relative importance of

adaptive and non-adaptive factors in the evolution of protein structure.

   MILKMAN: Thank you very much, Dr. Boyer.

STAGES IN CHROMOSOME EVOLUTION: THE CHROMATID TWINS AND HOW THEY GREW

E.J. DuPraw
Associate Professor
Department of Anatomy
Stanford University School of Medicine

INTRODUCTION

To judge from discussions of chromosome structure in several recent
textbooks (e.g., DeRobertis et al, 1970; Lehninger, 1970; Novikoff and
Holtzman, 1970; Cohn, 1969; Giese, 1968), the majority of cell biologists have
accepted the fact that higher chromosomes are composed of long folded fibers,
which vary around 100 Å to 500 Å in diameter, and that each fiber is based on
a long DNA molecule complexed with histone and non-histone proteins (Fig. 1A).
The existence of these fibers was completely unsuspected by chromosome
cytologists during the light microscope era (DeRobertis et al, 1965; Schrader,
1953; Kaufmann, 1936). With changing ideas about how chromosomes are built,
it has become necessary to revise our ideas about how chromosomes evolved. In
fact, this was the subject of a little book published recently (DuPraw, 1970),
from which four or five particularly important points deserve to be expanded.

CHROMOSOME STRANDEDNESS

In this volume, Sparrow et al (1971) have provided a set of graphs
documenting more than a 1000-fold increase in the amount of DNA per haploid
nucleus, as well as in DNA per chromosome, during the evolution of plants and
animals from micro-organisms.

Although much of this increase must reflect the greater genetic complexity
of multicellular species, a few traditional cytologists have repeatedly urged
that a 2- or 4-fold component of the overall 1000-fold increase is due to
multistrandedness (Stubblefield and Wray, 1971; Gay et al, 1970; Wolff, 1969;
Sparvoli, Gay and Kaufmann, 1965). Their thesis is that eukaryotic chromosomes
differ fundamentally from those of bacteria (and viruses) by containing a

minimum of 2 or 4 identical copies of the species' DNA at all times. Each
metaphase chromatid is thought to contain either two half-chromatids, or four
quarter-chromatids lying in parallel. At issue is not the occasional instance,
such as Drosophila salivary gland, in which polytene chromosomes are well
documented, but the general principle of eukaryotic chromosome structure.

In several earlier publications (DuPraw, 1966; 1968; 1970), I have
emphasized that no duplex or quadriplex organization can be seen in human
chromatids prepared as whole-mounts for the electron microscope (Fig. 1A).
This observation has been confirmed repeatedly for human cells (Abuelo and
Moore, 1969; Lampert, 1969; Golomb and Bahr, 1971), and has been extended to
at least seven other animal species (Comings and Okada, 1970a). In addition,
the great majority of living metaphase chromosomes, observed by phase contrast
or interference microscopy, reveal only pairs of chromatids without detectable
duplex or quadriplex substructure. The one constantly quoted example in which
living chromosomes of the plant Haemanthus, on rare occasions, seem to be
divided into half-chromatids (Bajer, 1965), is the exception that proves the
rule; here, even at metaphase "half-chromatid substructure often cannot be
recognized". Only after chromosomes have been digested with acid (Kaufmann,
1936), treated with ammonia in 50 percent ethanol (Manton, 1945), digested
with trypsin (Trosko and Wolff, 1965), or extracted with 2 M NaCl or 6 M
urea (Stubblefield and Wray, 1971) do the chromatids seem to show a duplex
organization. It is fair to ask whether the reagents are really uncovering
hidden substructure, or whether they are creating the duplex morphology as
a preparative artefact.

That the ultrastructure of mitotic chromosomes is prone to this type of
artefact is demonstrated by figure 1. Part A shows a well-preserved human
chromatid pair, in which the typical morphology of a D-group chromosome can
be recognized, the body of the chromatids consists entirely of 200-500 Å
chromatin fiber, and there is no sign of duplex or quadriplex organization
within each chromatid. By contrast, part B shows a chromatid pair in which
the gross morphology is poorly preserved, individual 200-500 Å fibers cannot
be seen in the body of the chromosome, but larger, obviously artefactual
subchromatid strands are conspicuous. Though these artefactual strands arise
by the collapse and aggregation of the original chromosome fibers, they display
a remarkable regularity. Seen at the lower resolution of the light microscope,
they would present exactly the appearance of half- and quarter-chromatids
(arrow), including the illusion that they are plectonemically coiled around
one another.

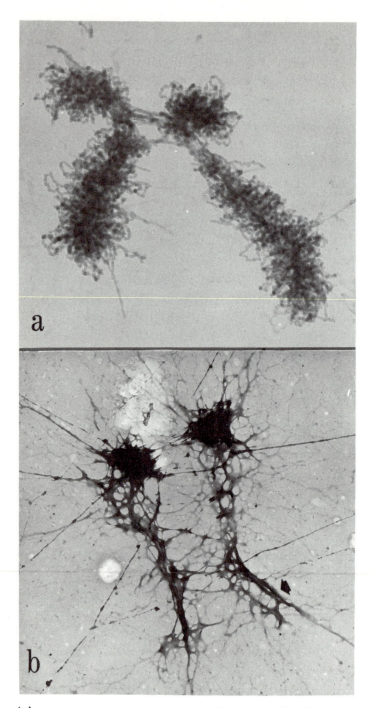

Fig. 1. (A) A human chromosome of the 13–15 group. Total dry mass for two chromatids is 10.3 X 10^{-13} gm. About 24,000 X. (B) Human chromatid pair that has undergone collapse and aggregation during whole-mount preparation.

The theory that mitotic chromosomes contain replicate half-chromatids was
introduced and accepted long before DNA was identified as the genetic material.
In this era of molecular genetics, however, the presence of 2, 4 or any other
number of identical DNA molecules cannot be tested, much less proved, by
observing the shape of a chromosome with a light microscope. When appropriate
techniques are used to determine the number of gene copies per chromatid,
either by quantitative mutagenesis (Wolff, 1963) or by quantitative DNA
hybridization (Laird, 1971), the results are consistent in showing that
nucleotide sequences not reiterated relative to the rest of the genome are
present as a single copy per chromatid. This is true in man, in mouse, in
Drosophila, in Tradescantia, and in a wide range of other species (Laird,
1971). Though there may be rare exceptions, chromosome evolution has involved
primarily a progressive modification of chromatid twins, rather than the
introduction of chromatid quadruplets and octuplets.

CHROMATID CORES

During the 1950's it was erroneously assumed that each gene would be a
separate DNA molecule, and that an indispensable part of any chromosome would
be a core or skeleton to hold these molecules together. This core, in turn,
would require its own special mechanism of replication. However, when it
became possible to study thin sections of chromosomes with the electron
microscope, the postulated cores could not be found (Moses and Coleman, 1964).
At the same time, it was discovered that genes are usually linked together by
DNA itself, forming single molecules that are millimeters or centimeters in
length (Cairns, 1963; 1966; Sasaki and Norman, 1966). These discoveries made
the existence of chromatid cores questionable both on theoretical and
empirical grounds.

In well-preserved chromosomes prepared as whole mounts, there is never
any sign of cores. Figure 1A, for example, is a kind of chromosome X-ray, in
which the darkness of the image is proportional to the dry mass at each point,
and in which the electron beam has passed entirely through each chromatid.
Cores, if they were present, would be as conspicuous as the bones in an X-ray
of an arm or a leg. However, not even the trace of a core is seen.

Figure 2, on the other hand, illustrates a human chromosome distorted in
a way that has recently been interpreted as revealing a core, supposedly after
the stripping away of an outer "epichromatin" (Stubblefield and Wray, 1971).
The latter authors also claim that such cores are duplex in structure,
demonstrating the presence of half-chromatids in Chinese hamster chromosomes.

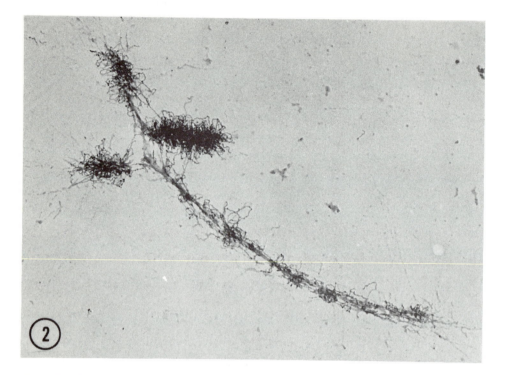

Fig. 2. Unidentified human chromosome prepared as a whole-mount by
surface spreading-critical point drying. The long arm of
one chromatid has been accidentally stretched and partially
aggregated to give the false impression of a chromatid "core".

Again the question arises: has the treatment with NaCl and urea, centrifugation at 1500 g, and air drying simply revealed the hidden cores, or has it in
fact created them? For the stretched chromatid arm shown in figure 2, it can
be said quite definitely that the apparent core is not an authentic structure,
but just a badly preserved chromosome. This is evident because no extracting
agents have been used, and consequently no epichromatin has been removed. One
sister chromatid is intact, while its supposedly core-containing twin differs
only in having been stretched and partially aggregated in the same manner as
the chromosome in figure 1B. Progressive aggregation is also evident in the
fact that the longitudinal "core" is morphologically single near the centromere
duplex farther out, and decidedly multiple near the end of the arm. Specimens
such as this do not represent duplex or quadriplex elements in native chromosomes, but they are striking examples of artefact at an ultrastructural level.

FIBER SUBSTRUCTURE

For some years, Ris (1959; see also Ris and Chandler, 1964) maintained that each 200 Å chromosome fiber consists of four identical DNA molecules loosely twisted around one another. This concept was surely the least well documented, and fortunately the shortest lived, of all the variants of the quarter-chromatid hypothesis. Almost immediately it was contradicted by enzyme digestion experiments, which showed only a single DNA molecule in each fiber (DuPraw, 1965a; Ris, 1966a). That this DNA molecule is tightly packed, probably by supercoiling, could be inferred from early estimates of the maximum fiber length per chromatid (DuPraw, 1965b; 1966), which were about 10 times smaller than the known lengths of DNA double helix in the same chromosomes.

During the last three years, these revised concepts of chromosome structure have been subjected to rigorous quantitative proof by applying an electron microscope technique for weighing individual chromosomes and parts of chromosomes, including single segments of chromosome fiber. In this method, electron micrographs are exposed in such a way that the darkness of each point in the image is linearly proportional to that point's dry mass in the specimen (Bahr and Zeitler, 1965); it is then possible to weigh any part of a chromosome by masking out the rest, measuring the integrated density of that part, and calibrating this measurement against standard specimens of known weight (small polystyrene spheres). Several authors have now made use of this technique to establish the quantitative properties of chromosomes, and their carefully confirmed values serve to rule out many concepts of chromosome structure that date from the era of light microscope cytology (DuPraw and Bahr, 1968; 1969; DuPraw, 1970; Lampert, 1969; Bahr, 1970; Bahr and Golomb, 1971).

For example, by quantitative electron microscopy the exact length of fiber per chromatid (L) can be determined from the formula

$$L = \frac{M}{\bar{m}}$$

where M is the total dry mass of the chromatid and $\bar{m}$ is the average mass per micron of the fiber. In the larger and smaller human chromosomes, these fiber lengths range from 723 microns down to 135 microns per chromatid (DuPraw and Bahr, 1969). Comparison of these exact fiber lengths with DNA helix lengths, as well as fiber dry masses with DNA helix mass, has established that each micron of fiber on the average must contain from 50 to over 100 microns of DNA molecule (DuPraw, 1970; Bahr, 1970). These measurements are incompatible with Ris's early (1959) 4-strand model of the fiber; in fact, Ris and Kubai

(1971) now agree that: "The basic nucleohistone fiber contains a single DNA
double helix. . . The DNA in the nucleohistone fiber is highly compacted."
They also take the position that "the only contradictory interpretation,"
in which "the DNH unit is composed of four parallel DNA strands", should be
ascribed to Luzzati and Nicolaieff (1963).

CHROMOSOME EVOLUTION

Once the tenuous superstructure of half-chromatids and chromatid cores
is stripped away, remarkable similarities can be seen between the chromosomes
of prokaryotes and eukaryotes. Figure 3 emphasizes diagrammatically that both
types of chromosome may be portrayed as single, long DNA molecules, which
replicate once and divide once during every cell cycle. Replication, as it
proceeds, generates Siamese chromatid twins that are held together by regions
not yet reproduced. In eukaryotic chromosomes, these late-replicating
segments have the appearance of typical chromosome fibers, which frequently
can be noticed passing between the sister chromatids at the centromere and
sometimes in other regions as well (Fig. 1A). Of course, the total amount

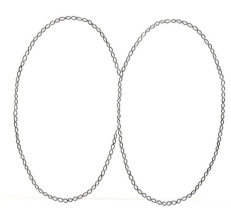

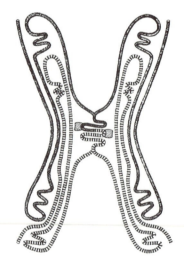

A.   **B.**

Fig. 3. (A) A circular bacterial chromosome shortly before replication
         is complete, with the DNA double helix exhibiting a modified
         "theta" form. (B) Folded fiber model of chromosome structure
         in eukaryotes. Each sister chromatid contains a single long
         DNA-protein fiber, with unreplicated segments holding the two
         together at the centromere. From DuPraw (1968).

of DNA in an average eukaryotic chromosome is much greater, but this DNA is
packed into a fiber that is often shorter than a bacterial DNA molecule.

What is the 3-dimensional packing configuration of DNA in a eukaryotic
chromosome? At least two separate problems are included in this question:
1) the packing of DNA into the chromatin fiber, which is a feature at all
stages of the cell cycle; and 2) the folding or condensation of the fiber into
a visible chromosome, which in higher plants and animals is a reversible
process occurring immediately before each cell division. Insights about both
these levels of packing have been obtained by using quantitative electron
microscopy to analyze intact human chromosomes.

Patterns of Fiber Folding

Figure 4A illustrates a human chromosome 13-15, which has been the subject
of previous investigations (DuPraw and Bahr, 1969a; DuPraw, 1970). Each
chromatid weighs 4.95×10^{-13} gm and contains 1.00×10^{-13} gm of DNA; this
is equivalent to 3.1 centimeters of double helix, packed into 218 microns of
chromosomal fiber.

In collaboration with Dr. G.F. Bahr, I have analyzed the arrangement of
fiber in each chromatid by scanning the chromosome with a recording densito-
meter at the telomeres, centromere and intermediate positions in the long
arms; for any given cross-scan, the area under the recorded curve was
proportional to the dry mass seen by the aperture during its scan. The
various areas were measured in arbitrary units with a planimeter and then
expressed in "fiber equivalents", i.e., some multiple of the average area
recorded in cross-scans of single fibers in the same chromosome. As shown
in figure 4B, the ends of the two short arms (cytological telomeres) have
cross-sectional dry masses equivalent respectively to 57 and 66 single fibers;
by contrast, the two long arm telomeres contain 85 and 81 fiber equivalents.
The centromere value of 100 fiber equivalents embraces both chromatids and is
comparable to values found for each chromatid at a mid-point along the long
arm (106 or 90 fiber equivalents).

These cross-sectional measurements tend to confirm the total fiber length
measurements, since 218 microns of fiber folded into a chromatid 2.5 microns
long should average about 87 side-by-side fiber segments (218/2.5). From the
fiber equivalents it is also possible to express the cross-sectional dry mass
at any point in the chromosome as grams per micron. In this chromosome 13-15,
for example, one fiber equivalent can be taken as 21.6×10^{-16} gm/μ; therefore,
the centromere value of 100 fiber equivalents corresponds to a dry mass of
approximately 2.16×10^{-13} gm per micron. Slight differences observed between

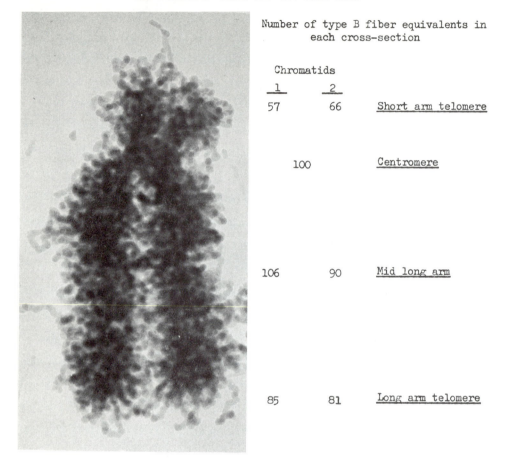

Number of type B fiber equivalents in
each cross-section

Chromatids		
1	2	
57	66	Short arm telomere
	100	Centromere
106	90	Mid long arm
85	81	Long arm telomere

Fig. 4. (A) Whole-mount specimen of human chromosome 13-15, prepared by
surface spreading-critical point drying. Each chromatid weighs
4.95×10^{-13} gm. About 50,000 X. (B) Fiber equivalents of dry
mass measured in cross-scans at different levels of the chromosome.
Each fiber equivalent equals 21.6×10^{-16} gm per micron.

sister chromatids when measured at comparable points may be attributed to
minor irregularities in the folding of fiber in each chromatid, as well as
to an instrumental error of approximately ± 2 percent.

The fact that the telomeres of these chromatids contain 57 to 85 fiber
equivalents is, of course, incompatible with the idea that each chromatid
might be composed of 2, 4, 8 or even 16 replicate copies of its DNA lying in
parallel and terminating at the two opposite telomeres. Rather, these high

values tend to confirm that the cytological telomeres are sites of folding or
looping of long fiber segments (DuPraw and Rae, 1966). In addition, the
discovery that the centromere, which is shared by two chromatids, contains
about the same number of fiber equivalents as a single chromatid arm, supports
the hypothesis that the centromere contains unreplicated fiber regions shared
jointly by the sister chromatids (Fig. 3; DuPraw, 1965b; 1970).

Chromosomes and the Nuclear Envelope

In many primitive eukaryotes, such as dinoflagellates, euglenoids, and
hypermastigid flagellates, the chromosomes remain condensed and visible by
light microscopy at all stages of the cell cycle; in these species, the nuclear
envelope is also a permanent structure that does not fragment during cell
division, but possesses a mechanism for constricting itself into two (or more)
daughter nuclei. Division of the nucleus is closely integrated with replication of the chromosomes, which are visibly attached to the envelope (Cleveland,
1938; Kubai and Ris, 1969).

These primitive features have great relevance for the chromosomes of
higher eukaryotes, which at interphase also exhibit direct attachments between
the dispersed chromosome fibers and the annuli of the nuclear envelope (DuPraw,
1965a; 1970; Beams and Mueller, 1970; Comings and Okada, 1970b). DNA polymerase
activity has also been demonstrated in nuclear envelopes isolated from HeLa
and Chinese hamster cells by Yoshikawa-Fukada and Ebert (1971). Since chromosome replication in eukaryotes requires close coordination between the
synthesis of histone proteins outside the nucleus and of DNA inside the nucleus
(Alfert et al, 1955; Robbins and Borun, 1967), attention focuses on the annuli
as organelles ideally located to coordinate assembly of both the protein and
DNA components of the chromatin fibers (Fig. 5A).

The combined molecular weights of all compounds composing a single annulus
in honeybee embryonic cells has been estimated by quantitative electron
microscopy (DuPraw and Bahr, 1969b). Figure 5B shows the total dry mass for
105 annuli measured in three different preparations (coded as stipples, slashes,
and vertical lines). The average dry mass per annulus was 5.23×10^{-16} gm,
and this value multiplied by Avogadro's number gives a total molecular weight
of 315 million per annulus. This is sufficient to accomodate nearly 3000
protein molecules equivalent to the size of bacterial DNA polymerase. Although
annuli floated on sucrose had slightly larger average diameters than annuli
floated on distilled water (919 Å compared with 834 Å), their weights were in
the same range. This indicates that an annulus can open and close in a
sphincter-like action, without any significant change in dry mass.

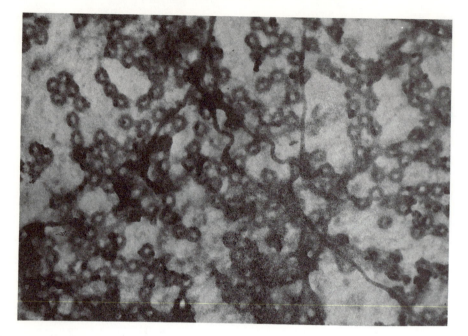

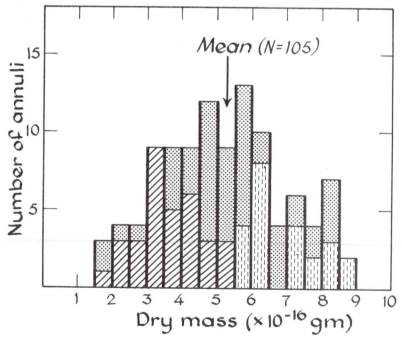

Fig. 5. (A) Doughnut-shaped annuli seen in the nuclear envelope of a honey
bee embryonic cell after surface spreading-critical point drying. About
50,000 X. (B) The dry weights of single annuli (N = 105) measured in three
different preparations (coded as stipples, slashes and vertical lines).

Several laboratories have published autoradiographic evidence that the nuclear envelope is a site for DNA synthesis in living tissue cells, at least during some phases of the S-period (Moses and Coleman, 1964; Comings and Kakefuda, 1968; O'Brien et al, 1971). Assuming that the envelope does function in this way, it must contain mechanisms for unpacking and repacking chromosomal DNA during replication. Condensation of the heterochromatic X-chromosome in mammalian females may also depend on the fact that at interphase this chromosome is firmly attached to the envelope (Beams and Mueller, 1970). Perhaps one of the nuclear envelope's earliest functions was to provide anchor points for the ends of linear DNA molecules, thereby preventing the spinning out of supercoils in the chromosomal fibers. This would be consistent with the cytology of primitive dinoflagellates, in which: 1) chromosomal DNA is visibly supercoiled throughout the cell cycle (Dodge, 1963; Haller et al, 1964); 2) the nuclear envelope remains intact throughout the cell cycle; 3) the chromosomes are attached to the nuclear envelope (Kubai and Ris, 1969); and 4) the chromosomes lack the histone proteins thought to maintain DNA packing in higher eukaryotes (Dodge, 1964; Mendiola et al, 1966).

Just before division in dinoflagellates and other primitive unicells, a microtubular apparatus assembles outside the envelope, but it tends to be small and does not establish direct attachments to the chromosomes. In more advanced flagellates, the mitotic spindle is relatively complex, and it does contact each chromosome at a mutual attachment site on the permanent nuclear envelope (Cleveland, 1938). Disruption of the envelope during cell division evidently became advantageous after the chromosomes had evolved kinetochores; at that stage the histone proteins, which were still relatively primitive in euglenoids (Netrawali, 1970), very likely became adapted as a supplementary mechanism to maintain supercoiling and control the enormously long DNA molecules. As the ciliates evolved, these proteins were able to differentiate into specific micronuclear and macronuclear histones, capable of fine-tuning the DNA for very different synthetic roles in the two kinds of nucleus (Gorovsky, 1970).

It should be pointed out that any replication mechanism in which DNA spins past a fixed point on the envelope is incompatible with back-to-back replication forks moving in opposite directions down the same double helix. That such back-to-back forks occur in Chinese hamster cells was first indicated by the DNA autoradiographs of Huberman and Riggs (1968), later supported by Schnös and Inman (1970), who detected a dual fork process in bacteriophage lambda. Replication sites almost certainly are not associated with the envelope in

macronuclei of the ciliate <u>Euplotes</u>, since DNA synthesis there is confined to
two "reorganization bands" that cut across the interior of the ribbon-shaped
nucleus (Gall, 1959; Ringertz et al, 1967). These examples merely emphasize
that eukaryotic cells, evolving through hundreds of millions of years, may
well have developed more than one way to replicate their genetic material.
Even if a particular mechanism is associated with the nuclear envelope,
another may be quite independent.

Chromosomal Proteins

However striking the increases in DNA during the evolution of multicellular plants and animals, even more astonishing increases occurred in the
relative amounts and variety of proteins attached to the DNA. For example,
the 2 meters of double helix in each human diploid nucleus constitute only
about 16 percent of the total metaphase chromosome mass (Huberman and Attardi,
1966); by contrast, over 70 percent of each chromosome is protein, held there
by a direct or indirect attachment to the DNA. Less than half this protein
consists of basic histones, while the remainder is a poorly defined "acid
protein" component. The biochemical properties of the many chromosomal
protein fractions, and their functions in eukaryotes, have recently been the
subject of a masterly review by James Bonner and his associates (Elgin et al,
1971).

Evolutionary increases in the total mass of proteins associated with DNA
naturally contributed to a vast growth in the total size and weight of the
chromatid twins. To a certain extent, however, the amount of this growth is
not species-constant, but varies from tissue to tissue and fluctuates during
the course of the cell cycle. For example, Huberman and Attardi (1966) found
that chromatin from HeLa cells at interphase contains about 31 percent DNA,
compared with the 16 percent DNA found when chromosomes were isolated in bulk
from the same cells at metaphase. Assuming that the absolute amount of DNA
per chromatid is constant, the drop in percent DNA means that metaphase
chromosomes weigh on the average about twice as much as interphase ones
(chromatid for chromatid).

This variation in chromosomal dry mass during the cell cycle is matched
by an unexpected variation in the weights of homologous chromosomes from one
metaphase plate to another. Within a given metaphase, the total dry mass of
each chromosome is very nearly proportional to its DNA content (DuPraw and
Bahr, 1969a); however, the proportionality constant (or chromosome index) is
much greater in some cells than in others. Evidently this is due to significant differences in the amount of non-DNA material bound to a constant DNA

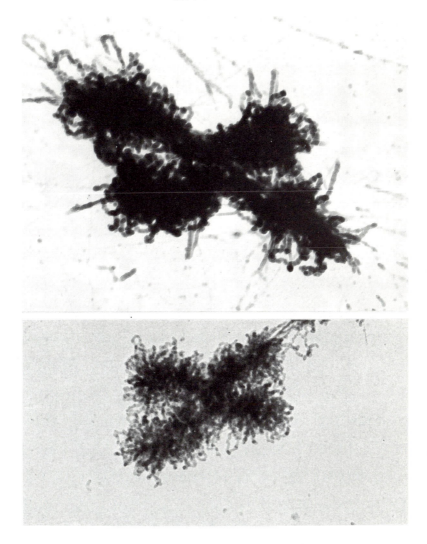

Fig. 6. Human chromosomes 19-20 isolated from two different metaphase
plates by surface spreading-critical point drying. The heavy
specimen weighs 17.0×10^{-13} gm, while the light one weighs
7.4×10^{-13} gm. Both have been printed to the same magnification
(about 30,000 X). Notice that the heavier chromosome is
proportionately longer and contains fiber of greater average
diameter. Despite the 2.3-fold difference in weight, the
heavier chromatids do not contain pairs of light chromatid
subunits.

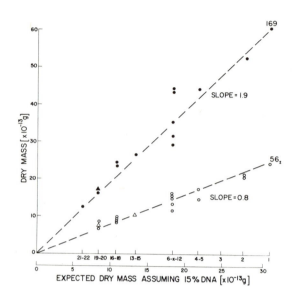

Fig. 7. Total weights of chromatid pairs from heavy and light metaphase
         plates grown in peripheral lymphocyte cultures. The values have
         been plotted against expected weights based on the DNA content
         of each karyotype group. Within a metaphase plate, there is a
         linear relationship between total dry mass and DNA content; however,
         the slope (chromosome index) is more than twice as great for the
         heavy set. Triangular points represent the chromosomes illustrated
         in figures 4 and 6.

component. What effect does such variation have on the 3-dimensional packing
of the DNA?

Figure 6 illustrates heavy and light chromosomes from the human F-group
(chromosomes 19-20), printed to identical magnifications. The total weight for
both chromatids is 17.0 X 10^{-13} gm for the larger specimen, but only 7.4 X 10^{-13}
gm for the smaller one. This remarkable difference is not due to misidentification of the chromosomes, or to accidental contamination of the heavier specimen.
The F-group identification is based not only on the chromosomes' characteristic
metacentric morphology, but also on their small size and exact dry masses
relative to other chromosomes in their respective metaphase plates. The weight
distribution of these other chromosomes is shown in figure 7, where the
triangular points mark the specimens illustrated in figure 6 (as well as the
chromosome 13-15 in figure 4). Electron micrographs showing heavy and light

chromosome 2's from the same metaphase plates have been published previously (DuPraw, 1970).

As seen in figure 6, the 2.3-fold difference in total dry mass between these two chromosomes is correlated with a proportional difference in their maximum lengths and the diameters of their constituent fibers. This rules out the possibility that the weight difference is the result of specimen contamination in the electron beam, or of adventitious binding of extraneous material during preparation; neither of these factors could produce so proportional an increase in the overall lengths of the chromosomes. Neither does the heavy chromosome show any indication that its chromatids are constructed of two "light" chromatid subunits. Rather, the data are most consistent with the interpretation that these chromosomes differ greatly in the amount of protein complexed with the F-group DNA component of 0.63×10^{-13} gm (Rudkin, 1967; Mendelsohn et al, 1969).

By substituting the appropriate numbers into equation (3) of DuPraw (1970), we can calculate the percent DNA in each of these two chromosomes; this turns out to be 17 percent for the smaller specimen, compared with only 7.4 percent for the heavier specimen. Associated with this difference is a visible change in the size and folding pattern of the chromosome fiber, suggesting that the 3-dimensional arrangement of DNA must be different in the two, even though both chromosomes are metacentric in their gross anatomy.

That the same DNA helix can be packed in a wide variety of ways is not really surprising. Configurational changes of this type have long been known to distinguish meiotic from mitotic chromosomes, and also occur universally during the differentiation of sperm nuclei. Evidence that human cancer cells develop abnormalities in the packing configuration of their chromosomal DNA has been reported by Lampert, Bahr and DuPraw (1969; see also Lampert, 1971). Undoubtedly repacking of the same DNA into new configurations has occurred frequently during evolution, and accounts for various examples in which chromosome number and karyotype differ strikingly between closely related species.

Is all the DNA in a chromatid part of a single, long double helix? For a human F-group chromatid, the length of such a helix would be about 1.9 centimeters. However, this value is actually _less_ than the 2.2 centimeter DNA molecules isolated from human lymphocytes by Sasaki and Norman (1966). Provided that their DNA molecules came from chromosomes such as those shown in figure 6, there seems to be little choice but to conclude that the basic unit chromatid contains only one continuous length of DNA.

REFERENCES

1. Abuelo, J.G., and Moore, D.E. (1969). J. Cell Biol. 41: 73.

2. Alfert, M., Bern, H., and Kahn, R. (1955). Acta Anat. 23: 185.

3. Bahr, G.F. (1970). Expt. Cell Res. 62: 39.

4. Bahr, G.F. and Golomb, H.M. (1971). Proc. Nat. Acad. Sci. U.S. 68: 726.

5. Bahr, G.F. and Zeitler, E. (1965). In "Symposium on Quantitative
   Electron Microscopy" (G.F. Bahr and E. Zeitler, eds.), p. 217.
   Williams and Wilkins, Baltimore, Maryland.

6. Bajer, A. (1965). Chromosoma 17: 291.

7. Beams, H.W. and Mueller, S. (1970). Z. Zellforsch. 108: 297.

8. Cairns, J. (1963). J. Mol. Biol. 6: 208.

9. Cairns, J. (1966). J. Mol. Biol. 15: 372.

10. Cleveland, L.R. (1938). Biol. Bull. 74: 1, 51.

11. Cohn, N.S. (1969). "Elements of Cytology." 2nd edition. Harcourt,
    Brace and World, Inc., New York.

12. Comings, D.E. and Kakefuda, T. (1968). J. Mol. Biol 33: 225.

13. Comings, D.E. and Okada, T.A. (1970a). Cytogenetics 9: 450.

14. Comings, D.E. and Okada, T.A. (1970b). Expt. Cell Res. 62: 293.

15. DeRobertis, E., Nowinski, W. and Saez, F. (1965, 1970). "Cell Biology."
    Saunders, Philadelphia. 4th and 5th ed. 555 pp.

16. Dodge, J.D. (1963). Arch. Mikrobiol. 45: 46.

17. Dodge, J.D. (1964). Arch. Mikrobiol. 48: 66.

18. DuPraw, E.J. (1965a). Proc. Nat. Acad. Sci. U.S. 53: 161.

19. DuPraw, E.J. (1965b). Nature 206: 338.

20. DuPraw, E.J. (1966). Nature 209: 577.

21. DuPraw, E.J. (1968). "Cell and Molecular Biology." Academic Press,
    New York. 739 pp.

22. DuPraw, E.J. (1970). "DNA and Chromosomes." Holt, Rinehart, and
    Winston, New York. 340 pp.

23. DuPraw, E.J. and Bahr, G.F. (1968). J. Cell Biol. 39: 38a.

24. DuPraw, E.J. and Bahr, G.F. (1969a). Acta Cytol. 13: 188.

25. DuPraw, E.J. and Bahr, G.F. (1969b). J. Cell Biol. 43: 32a.

26. DuPraw, E.J. and Rae, P.M.M. (1966). Nature 212: 598.

27. Elgin, S., Froehner, S., Smart, J. and Bonner, J. (1971). Adv. Cell
    Mol. Biol. 1: 1.

28. Gall, J.G. (1959). J. Biophys. Biochem. Cytol. 5: 295.

29. Gay, H., Das, C., Forward, K. and Kaufmann, B.P. (1970). Chromosoma
    32: 213.

30. Giese, A. (1968). "Cell Physiology." Saunders, Philadelphia. 3rd Ed.

31. Golomb, H.M. and Bahr, G.F. (1971). Science 171: 1024.

32. Gorovsky, M.A. (1970). J. Cell Biol. 47: 619, 631.

33. Haller, G., Kellenberger, E., and Rouiller, C. (1964). J. Microscopie
    3: 627.

34. Huberman, J.A. and Attardi, G. (1966). J. Cell Biol. 31: 95.

35. Huberman, J.A. and Riggs, A.D. (1968). J. Mol. Biol. 32: 327.

36. Kaufmann, B.P. (1936). Botan. Rev. 2: 529.

37. Kihlman, B.A. (1971). Adv. Cell Mol. Biol. 1: 59.

38. Kubai, D.F. and Ris, H. (1969). J. Cell Biol. 40: 508.

39. Laird, C.D. (1971). Chromosoma 32: 378.

40. Lampert, F. (1969). Naturwissen. 56: 629.

41. Lampert, F. (1971). Adv. Cell Mol. Biol. 1: 185.

42. Lampert, F., Bahr, G.F. and DuPraw, E.J. (1969). Cancer 24: 367.

43. Lehninger, A.L. (1970). "Biochemistry." Worth Co., New York.

44. Luzzati, V. and Nicolaieff, A. (1963). J. Mol. Biol. 7: 142.

45. Manton, I. (1945). Amer. J. Bot. 32: 342.

46. Mendelsohn, M., Hungerford, D., Mayall, B., Perry, B., Conway, T., and
    Prewitt, J. (1969). Annals N.Y. Acad. Sci. 157: 376.

47. Mendiola, L.R., Price, C. and Guillard, R. (1966). Science 153: 1661.

48. Moses, M.J. and Coleman, J.R. (1964). In "The Role of Chromosomes in
    Development." (M. Locke, ed.), p. 14. Academic Press, New York.

49. Netrawali, M.S. (1970). Expt. Cell Res. 63: 422.

50. Novikoff, A.B. and Holtzman, E. (1970). "Cells and Organelles." Holt,
    Rinehart and Winston, New York. 337 pp.

51. O'Brien, R.L., Sanyal, A. and Stanton, R. (1971). Expt. Cell Res.
    (in press).

52. Ringertz, N., Ericsson, J. and Nilsson, D. (1967). Expt. Cell Res.
    48: 97.

53. Ris, H. (1959). Colloq. Ges. Physiol. Chem. 9: 1.

54. Ris, H. (1966a). Proc. 6th Intern. Conf. Electron Microscopy, Kyoto.
    p. 339

55. Ris, H. (1966b). Proc. Roy. Soc. B164: 246.

56. Ris, H. and Chandler, B.L. (1964). Cold Spring Harbor Symp. Quant.
    Biol. 28: 1.

57. Ris, H. and Kubai, D.F. (1970). Ann. Rev. Genet. 4: 263.

58. Robbins, E. and Borun, T. (1967). Proc. Nat. Acad. Sci. U.S. 57: 409.

59. Rudkin, G.T. (1967). In "The Chromosome: Structural and Functional
    Aspects." (G. Yerganian, ed.), p. 12. Williams & Wilkins, Baltimore.

60. Sasaki, M. and Norman, A. (1966). Expt. Cell Res. 44: 642.

61. Schnös, M. and Inman, R. (1970). J. Mol. Biol. 51: 61.

62. Schrader, F. (1953). "Mitosis." Columbia Univ. Press, New York.

63. Sparrow, A.H., Price, H. and Underbrink, A. (1971). Brookhaven Symp. in Biol. No. 23 (in press).

64. Sparvoli, E., Gay, H. and Kaufmann, B.P. (1965). Chromosoma 16: 415.

65. Stubblefield, E. and Wray, W. (1971). Chromosoma 32: 262.

66. Trosko, J.E. and Wolff, S. (1965). J. Cell Biol. 26: 125.

67. Wolff, S. (1963). "Radiation-Induced Chromosome Aberrations." Columbia Univ. Press, New York.

68. Wolff, S. (1969). Internat. Rev. Cytol. 25: 279.

69. Yoshikawa-Fukada, M. and Ebert, J.D. (1971). Biochem. Biophys. Res. Comm. 43: 133.

70. Yunis, J.J. (1965). "Human Chromosome Methodology." Academic Press, New York.

DISCUSSION

SPARROW: You implied, I believe, that you are skeptical about the existence of 1/2 chromatids. If there are no 1/2 chromatids, how do you explain radiation-induced side-arm bridges, as reported by several investigators?

DU PRAW: Side-arm bridges certainly occur, and tell us that recombination can take place between elements within a chromatid that are much smaller than the chromatid itself. However, these elements do not have to be half-chromatids; they might just as well be the side-by-side segments of 200-500 Å fiber that we see with the electron microscope. Prof. B. A. Kihlman has recently devoted much of a review article to making this important distinction (Kihlman, Adv. Cell Mol. Biol. 1, 59 (1971)). He and his collaborators have repeated and confirmed earlier experiments by Östergren and Wakonig, who were the first to suspect that induced aberrations such as side-arm bridges do not necessarily mean that the chromatids contain half-chromatids. The reasoning was that if these are half-chromatid anomalies, then after one round of replication, they should appear as single chromatid anomalies. On the other hand, if the aberrations involve side-by-side loops of a folded chromatid axis, then after one round of replication they should be whole chromosome aberrations, i.e. affecting both sister chromatids.

When the experiment was done, the results were inconsistent with the half-chromatid hypothesis.

GALINSKY: You showed photographs with non-homologous chromosomes connected by fine strands of chromosomal material. How do sticky chromosomes produced by chemicals or radiation relate to such connections and to your interpretation of them?

DU PRAW: The fact that non-homologous chromosomes are often connected by fine chromatin fibers has been observed by many electron microscopists working with whole-mount preparations. How to interpret these is still a problem, and an entire chapter of my book "DNA and Chromosomes" was devoted to exploring various possibilities. We generally assume that non-homologous chromosomes behave independently at mitosis, but many exceptions are known (e.g., Priest and Shikes, J. Cell Biol. 47, 99 (1970)). "Stickiness", as applied to chromosomes, is not a very precise term. However, it brings to mind the so-called "sticky" ends of phage lambda chromosomes. These represent short single-stranded regions that protrude from opposite ends of a linear DNA molecule and which have complementary base sequences capable of hydrogen bonding together into a circular molecule. Perhaps the stickiness produced in higher chromosomes by radiation is due to breakage of the connecting fibers, and the formation of "frayed" DNA ends. If this inter-chromosome DNA contains fairly simple, repetitive base sequences (as in chromosomal satellite DNA), then it might readily hydrogen bond with other frayed ends. To the light microscopist, the chromosomes would appear to be sticky. In this connection it is interesting that the stickiness is sometimes reported to disappear after an intervening cycle of DNA replications. This, of course, would abolish any single-stranded DNA at the frayed ends. All of these ideas, I hasten to add, are still highly speculative.

HIS4 A GENE COMPLEX OF SACCHAROMYCES CEREVISIAE

Barbara Shaffer, Stuart Edelstein, and Gerald R. Fink

From the Section of Genetics, Development and Physiology
and the Department of Biochemistry
Cornell University
Ithaca, New York   14850

Abstract. The protein encoded by the his4 region of S. cerevisiae has
been purified. The purified protein has a molecular weight of 80,000
and consists of two subunits of MW 40-50,000, held together by disulfide
bonds. This evidence on the subunit structure places restrictions on the
interpretation of this fungal gene cluster.

INTRODUCTION

In yeast and other fungi, the genes coding for the individual enzymes
in most biosynthetic pathways are scattered throughout the genome, in contrast to the situation which exists in the enteric bacteria (1). A few
instances of gene clustering have been described in the fungi (2,3,4).
These genetic regions encode the information for two or more enzymic
activities in the same biochemical pathway. A common feature of the known
gene clusters is that the enzymic activities which they specify are physically associated. Recent reviews (2,5) have centered on a number of
paradoxes in the biochemical and genetic data which have so far prevented
a decision as to whether the fungal gene cluster is equivalent to a bacterial operon or to a single cistron. These genetic regions in some
respects resemble fused genes, analogous to the fusion between the hisC
and hisD gene in the Salmonella histidine operon (6). One step towards an
understanding of the gene cluster is the isolation and characterization of
the protein product, the enzyme complex.

250

The his4 region in Saccharomyces cerevisiae (Fig 1) codes for an enzyme complex which catalyzes three nonsequential steps in the pathway of histidine biosynthesis; the second, third and tenth. The associated enzymic activities are PR-AMP[1] cyclohydrolase (his4A), PR-AMP pyrophos- phohydrolase (his4B) and histidinol dehydrogenase (his4C). The purification of the his4 enzyme complex and the physical nature of the purified protein are described in this report.

PURIFICATION PROCEDURE

An outline of the procedure used to isolate the his4 complex and a summary of the results are given in Table I. Cyclohydrolase activity was assayed at the end of each purification step as indicated. Histidinol dehydrogenase activity was used when monitoring chromatographic procedures and for the calculation of specific activities, because this enzymic activity is the most stable of the three and the assay is the most accurate. Specific activities of histidinol dehydrogenase cannot be accurately com- puted prior to the Sephadex G50 step because of interference by NADH oxidase and a high rate of endogenous reduction without the addition of substrate. Adams (7) reached similar conclusions in his early studies with this enzyme. Specific activities were calculated only after step 4, which separates many of these interfering activities from the his4 complex. Only after step 5 is the rate strictly proportional to the enzyme concentration and is there no reduction of NAD^{+} without the addition of histidinol. The results from two large scale purifications are reported in the following sections. Unless otherwise indicated, the data are from the second of these.

---

[1] The abbreviations used are: PR-AMP, PR-ATP, N-1-(5-phosphoribosyl) adenosine mono-and triphosphate; PMSF, phenylmethylsulfonylfluoride.

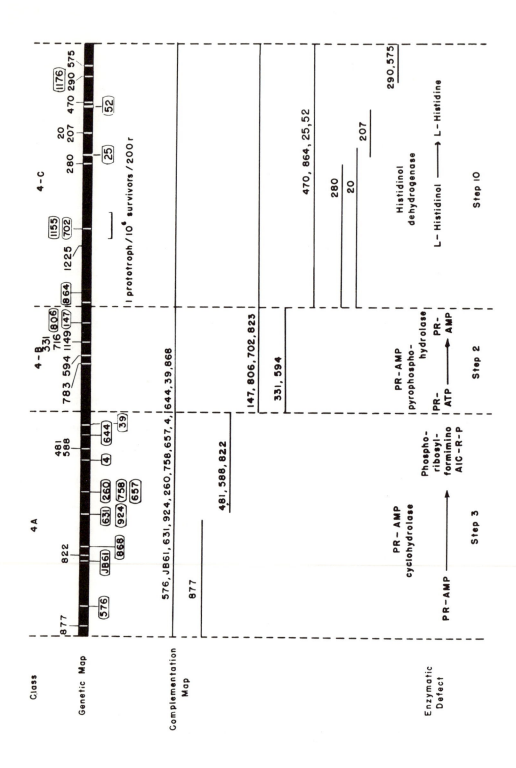

FIGURE LEGENDS

Fig. 1. The organization of the his4 region of yeast. At the top of
the figure is a genetic map of the region. The numbers circled
on the map designate nonsense mutations. Noncomplementing
mutants are placed below the line indicating the chromosome. The
regions A, B, and C are based on the complementation map. Allelic
complementation occurs within A and C, but that in C is much more
detailed than is shown here. Reactions in the pathway of histidine
biosynthesis which are affected by mutations in the A, B, and C
regions are shown below the appropriate region of the complementation map.

TABLE I

PURIFICATION OF THE HIS4 COMPLEX FROM SACCHAROMYCES CEREVISIAE

Purification Step	Volume (ml)	Cyclohydrolase Activity	Total Protein (mg)	Total Activity* (units)	Specific Activity (units/mg)
1. Crude Extract	700	+	12000	-	-
2. MnCl$_2$ super-natant	700	+	11200	-	-
3. Ammonium sul-fate concentrate	150	+	3300	-	-
4. G50	285	+	2400	180	0.075
5. DEAE-cellulose concentrate	80	+	376	140	0.37
6. DEAE-Sephadex concentrate	42	-	22.3	30	1.34
7. CaPO$_4$ concen-trate	7.6	-	6.8	14	2.06
8. G150 concen-trate	5.3	-	2.1	5.8	2.76

* 1 unit is equivalent to one micromole of NAD$^+$ reduced per milligram of protein per ml.

STABILITY OF THE ENZYME COMPLEX

Two of the his4 enzymic activities, cyclohydrolase and pyrophosphohydrolase, were found to be extremely labile, while the third, histidinol
dehydrogenase is relatively more stable. Crude extracts lose the majority
of their cyclohydrolase and pyrophosphohydrolase activity within three days,
when stored at 4°. These two activities are always lost simultaneously.
Under no conditions have we seen a preparation of "wild-type" enzyme lose
only one of these two enzymic activities.

There are at least two different sources of instability. The lability
of the cyclohydrolase and pyrophosphohydrolase activities in crude extracts
is in part due to the action of a proteolytic enzyme(s). The action of the
proteolytic enzyme(s) on the his4 complex is specific. Experiments, using
sucrose gradient ultracentrifugation, show that the his4 complex loses a
fragment of molecular weight ca. 30,000 upon standing; this loss of molecular weight is concomitant with the loss of the cyclohydrolase and pyrophosphohydrolase activities. Addition of PMSF (1×10^{-4}M) to crude extracts
prevents this shift in molecular weight, as well as preserving cyclohydrolase and pyrophosphohydrolase activities. Our data indicate that the
protease(s) is not randomly cleaving the complex; the same molecular weight
reduction is always seen and the remaining activity, histidinol dehydrogenase, appears as a symmetrical peak in sucrose gradients. In large scale
purifications, PMSF was used in the buffers for the purpose of preserving
cyclohydrolase and pyrophosphohydrolase activities.

Even in the presence of PMSF, however, cyclohydrolase and pyrophosphohydrolase activities are considerably more unstable than histidinol dehydrogenase activity. Previously reported experiments (8) showed that the three
enzymic activities coded for by the his4 genetic region remain associated

through step 6 of the purification procedure. However, in this earlier
work less than 50 g of cells were processed, in order that the purification procedures could be rapidly completed in one or two days. In large
scale preparations (ca. 800 g cells), requiring eight to ten days,
significant losses of the cyclohydrolase and pyrophosphohydrolase
occurred. This loss is considered different from that which is prevented
by PMSF since this second type of inactivation results in no change in
molecular weight, in electrophoretic mobility on acrylamide gels, or in
isoelectric point (see below).

AMPHOLINE ELECTROFOCUSING

The purified protein was subjected to electrophoresis in the presence
of amphoteric compounds pH3-6. Under these conditions the protein is quite
stable and does not precipitate at its isoelectric point. A single protein
peak coincident with the peak of histidinol dehydrogenase activity was
found (Figure 2). The isoelectric point of the protein is pH 4.7. The
identical isoelectric point was determined for the trifunctional enzyme
examined in preparations taken through Step 5 of the purification procedure.
The active fractions were combined, concentrated by ultrafiltration, and the
carrier ampholytes removed from the concentrate by chromatography on
Sephadex G-150.

POLYACRYLAMIDE GEL ELECTROPHORESIS

The concentrates after the DEAE-cellulose, DEAE-Sephadex, $CaPO_4$
and Sephadex G-150 steps were subjected to polyacrylamide slab gel electrophoresis, the gel was stained for histidinol dehydrogenase activity
(Figure 3).

At each stage of the purification, the samples showed a single band
of enzyme activity. If the loss of cyclohydrolase activity resulted from
the dissociation of the <u>his4</u> complex into smaller, dissimilar polypeptide
chains, this dissociation might well have been reflected in an altered

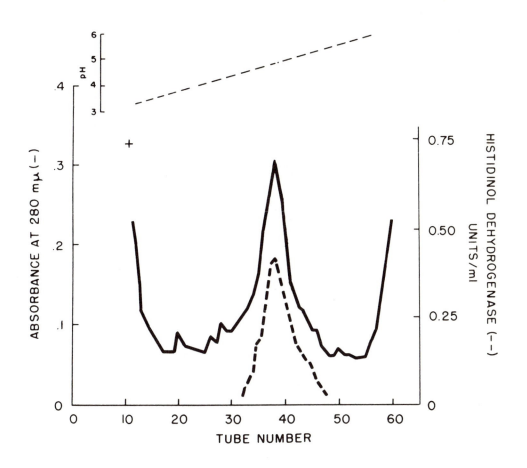

Fig. 2.   Ampholine electrofocusing of the purified his4 protein. The concentrate from Step 8 of the purification procedure (April, 1970
preparation, see text) was applied to a 110 ml ampholine column,
pH 3-6, and electrofocused for 48 hr at 700 volts and 4°. The
gradient was pumped out of the column at 2 ml/min and 7.8 ml fractions collected. The fractions were assayed for histidinol
dehydrogenase activity and protein content (absorbance at 280 mu).
Fractions 32-46 were combined, concentrated and chromatographed
on a Sephadex G150 column to remove carrier ampholytes.

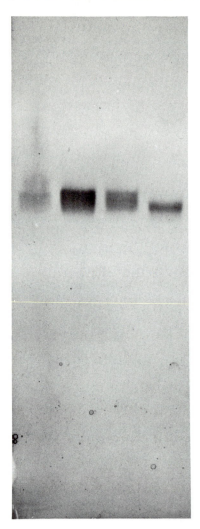

Fig. 3. Polyacrylamide slab gel monitoring of the purification of the

his4 protein. Samples (50 µl in 20% sucrose) were subjected to

electrophoresis in a 6½% vertical gel at 400 volts at 4° for 2 hr.

The gel was stained for histidinol dehydrogenase activity by the

method of Loper and Adams (10). From left to right, the samples

were from the concentrates after the DEAE-cellulose, DEAE-Sephadex,

$CaPO_4$ and Sephadex G150 steps in the purification procedure.

migration pattern for the dehydrogenase activity. However, the position

of the histidinol dehydrogenase activity on the gel was not altered by the

DEAE Sephadex procedure, during which the cyclohydrolase activity was lost,

or by any subsequent procedure.

To estimate the purity of the protein, samples from Step 8 of the puri-

fication procedure were analyzed by polyacrylamide disc gel electrophoresis.

The protein patterns obtained from two different purifications are shown in

Figure 4. Each gel showed a single dark band when stained with Coomassie

blue.

ULTRACENTRIFUGATION

In order to study the molecular properties of the enzyme, molecular

weight measurements were performed under a variety of conditions. The

first measurements were concerned with the enzyme in solvents which main-

tained the native structure to obtain an estimate of its size and its degree

of homogeneity. Results of a sedimentation equilibrium experiment in

dilute, neutral buffer are presented in Figure 5. The data obtained with

the absorption optical system and scanner indicate that to within the limits

of the experimental methods, the protein behaves as a single component. The

distribution corresponds to a molecular weight of 80,000 on the basis of the

assumed partial specific volume. This molecular weight value is not sub-

stantially different from that found by sucrose gradient centrifugation of

either crude extracts or earlier stages of the purification where all three

activities are present. This leads us to believe that the molecular weight

of the protein, although it lacks cyclohydrolase and pyrophosphorylase

activities, has not been radically altered by the purification procedure.

Since earlier work (8) had led to the supposition that the purified

protein would contain more than one polypeptide chain, experiments were

conducted with strong solutions of guanidine-HCl, (Gu-HCl), a dissociating

agent. However, the initial experiments showed no evidence of subunits.

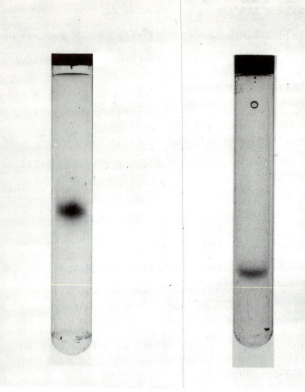

Fig. 4. Polyacrylamide disc gel electrophoresis of the purified his4

protein. The gels represent the purified protein from two

different preparations. Samples containing approximately 20 μg

of protein from the Step 8 concentrate were subjected to electro-

phoresis at 4° in 7½% acrylamide gels at 3 ma per tube. Migration

was toward the anode, which is at the bottom of the photograph.

(The samples were subjected to electrophoresis for different

periods of time).

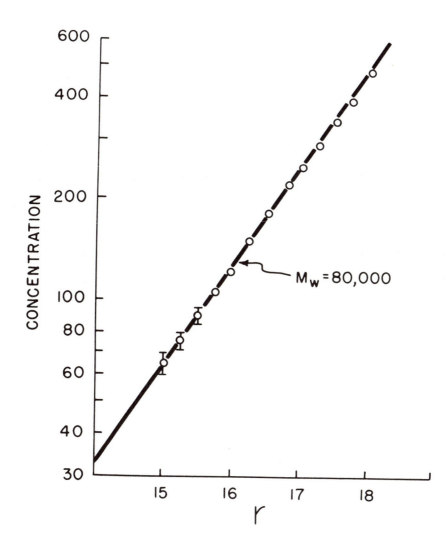

Fig. 5. Sedimentation equilibrium of the native <u>his4</u> protein. Concentration is mm scanner recorder trace versus distance in cm on the trace. Data was recorded with the absorption optical system scanner (10, 11) with light of 280 mμ after 18 hours of centrifugation at 26,000 rpm in a Model E ultracentrifuge (20°). The experiment was conducted with 0.1 ml of a solution containing approximately 0.3 mg/ml protein in 0.1 Tris buffer, pH 7.5. For convenience, the data were examined in the form of concentration versus distance on the photographic plate or trace. The use of r on the abscissa of the plot instead of r^2 introduces only insignificant errors since the range of r is so small (12). The value of the partial specific volume, $\bar{v}$, was assumed to be 0.74 μ/gm, since no amino acid composition data is known.

As seen in Figure 6, the distribution of protein at sedimentation equilibrium in a solution of 5 M Gu-HCl is consistent with a homogeneous material corresponding to a molecular weight of approximately 80,000.

Since the surprising result was obtained that the protein is not resolved into smaller units in Gu-HCl in spite of earlier indications of subunits, the possibility that subunits are present but held together by disulfide linkages was considered. To test this possibility, measurements on the protein in solutions of Gu-HCl were repeated, but with the addition of mercaptoethanol. As seen in Figure 7, under these conditions breakdown of the protein is observed, with a molecular weight of 45,000 obtained. A high degree of apparent homogeneity is still retained. It should be noted that if the assumed $\bar{v}$ is in error, the value of the molecular weight in Gu-HCl could require significant adjustment due to the high solution density and the term $(1-\bar{v})$ which magnifies error in $\bar{v}$. Therefore, the value of 45,000 must be considered approximate.

With the appearance of subunits under the reducing and dissociating conditions, the possibility that reducing conditions alone would lead to subunits was tested. However, as seen in Figure 8, the addition of mercaptoethanol without Gu-HCl has no significant effect on the behavior of the protein.

DISCUSSION

The existence of genetic regions specifying multiple enzymic activities has focused on shortcomings in the use of genetic analysis to define gene boundaries. In practice, geneticists rely on the complementation test as well as genetic mapping to determine the number of genes. The basic assumption of the complementation test, that a mutation in one gene does not affect the activity of another gene, is frequently not valid. In an operon, the unit of function is not only the polypeptide chain but also the polycistronic message. Mutants causing defects in the translation or

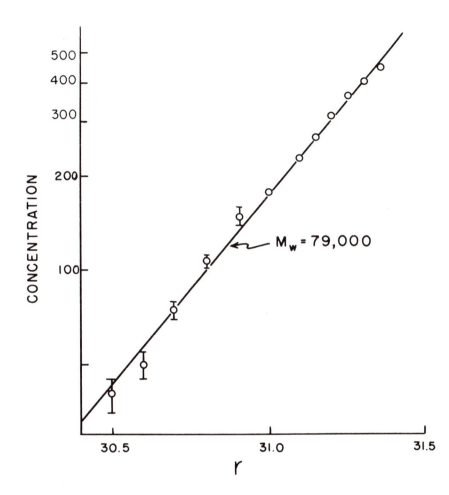

Fig. 6.   Sedimentation equilibrium of the his4 protein in a solution containing Gu-HCl. Concentration is in microns displacement versus
distance in units of the microcomparator. Data were recorded with
the interference optical system after 18 hours of centrifugation
at 36,000 rpm. The solution (0.1 ml) contained approximately 0.3
mg/ml protein in 0.1 M Tris buffer, pH 7.5 and 5 M Gu-HCl.

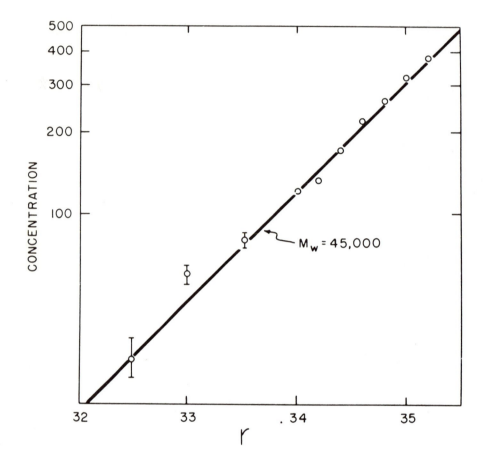

Fig. 7. Sedimentation equilibrium of the <u>his4</u> protein in a solution of
         Gu-HCl and mercaptoethanol. Details as in Fig. 6 with rotor
         speed 26,000 rpm, Gu-HCl concentration 6 M and mercaptoethanol
         added to a level of 0.1 M.

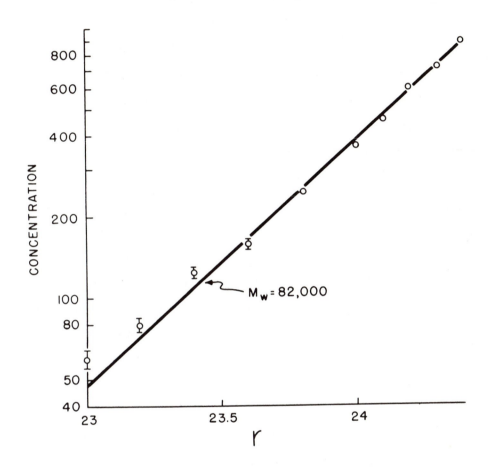

Fig. 8. Sedimentation equilibrium of the his4 protein in the presence of

mercaptoethanol. Experimental details as in Fig. 6 with the

ommission of Gu-HCl.

transcription of this message (nonsense, promoter, and regulatory mutations)
can simultaneously affect the function of some or all the genes in the
operon. Strains carrying these pleiotropic mutations frequently fail to
complement strains with a mutation in any of the other genes within the
operon; a result which catagorizes the whole operon as a single functional
unit. Indeed, once it is known that a genetic region specifies different
proteins, the presence of mutations causing pleiotropic effects is taken as
<u>prima</u> <u>facie</u> evidence of operon-like organization.

A region like <u>his4</u> which specifies a multifunctional protein is
especially difficult to analyze genetically. The protein encoded by <u>his4</u>
might be a polymer of identical or different polypeptide chains. In the
former case a single gene is implicated and in the latter two. The
pleiotropic effects of nonsense, promoter, and regulatory mutations on
the activity of a multifunctional protein can easily be mimicked by mutational changes affecting protein-protein interactions. Consequently,
genetic analysis does not provide the resolution necessary to distinguish
between two contiguous genes in an operon and a single gene specifying a
multifunctional protein.

Any model of the structure of the <u>his4</u> region must explain the evidence
of previous studies on nonsense mutations in this region. 1) Nonsense
mutations in <u>his4</u> show complete polarity (e.g. 260, 644, 39 etc. Fig. 1).
No residual <u>his4C</u> activity has been found either by complementation or
enzyme assay. 2) Nonsense mutations 864, 1155, 25, 52, and 1176, which
map at various positions throughout the <u>his4C</u> region, result in partial
complexes with residual A and B activities of MW 45,000 despite the fact
that the distance between 864 and 1176 is equivalent to half the size of
the <u>his4</u> region (8, 13).

We suggest two models (Figure 9) to explain the biochemical and genetic evidence accrued so far. The first is the operon model. According to this hypothesis, the region specifies two polypeptide chains; one coded for by the A and B regions and the other coded for by the C region. The two different polypeptides would be associated by disulfide bonds into a hetero-polymeric aggregate. The genetic and biochemical data correlate well with this hypothesis. Based on map distances, the length of the his4 region has been estimated to be 2250 nucleotides and would therefore code for a protein of molecular weight 86,000 (13). The C region is about half the genetic map and should code for a protein of MW 40-50,000. This size for the C product agrees well with our finding that the nonsense mutations in his4C result in partial complexes of MW 45,000 with A and B activity. According to the operon model, these 45,000 MW proteins with A and B activity are specified by the his4AB region and fail to form disulfide bonds with the prematurely terminated his4C nonsense peptide. One complication with this model is that it predicts that an AB subunit in a C nonsense mutant should hybridize with C subunits provided by a normal complex. Experiments with heterozygous diploids (i.e. 588 x 864) and mixtures of extracts have failed to detect any hybridization between the subunits (2).

The second model envisions the his4 region as a single cistron encoding a single multifunctional polypeptide chain. Based on the biochemical evidence, the active protein would have to be a dimer of MW 80,000 composed of two identical subunits of MW 45,000. If the primary gene product is a protein of MW 45,000 there is no simple hypothesis by which the 45,000 MW fragments produced by his4C nonsense mutants can be explained. An alterna-tive to this hypothesis (Figure 9) is that a single chain of 80,000 MW is produced, disulfide bonds form, and the correctly folded protein is sub-sequently cleaved by proteolytic enzymes to form the active complex. This could occur by a process similar to that described by Jacobson and Baltimore

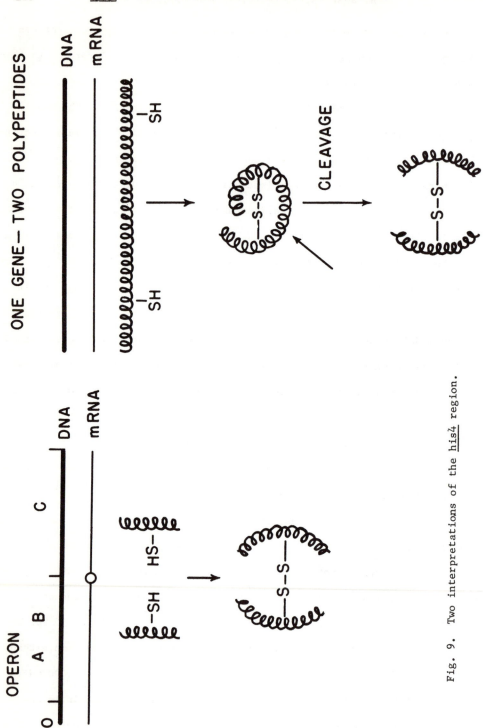

Fig. 9. Two interpretations of the <u>his4</u> region.

for the maturation of polio virus proteins (14). This "one gene-two polypeptide model" would explain the absolute polarity of his4 nonsense mutants
(e.g. his4-39, -260, -644, etc.) but provides no explanation for the uniformity of size of his4C nonsense fragments. Perhaps, prematurely terminated
proteins are degraded to yet smaller proteins. Fragments of MW 45,000 may
be the smallest fragments with activity and, therefore, the only ones
detectable by the enzyme assay. Both an analysis of the subunit structure
of the his4 protein as well as a study of the role of proteolytic enzymes in
protein maturation appear to be necessary in order to gain an understanding
of this fungal gene cluster.

Acknowledgements: This work represents experimentation supported by
N.I.H. grant GM 15408 to G.R.F., N.S.F. grant GB 8773 to S.E. and N.I.H.
training post-doctoral fellowship GM 01035-09 to B.S.

REFERENCES

1. Demerec, M., Proc. Natl. Acad. Sci. U.S., 51, 1057 (1964).

2. Fink, G. R., Metabolic Pathways, Chap. 7, p. 200, Greenberg and Vogel editors; Academic Press, New York (1970).

3. Rines, H. W., Case, M. E., and Giles, N. H., Genetics, 61, 789 (1969).

4. DeMoss, J. A., and Wegman, J., Proc. Natl. Acad. Sci. U.S., 54, 241 (1965).

5. Calvo, J. and Fink, G. R., Annual Rev. Biochem. (in press).

6. Yourno, J., Kohno, T., Roth, J. R., Nature, 228, 820 (1970).

7. Adams, E., J. Biol. Chem., 217, 325 (1955).

8. Shaffer, B., Rytka, J. and Fink, G. R., Proc. Natl. Acad. Sci. U.S., 63, 1198 (1969).

9. Loper, J., and Adams, E., J. Biol. Chem., 240, 788 (1965).

10. Goldberg, M. E., and Edelstein, S. J., J. Mol. Biol., 46, 431 (1969).

11. Edelstein, S. J., Rehmar, M. J., Olson, J. S., and Gibson, Q. H., J. Biol. Chem., 245 (1970) in press.

12. Yphantis, D. A., Biochemistry, 3, 297 (1964).

13. Shaffer, B., and Fink, G. R., Genetics, 61, 54 (1969).

14. Jacobson, M. F. and Baltimore, D., P.N.A.S. 61, 77 (1968).

CHROMOSOMAL LOCALIZATION OF REPETITIVE DNA

Ronald A. Eckhardt
Department of Biology, Yale University
New Haven, Connecticut

For about seven years now, the DNA of eukaryotic organisms has been known to include families of nucleotide sequences which are repeated to varying degrees.[1,2] This observation has provided many new insights into the molecular organization of the eukaryotic genome. As discussed in the opening sessions of this symposium, substantial progress has been made recently in understanding the evolutionary significance of repetitive DNA. In addition, several papers were presented dealing with the evolution of genes and gene groups. The studies on which I shall report have taken much of this information at the molecular and genic levels of genetic organization and have used it to help interpret problems at a relatively higher level of organization, that of the eukaryotic chromosome.

Briefly stated, eukaryotic DNA can be divided into three arbitrary classes of nucleotide sequences delineated by the degree of repetition within each class. These are: (1) a highly repetitive class, (2) an "intermediately," or moderately repetitive class, and (3) an essentially non-repetitive, or "unique" class of nucleotide sequences. Much of the evidence for this division is derived from studies of DNA-DNA reassociation kinetics.[1,2]

Until about four years ago, very little was known concerning the spatial and structural distribution of repetitive DNA in the eukaryotic genome. Since then, various approaches have been used to study this

271

distribution. Many of these investigations have involved fractionating either nuclear components, size classes of chromosomes, or different types of chromatin and then examining the DNA isolated from each fraction.[3,4,5,6] Through these experiments it has been shown that certain repetitive DNAs are distributed throughout the chromosomal set.

It was not until 1969, however, that a method was available for the direct examination of the distribution of specific nucleotide sequences in eukaryotic chromosomes. The technique of *in situ*, or cytological hybridization involves the molecular hybridization of radioactively labeled nucleic acids to the DNA present in cytological structures and the subsequent visualization of this hybridization by means of autoradiography.[7,8,9,10,11] This technique has proven to be a useful tool for cytogenetic analysis combining the specificity of molecular hybridization with the precision of cytological localization.

In the following report, I shall discuss some of the ways in which the technique of cytological hybridization has been modified to extend its general applicability and then briefly review the progress made during the past two years in determining the chromosomal localization of several repetitive DNAs. Some of these repetitive DNAs have well defined cellular functions such as the DNAs coding for ribosomal RNA, 5S RNA, and transfer RNA. Others, such as highly repetitive DNA and the ill-defined bulk of the moderately repetitive DNAs, have no known cellular functions. In addition, I shall present some new data from our laboratory pertaining to the under-replication of repetitive DNA in the polytene cells of various species of *Drosophila*.

IN SITU MOLECULAR HYBRIDIZATION

The technique of *in situ*, or cytological hybridization was developed independently about three years ago in three laboratories.[7,8,9,11] Since these early reports appeared, several additions and modifications have been made to the basic procedure which have increased its sensitivity and selectivity thus extending its general usefulness for studying the chromo- somal distribution of repetitive DNA.

In the original procedure, *in vivo* labeled radioactive RNA was used for the hybridization to the DNA of cytological structures. The use of RNA from living cells as the hybridizing species imposed several serious limitations. First of all, it was difficult to obtain RNA above certain practical levels of specific activity because of the inherent limitations of *in vivo* labeling. Secondly, the number of single, specific RNAs separable from the great diversity of sequences within intact cells was also limited. Both of these factors are of paramount importance in achieving the maximum sensitivity of detection in cytological hybrids.[10]

A later modification which eliminated the necessity of fractionating pure samples of specific RNAs was the demonstration that radioactively labeled single-stranded DNA could also be used as the hybridizing species.[12,13,14] Since DNA can be fractionated by physical techniques more selective than those available for most RNAs, this modification had the additional benefit of providing a greater variety of specific nucleotide sequences to be examined. However, like the RNA in the original procedure, the DNA was labeled *in vivo* and therefore it was difficult to obtain samples of high specific activity.

One recent development for obtaining nucleotide sequences of ex- tremely high specific activity involves their synthesis *in vitro* using DNA-dependent RNA polymerase transcription systems.[15,16,17] By this method, the complementary RNAs (cRNAs) thus produced are as "hot" as the precursors introduced into the reaction mixture. When used in conjunction with *in situ* hybridization, the sensitivity of detection of specific nucleotide sequences is greatly increased.

While *in vitro* transcription systems can provide nucleotide sequences

of extremely high specific activity, there are practical limits to the
number of single DNA sequences that can be isolated as pure samples and
then copied. It is possible, however, to investigate the distribution
of mixtures of nucleotide sequences using "pulse-labeled" RNA from intact
cells,[18] or by using either cRNA to total DNA[19,20] or to some separable
fraction of the total DNA.[13,19,21,22,23,24,25,26]

Another possibility would be to use chromatin preparations as
template material in transcription systems. Doing this would allow one
to assess the overall genetic activity of cells in particular states of
development or differentiation.[27,28,29] A major drawback here, as in any
cytological hybridization, is that it is not possible to detect the
localization of sequences below a certain degree of repetition. Gall
and Pardue[10] have estimated that the smallest amount of a specific DNA
sequence detectable by *in situ* hybridization to be about 10^8 daltons.
Assuming most "unique" sequences to be about 10^5-10^6 daltons of DNA, their
occurrence could not be examined except for special situations such as in
some polytene cells.

Other possible ways in which nucleotide sequences of high specific
activity could be obtained for cytological hybridization include the use
of DNA-dependent DNA polymerase,[30] the newly characterized RNA-dependent
DNA polymerases,[31,32,33,34] or some of the restrictive DNA-dependent
RNA polymerases found in eukaryotic organisms.[35,36,37,38] None of these
enzyme systems has as yet been extensively exploited for this purpose but
their application could permit investigation of problems which are otherwise currently unapproachable.

THE LOCALIZATION OF DNA CODING FOR RIBOSOMAL RNA

One of the earliest applications of *in situ* molecular hybridization
was the demonstration of the presence of ribosomal DNA (*ie:* that DNA coding
for ribosomal RNA) in the extra-chromosomal DNA forming the "nuclear cap"

during late pachytene stages of oogenesis in *Xenopus laevis*.[8,9,11] Other

investigations had previously established that during pachytene there is

a selective amplification of the ribosomal cistrons in the oocytes.[39,40]

[41,42] Using tritium-labeled rRNA produced *in vivo* it was possible to

demonstrate the selective binding of rRNA to the nuclear cap DNA whereas

the chromosomal DNA remained essentially unlabeled (Figure 1).

These observations were later extended by Pardue and Gall[13] who

hybridized pachytene stage nuclei with radioactively labeled DNA. As

shown by earlier investigations, *Xenopus* DNA can be separated on CsCl

gradients into a high density satellite DNA which contains most of the

ribosomal cistrons.[43] Through a series of successive isopycnic centri-

fugations, the rDNA can be removed from the main band DNA. When Pardue

and Gall[13] hybridized both the pachytene chromosomes and the nuclear cap

DNA with tritium-labeled main band DNA, the converse cytological picture

of the results obtained with rRNA was observed. In this case only the

chromosomes were labeled while the nuclear cap remained essentially

unlabeled (Figure 2). Together these investigations gave clear evidence

of the specificity of cytological hybridization.

The distribution of ribosomal cistrons has been examined in other

eukaryotes. Pardue, Gerbi, Eckhardt, and Gall have determined the

cytological distribution of rDNA in three genera of flies.[25] These

organisms were choosen for study because of the presence of polytene

chromosomes in various tissues and therefore, greater resolution of

chromosomal sub-structure was possible. One unusual feature of this

study was that the hybridizing species was cRNA copied from isolated

rDNA of *Xenopus*. Brown, Weber, and Sinclair had earlier shown that

Xenopus rRNA would specifically hybridize to the rDNA of other eukaryotes.

[44] **Pardue** *et al.*[25] showed that cRNA copied from *Xenopus* rDNA would like-

wise selectively hybridize to Dipteran rDNA. Because of problems in

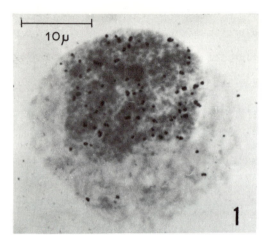

Figure 1. Slide hybrid of *Xenopus* pachytene oocyte nucleus hybridized
with *in vivo* labeled *Xenopus* ribosomal RNA-H^3 [9] Note the localization
of silver grains over the nuclear cap of amplified rDNA while the chromosomes remain essentially unlabeled.

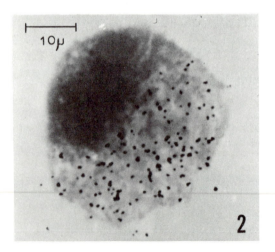

Figure 2. Slide hybrid of *Xenopus* pachytene oocyte nucleus hybridized
with *in vivo* labeled *Xenopus* DNA-H^3 lacking the ribosomal cistrons.[13]
There is significant labeling over the chromosomes but not over the cap
of amplified rDNA.

obtaining Dipteran rRNA of high specific activity and because of the

convenience of obtaining *Xenopus* rDNA for use in RNA polymerase transcrip-

tion systems, heterologous hybridizations were used to study the cytolog-

ical distribution of Dipteran rDNA.

Briefly summarized, it was found that the ribosomal cistrons of

Drosophila hydei could be demonstrated only within the body of the nucleo-

lus in polytene nuclei rather than structurally confined to distinct

chromosomal bands (Figure 3). Previous studies had shown that the nucleo-

lus organizer regions of *D. hydei* are located in the heterochromatin of

the X and Y chromosomes.[45] These regions become part of the chromocenter

in polytene cells. From the results of the cytological hybridizations,

it would appear that the rDNA is spatially arranged in polytene cells in

such a way that it extends outward from the chromocenter to form the

nucleolus while the rest of the heterochromatin remains compacted.

In the same study, the rDNA of the Sciarid fly, *Rhynchosciara*

hollaenderi, was shown to be present in three structures: (1) the known

nucleolus organizer region (NOR) on the X chromosome, (2) an unsuspected

NOR on the C chromosome, and (3) micronucleoli scattered throughout the

nucleus (Figure 4). The ribosomal cistrons of *Sciara coprophila* were

found on the X chromosome, which contains a prominent NOR, and in certain

micronucleoli in the nucleus. No association of rDNA could be found with

the DNA puffs of the two Sciarid flies. Beyond the actual findings, this

study conclusively demonstrated that it is possible to determine the

localization of specific genes at the chromosomal level using *in situ*

hybridization.

The distribution of rDNA has also been studied in another organism.

In a recent investigation, Buongiorno-Nardelli and Amaldi have shown that

the ribosomal genes of the golden Chinese hamster are localized in

association with the nucleolus in brain and liver cells.[7] Their study

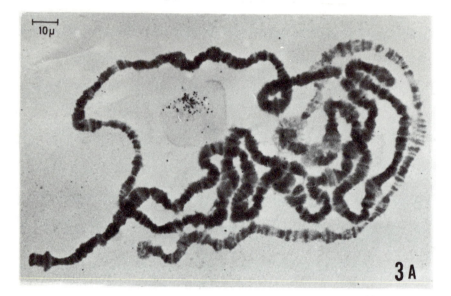

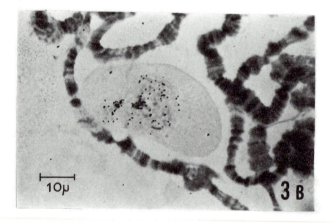

Figure 3. Slide hybrids of the polytene chromosomes and nucleolus from the salivary gland cells of *Drosophila hydei* hybridized with *Xenopus* ribosomal RNA-H[3] synthesized *in vivo*.[25] A: Entire *D. hydei* genome. Ribosomal cistrons were detected only within the body of the nucleolus. B: An enlargement of a nucleolus showing the localization of silver grains over the nucleolar DNA.

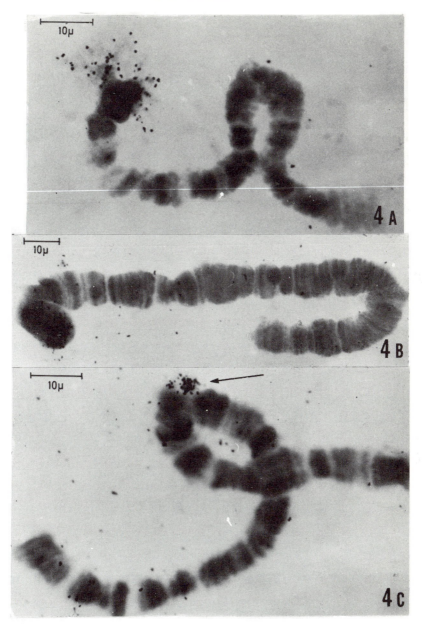

Figure 4. Slide hybrids of *Rhynchosciara* salivary gland chromosomes
hybridized with complementary RNA transcribed *in vitro* from isolated
Xenopus rDNA.[25] A: The X chromosome showing labeling over the nucleolar DNA. B: The C chromosome showing labeling over the terminal heterochromatin. C: Portions of the A chromosome with an associated micro-
nucleolus (arrow) showing heavy labeling. The B chromosome (not illustrated) does not possess rDNA nor does the A chromosome.

was the first to use tissue sections for cytological hybridizations.
Using their procedure, it should be possible to examine the distribution
of nucleotide sequences in tissues which are not easily handled as squash
or smear preparations. In addition, Jacob, Todd, Birnstiel, and Bird
have reported success in examining the distribution of rDNA in *Xenopus*
oocytes at the ultrastructural level by E. M. autoradiography.[46] This
modification obviously extends the limits of cytological resolution
possible.

Many biologically interesting problems remain to be examined, for
example, the distribution of rDNA in the extra-chromosomal DNA bodies of
various insects and in other situations where rDNA amplification is thought
to occur.[8] It should be relatively easy to investigate these problems
using the cytological hybridization procedure.

THE LOCALIZATION OF DNA CODING FOR 5S RNA AND TRANSFER RNA

Comparatively fewer studies are available which have investigated
the distribution of DNAs coding for 5S RNA and transfer RNA than those
which have examined the localization of ribosomal DNA. This is partially
due to difficulties in isolating these DNAs in workable quantities for
in vitro transcription. The cytological localization of 5S DNA and 4S
DNA has as yet been reported in the literature in only two organisms.

Amaldi and Buongiorno-Nardelli have studied the distribution of 5S
and 4S DNAs in tissue sections of liver and brain cells from golden
Chinese hamsters.[18] For the most part, the labeling was preferentially
localized in the perinucleolar dense chromatin. This is quite interesting
in view of their earlier observations which showed ribosomal DNA also to
be found primarily in this region.[7] Previous work in *Xenopus*[47,48] and
in HeLa cells[49] have shown that the 5S genes are not linked molecularly
to the nucleolus organizer. But as suggested by Amaldi and Buongiorn-
Nardelli, the spatial association in the cell of 5S genes with ribosomal

genes might facilitate ribosome assembly.

Wimber and Steffensen have investigated the distribution of 5S DNA in the polytene chromosomes of *Drosophila melanogaster*.[50] It was their finding that the 5S genes are primarily localized within region 56E-F of the right arm of the second chromosome. The ribosomal cistrons of *D. melanogaster* have been shown to be associated with the *bobbed* loci of the X and Y chromosomes.[51,52,53]

In a brief abstract, Steffensen and Wimber have also reported that the DNA coding for transfer RNAs in *D. melanogaster* was found in a number of sites throughout the chromosomal set.[54] These sites were indicated to correspond to loci associated with the *Minute* mutations of *Drosophila*. The involvement of 4S DNA in the *Minute* mutations has been suggested on other grounds for about five years.[55] The results of the cytological hybridizations seem compatible with this hypothesis. No information is as yet available concerning the location of genes for specific transfer RNAs.

LOCALIZATION OF OTHER REPETITIVE DNA'S

Considerable progress has been made during the past two years in investigating the chromosomal distribution of repetitive DNAs other than those for which a cellular function is known. Numerous reports of findings in a variety of eukaryotes have been published. The initial success in localizing highly repetitive DNA was reported in the mouse.[12,13,14,20] In this organism almost all of the highly repetitive DNA bands in CsCl gradients as a low density satellite. This situation obviously facilitates its isolation.

The distribution of mouse satellite DNA was examined by the *in situ* hybridization of *in vivo* labeled satellite DNA, and by the hybridization of cRNA copied from satellite DNA, to mouse metaphase chromosome preparations. Both approaches yielded comparable results. Highly repetitive DNA was found to be localized in the centromeric heterochromatin of each of the chromosomes with the possible exception of the Y chromosome (Figure 5). This result is consistent with earlier observations obtained using cell fractionation techniques.[3,4,5,6]

Recently Eckhardt and Gall have shown that the highly repetitive DNA (occuring as a low density satellite) of the fly, *Rhynchosciara hollaenderi*, is likewise found in association with the centromeric heterochromatin of each of the four chromosomes.[22,23] However, in this organism there were several other chromosomal regions which also contained highly repetitive DNA. In addition, Eckhardt and Gall have also examined the distribution of total repetitive DNA minus the satellite sequences. It was found that main band repetitive DNAs had a rather generalized distribution throughout the chromosomes including those areas shown to contain highly repetitive DNA (Figure 6). There was not, however, a consistent relationship between the amount of DNA present in a band (as judged by the intensity of staining in a band) and the density of silver grains over the band. Certain densely staining bands had fewer silver grains than lighter staining bands.

Other work by Rae has demonstrated that cRNA synthesized *in vitro* from repetitive DNA fractions of *Drosophila melanogaster* hybridizes with the chromocenter in salivary gland preparations as well as with certain bands in the chromosome arms.[26] Jones and Robertson have made similar observations using cRNA transcribed *in vitro* from total *D. melanogaster* DNA.[20] In a study of several species of the *Drosophila hydei* group, Hennig, Hennig, and Stein have demonstrated that certain repetitive satellite DNAs are localized in the centromeric heterochromatin while others may be more widely spread in the genome.[24] Arrighi, Hsu, Saunders, and Saunders have found that the heterochromatic portions of the sex chromosomes of *Microtis agrestis* are significantly enriched in repetitive

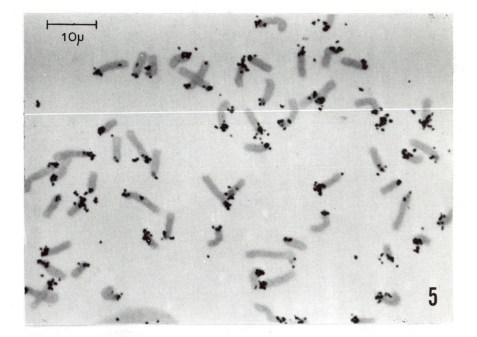

Figure 5. Slide hybrid of mouse metaphase chromosomes hybridized with complementary RNA transcribed *in vitro* from isolated mouse satellite DNA.[14] The cRNA bound to the centromeric heterochromatin of each of the mouse chromosomes with the exception of the Y chromosome which consistently appeared unlabeled.

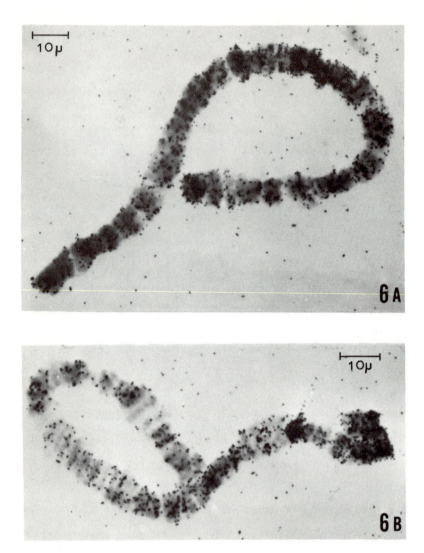

Figure 6. Slide hybrid of *Rhynchosciara* salivary gland chromosomes hybridized with complementary RNA transcribed *in vitro* from total DNA lacking the
highly repetitive satellite sequences.[23] The main band repetitive DNAs had
a generalized distribution throughout the chromosomal set. A: The A chromosome of *Rhynchosciara*. B: The B chromosome of *Rhynchosciara*. The distribution of silver grains over the C and X chromosomes (not illustrated)
was similar.

DNA.[21]

It appears then that repetitive DNAs can have a wide variety of distributions in different organisms. There seems to be a preference for the localization of highly repetitive sequences in the centromeric heterochromatin of many organisms but in others, there are multiple sites where highly repetitive DNA can be found. Other repetitive DNA's appear to have a more generalized distribution throughout the total genome.

UNDER-REPLICATION OF SATELLITE DNA IN *DROSOPHILA* POLYTENE CELLS

In the remainder of this report, I would like to consider some special aspects of the distribution of repetitive DNA in the polytene cells of *Drosophila*. As already mentioned, certain repetitive DNAs have been shown to be primarily localized in the centromeric heterochromatin of various Dipterans. The centromeric heterochromatin is included in the chromocenter of polytene nuclei in many species of *Drosophila*. In the mitotic chromosomes, however, the conspicuous centromeric heterochromatin comprises a relatively greater percentage of the total length of the chromosomes. This observation has led to the hypothesis that the centromeric heterochromatin fails to replicate during the formation of polytene chromosomes or at least replicates more slowly than the euchromatin.[45,56] Strong support for this hypothesis has been provided by the microspectrophotometric measurements of Rudkin,[57,58] of Berendes and Keyl,[59] and of Mulder, van Duijn, and Gloor.[60]

Since centromeric heterochromatin is thought to be under-replicated in polytene cells and since repetitive satellite DNA is found in this region, an important question can be asked: *is satellite DNA under-replicated during the formation of polytene cells?* This question has been examined by Gall, Cohen, and Polan from our laboratory.[19]

These workers have investigated the occurrence of satellite DNAs in two species of *Drosophila*, *D. melanogaster* and *D. virilis*. This was

done by the analytical isopycnic CsCl centrifugation of DNA from direct
lysates of diploid tissues. A satellite DNA found in *D. melanogaster*
diploid tissues constitutes about 8% of the total DNA while in *D. virilis*
diploid tissues, three large satellites were observed together constituting
about 41% of the total DNA. However, when the DNA of salivary glands
was examined under similar conditions, a remarkable difference was seen!
In both species the satellite DNAs were either undetectable or present in
greatly reduced amounts. These observations are summarized in Figure 7.

The chromosomal distribution of several of the satellite DNAs has
been examined in both polytene chromosome preparations and in preparations
of mitotic chromosomes. I shall describe only the results seen for one
of these, the distribution being similar for the others. Complementary
RNA was prepared from the heavy strand of satellite I of *D. virilis*
(see Fig. 7) which had been isolated from alkaline CsCl gradients. In
polytene nuclei, labeling was detected only in the chromocenter region
especially over a densely staining mass designated the α-heterochromatin
by Heitz.[45] A few silver grains were seen elsewhere in the chromocenter
particularly around the base of the X chromosome but generalized labeling
of the rest of the chromocenter (termed β-heterochromatin[45]) was not
observed.

In mitotic chromosomes, heavy labeling was observed over the entire
centromeric heterochromatin of the X chromosome and of the autosomes,
but appeared relatively lighter over the Y chromosome. The euchromatic
areas were not significantly labeled.

One further important observation was that the amount of labeling
over polytene and diploid nuclei was similar even though the polytene
nuclei contain a vastly greater amount of total DNA. The results of the
cytological hybridizations are shown in Figure 8.

Together, the results of the centrifugation studies showing a

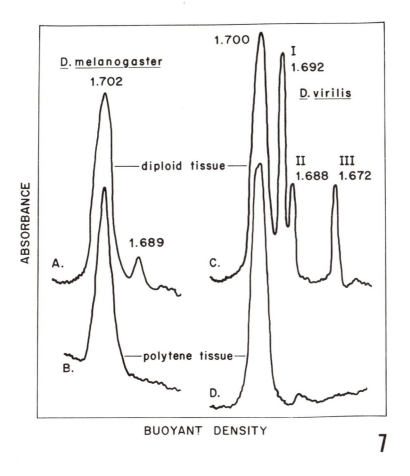

Figure 7. Banding profiles of *Drosophila* DNA after centrifugation to
equilibrium in neutral CsCl.[19] In both *D. melanogaster* and *D. virilis*,
satellite DNA is observed in diploid tissues (brains and imaginal discs)
but is not detected in polytene tissue (salivary glands). The tissues
were dissected from late third instar larvae, lysed in Sarkosyl detergent,
and were placed directly into CsCl for isopycnic centrifugation. A: Im-
maginal discs and brains from *D. melanogaster*. B: Salivary glands from
D. melanogaster. C: Imaginal discs and brains from *D. virilis*. D: Salivary glands from *D. virilis*.

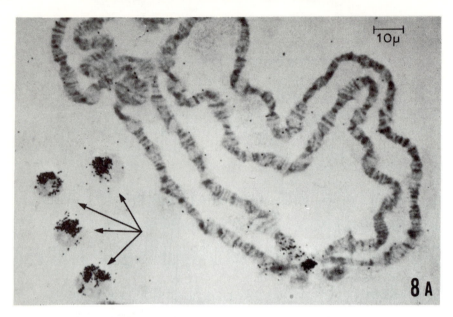

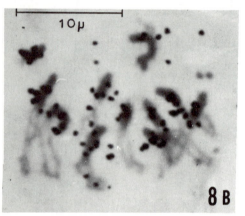

Figure 8. Slide hybrids of *Drosophila virilis* chromosomes hybridized with cRNA copied from the heavy strand of satellite I (see Fig. 7).[19] A: Polytene chromosomes from the salivary gland and a few diploid cells (arrows). Labeling was restricted to the α-heterochromatin of the chromocenter and to the base of the X chromosome. Note the similarity in the amount of label over the chromocenter and over the diploid cells. B: Diploid cells. Silver grains are primarily localized over the centromeric heterochromatin. The labeling over the Y chromosome was usually less intense than over the other chromosomes.

relative lack of satellite in polytene tissues, and the results of the cytological hybridizations showing similar amounts of satellite cRNA labeling over the centromeric heterochromatin of both polytene and diploid nuclei, provide strong evidence in support of the under-replication hypothesis. When the function of repetitive DNA is better understood, it should be interesting to see why during the formation of polytene cells repetitive satellite DNA is not replicated proportionally.

Acknowledgments: I wish to express my sincere gratitude to my colleagues at Yale, both past and present: Lesley W. Coggins, Edward H. Cohen, Susan A. Gerbi, Martin A. Gorovsky, Werner Kunz, Mary Lou Pardue, Mary Lake Polan, Brian B. Spear, and particularly Prof. Joseph G. Gall for the use of published and unpublished data in the preparation of this paper, and for many helpful and informative discussions. Portions of the research presented in this report were supported by funds from U. S. Public Health Service grants GM 12427 from the National Institute of General Medical Sciences and 5 T01 HD 32 from the National Institute of Child Health and Human Development. The author's present address is: Department of Biology, Brooklyn College of the City University of New York, Brooklyn, New York, 11210.

REFERENCES

1. Waring, M., and Britten, R.J., *Science 154*, 791 (1966).

2. Britten, R.J., and Kohne, D.E., *Science 161*, 529 (1968).

3. Maio, J.J., and Schildkraut, C.L., *J. Mol.Biol. 24*, 29 (1967).

4. Maio, J.J., and Schildkraut, C.L., *J. Mol.Biol. 40*, 203 (1969).

5. Schildkraut, C.L., and Maio, J.J., *Biochim. Biophys. Acta 161*, 76 (1968).

6. Yasmineh, W.G., and Yunis, J.J., *Exptl. Cell Res. 59*, 69 (1970).

7. Buongiorno-Nardelli, M., and Amaldi, F., *Nature 225*, 946 (1970).

8. Gall, J.G., *Genetics (Suppl.) 61*, 121 (1969).

9. Gall, J.G., and Pardue, M.L., *Proc. Natl. Acad. Sci. U.S. 63*, 378 (1969).

10. Gall, J.G., and Pardue, M.L., in *Methods in Enzymology*, L. Grossman and K. Moldave, Editors, in press, Academic Press, New York, 1971.

11. John, H.A., Birnstiel, M.L., and Jones, K.W., *Nature 223*, 582 (1969).

12. Jones, K.W., *Nature 225*, 912 (1970).

13. Pardue, M.L., and Gall, J.G., *Proc. Natl. Acad. Sci. U.S. 64*, 600 (1969).

14. Pardue, M.L., and Gall, J.G., *Science 168*, 1356 (1970).

15. Burgess, R.R., *J. Biol. Chem. 244*, 6160 (1969).

16. Chamberlin, M., and Berg, P., *Proc. Natl. Acad. Sci. U.S. 48*, 81 (1962).

17. Nakamoto, T., Fox, C.F., and Weiss, S.B., *J. Biol. Chem. 239*, 167 (1964).

18. Amaldi, F., and Buongiorno-Nardelli, M., *Exptl. Cell Res. 65*, 329 (1971).

19. Gall, J.G., Cohen, E.H., and Polan, M.L., *Chromosoma*, in press, (1971).

20. Jones, K.W., and Robertson, F.W., *Chromosoma 31*, 331 (1970).

21 Arrighi, F.E., Hsu, T.C., Saunders, P., and Saunders, G.F., *Chromosoma 32*, 224 (1970).

22. Eckhardt, R.A., *J. Cell Biol. 47*, 55a (1970).

23. Eckhardt, R.A., and Gall, J.G., *Chromosoma 32*, 407 (1971).

24. Hennig, W., Hennig, I., and Stein, H., *Chromosoma 32*, 31 (1970).

25. Pardue, M.L., Gerbi, S.A., Eckhardt, R.A., and Gall, J.G., *Chromosoma 29*, 268 (1970).

26. Rae, P.M.M., *Proc. Natl. Acad. Sci. U.S. 67*, 1018 (1970).

27. Marushige, K., and Bonner, J., *J. Mol.Biol. 15*, 169 (1966).

28. Paul, J., and Gilmour, R.S., *J. Mol. Biol. 34*, 305 (1968).

29. Smith, K.D., Church, R.B., and McCarthy, B.J., *Biochemistry 8*, 4271 (1969).

30. Lehman, I.R., Zimmerman, S., Adler, J., Bessman, M., Simms, E., and Kornberg, A., *Proc. Natl. Acad. Sci. U.S. 44*, 1191 (1958).

31. Baltimore, D., *Nature 226*, 1209 (1970).

32. Schlom, J., Hartner, D.H., Burney, A., and Spiegelman, S., *Proc. Natl.*

Acad. Sci. U.S. 68, 182 (1971).

33. Spiegelman, S., Burny, A., Das, M.R., Keydar, J., Schlom, J., Travnicek, M., and Watson, K., *Nature 227,* 563 (1970).

34. Temin, H.M., and Mizutani, S., *Nature 226,* 1211 (1970).

35. Jacob, S.T., Sajdel, E.M., and Munro, H.N., *Biochem. Biophys. Res. Commun. 32,* 831 (1968).

36. Kedinger, C., Gniazdowski, M., Mandel, J.L., Gissinger, F., and Chambon, P., *Biochem. Biophys. Res. Commun. 38,* 165 (1970).

37. Roeder, R.G., and Rutter, W.G., *Nature 224,* 234 (1969).

38. Tocchini-Valentini, G.P., and Crippa, M., *Nature 228,* 993 (1970).

39. Brown, D.D., and Dawid, I.B., *Science 160,* 272 (1968).

40. Evans, D., and Birnstiel, M., *Biochim. Biophys. Acta 166,* 274 (1968).

41. Gall, J.G., *Proc. Natl. Acad. Sci. U.S. 60,* 553 (1968).

42. Macgregor, H.C., *J.Cell Sci. 3,* 437 (1968).

43. Wallace, H., and Birnstiel, M.L., *Biochim. Biophys. Acta 114,*296 (1966).

44. Brown, D.D., Weber, C.S., and Sinclair, J.H., *Carnegie Inst. Year Book 66,* 580 (1967).

45. Heitz, E., *Biol. Zbl. 54,* 588 (1934).

46. Jacob, J., Todd, K., Birnstiel, M.L., and Bird, A., *Biochim. Biophys. Acta 228,* 761 (1971).

47. Brown, D.D., and Weber, C.S., *J. Mol. Biol. 34,* 661 (1968).

48. Brown, D.D., and Weber, C.S., *J. Mol. Biol. 34,* 681 (1968).

49. Aloni, Y., Hatlen, L.E., and Attardi, G., *J. Mol. Biol.,* in press (1971).

50. Wimber, D.E., and Steffensen, D.M., *Science 170,* 639 (1970).

51. Ritossa, F.M., Atwood, K.C., Lindsley, D.L., and Spiegelman, S., *Natl. Cancer Inst. Monograph 23,* 449 (1966).

52. Ritossa, F.M., Atwood, K.C., and Spiegelman, S., *Genetics 54,* 819 (1966).

53. Ritossa, F.M., and Spiegelman, S., *Proc. Natl. Acad. Sci. U.S. 53,* 737 (1965).

54. Steffensen, D.M., and Wimber, D.E., *J. Cell Biol. 47,* 202a (1970).

55. Ritossa, F.M., Atwood, K.C., and Spiegelman, S., *Genetics 54,* 663 (1966).

56. Heitz, E., *Z. Zellforsch. 20,* 237 (1934).

57. Rudkin, G.T., in *Genetics Today (Proc. XI Int. Cong. Gentics, The Hague, The Netherlands),* p. 359, Pergamon Press, New York, 1964.

58. Rudkin, G.T., *Genetics (Suppl.) 61,* 227 (1969).

59. Berendes, H.D., and Keyl, H.G., *Genetics 57,* 1 (1967).

60. Mulder, M.P., van Duijn, P., and Gloor, H.J., *Genetica 39,* 385 (1968).

DISCUSSION

RICHMOND: Have you had an opportunity to examine the polytene chromosomes of other organs of <u>Drosophila</u>, for example, from the malpighean tubules?

ECKHARDT: No, not as yet.

CHROMOSOMES AND RECOMBINATION

H.L.K. Whitehouse
Botany School, University of Cambridge

Abstract. The following inferences from eukaryote recombination data
appear well established. (1) The axis of the chromosome consists of one
duplex DNA molecule. (2) Recombination is initiated from predetermined
points. (3) Hybrid DNA extends for a variable distance from these points
and may pass through one gene into another, and may pass another opening
point. (4) Excision in the correction of mispairing may also extend from
one gene to another and may pass an opening point. The following are
tentative conclusions. (1) Recombination is initiated at open replicon
junctions controlled by promoter and operator sites. (2) Hybrid DNA is
often of unequal extent in the two recombining chromatids. (3) The
chromomeres are temporarily detached. (4) Recombination associated with
parental flanking markers represents crossovers that have been cancelled,
as part of a control mechanism.

INTRODUCTION

Experimental data on recombination in eukaryotes have provided

information in a number of different ways about chromosome structure and

organization. I shall consider in turn recombination data of various

kinds, and shall suggest inferences to be drawn from them about the

chromosome and the mechanism of recombination.

ABERRANT RATIOS IN TETRADS OR OCTADS OF SPORES

Postmeiotic segregation and conversion are known in a number of

eukaryotes and are most readily observed with spore colour mutants of

Ascomycetes, as in Ascobolus immersus, Sordaria fimicola and

S. brevicollis. In crosses between wild-type and mutant strains,

aberrant 4:4, 5:3, 3:5, 6:2 and 2:6 ratios of wild-type:mutant spores

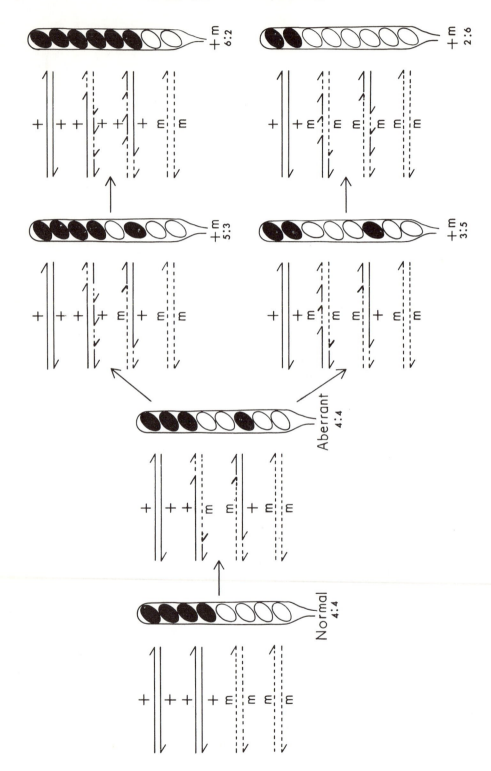

Fig. 1. Diagram to show how the formation of hybrid DNA will account for
         postmeiotic segregation, and how the subsequent correction of
         mispairing will explain conversion. The lines represent nucleotide
         chains at the pachytene stage of meiosis, broken lines distinguishing
         those of one parental genotype. Each chromatid is represented as a
         DNA duplex. The arrows indicate possible pathways of hybrid DNA
         formation and correction of mispairing, and the asci show the
         consequence for a spore colour mutant if the pathway stopped at
         that point.

are found (Fig. 1). Dissection of such asci and scoring for flanking markers, notably by Kitani and Olive (24) using mutants at the g locus in S. fimicola, has shown that the aberrant asci are associated with a greatly increased frequency of crossing-over between the flanking markers, compared with a random sample of asci. Furthermore, the aberrant segregation involves the same chromatids as took part in crossing-over.

These findings can be accounted for by making the following three postulates about the chromosome.

1.   The axis of the chromosome consists of only one DNA molecule, that
      is, two complementary nucleotide chains.

2.   Crossing-over involves breakage of these two strands in each
      recombining chromatid, the break-points being so placed that
      rejoining can occur by annealing of complementary nucleotide chains
      from the two molecules to form segments of hybrid DNA (17, 52)
      (Fig. 1).

3.   If the recombining chromatids differ by a mutation situated in the
      hybrid DNA segments, an enzyme system may recognise the resulting
      distortion in the molecule, excise from one strand a segment of
      nucleotide chain including the mispaired nucleotide or nucleotides,
      and replace the segment by nucleotides complementary to those in
      the other strand.

Holliday (17, 18) has questioned the need to postulate only one DNA duplex for the chromosome axis, and has argued that matching of complementary chains could occur between two or more DNA molecules lying parallel, thereby transferring nucleotide sequence to a second or further molecules. Although theoretically possible, there is no experimental evidence for the occurrence of such matching. Moreover, since postmeiotic segregation is attributed to failure of correction of mispairing, such failure would also be expected in the matching process.

This would lead to segregation of character differences at the second or later mitoses after meiosis, but this has not been recorded. In any case, strong support for the one-duplex chromosome axis is provided by the work of Brewen and Peacock (4).

The second postulate to account for aberrant asci, namely, that rejoining involves hybrid DNA formation, is supported by labelling experiments with prokaryotes, which have shown that hydrogen bonding precedes covalent joining of nucleotide chains, for example, in the recombination following conjugation in Escherichia coli (33) and in recombination in phage T4 (48).

The third postulate, correction of mispairing, gains strong support from the discovery by Leblon and Rossignol (27) that the relative frequencies of the various kinds of aberrant asci in Ascobolus immersus are influenced markedly by the mutagen used to obtain the spore colour mutant, and hence presumably by the molecular nature of the mutation. Furthermore, the phenomenon of map expansion is accounted for if correction of mispairing takes place and is extensive (13). Extensive excision in the correction of mispairing also explains the linked conversion of closely linked mutants, the frequency of which Fogel and Mortimer (15) have shown to be inversely related to the distance between the mutant sites. Correction of mispairing also accounts for the phenomenon of negative interference. When two closely linked auxotrophic mutants are crossed and prototrophs selected, they are usually found to comprise all four possible genotypes for flanking markers. Correction of mispairing in hybrid DNA offers an explanation for this additional recombination on one or both sides of the auxotrophic mutants (Table 1).

Although aberrant asci and their association with crossing-over can be explained, at least qualitatively, on the above three postulates, aberrant asci also occur regularly in the absence of crossing-over of

Table 1.   Correction of mispairing as an explanation of negative interference

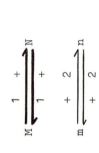

Parents:

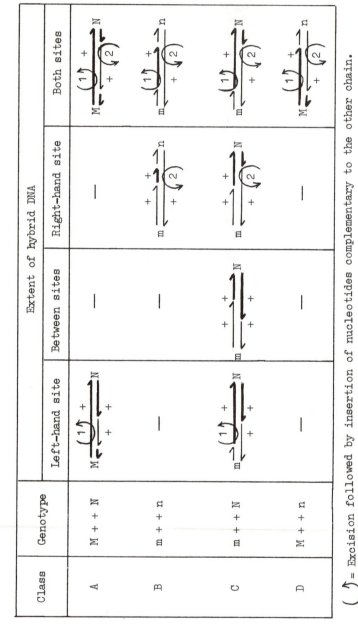

= Excision followed by insertion of nucleotides complementary to the other chain.

flanking markers. This is discussed below.

POLARITY IN RECOMBINATION

Systematic differences in the frequency of conversion of allelic mutants have been observed in many genes. This phenomenon is spoken of as polarity in recombination. Mutants situated near the ends of a gene usually show a relatively high conversion frequency compared with those near the middle. Information about relative conversion frequencies has generally been obtained by crossing allelic auxotrophic mutants carrying flanking marker genes, and selecting progeny which were wild-type for the allelic character. Those prototrophs with parental combinations of flanking markers provide information about polarity in recombination, because conversion to wild-type of the left-hand mutant of the pair will give a prototroph with recombination to the left of the alleles as well as between them (class A in Table 1), while conversion to wild-type of the right-hand mutant will give a prototroph with extra recombination to the right (class B). Pees (34) crossed 12 mutants at the lysine-51 locus in Aspergillus nidulans in almost all the pairwise combinations, and selected lysine prototrophs. Her data for those with parental flanking markers are shown graphically in Fig. 2. It is evident that conversion is about equally frequent at the two ends of the gene, but much less frequent in mid-gene. A more usual pattern is an asymmetric one, with conversion more frequent at one end of the gene than the other, giving a J-shaped rather than a U-shaped curve when conversion frequency is plotted against the position of the mutant site.

The following two postulates about the chromosome provide an explanation for the data on polarity in recombination.

1.   There are fixed points in the chromosome from which the dissociation of the nucleotide chains of the two recombining chromatids is initiated, to be followed by cross-annealing to give hybrid DNA.

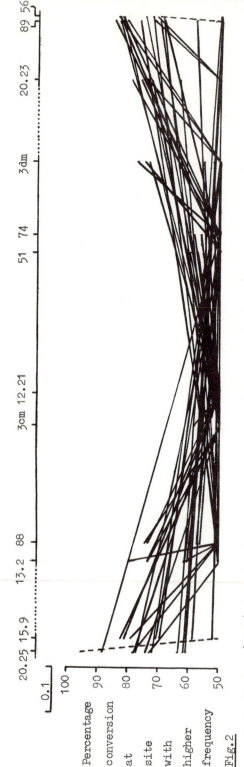

<u>Fig.2</u>

Data of Pees (1968) for 9712 <u>lysine-51</u> prototrophs with parental flanking markers from 63 of the 66 possible pairwise crosses between 12 <u>lys-51</u> mutants in <u>Aspergillus nidulans</u>. For each line the abscissa shows the mutants crossed and the ordinate the percentage of conversion at the site with the higher frequency. Two lines are shown for the cross of mutants 13.2 and 88 as the data from two crosses were heterogeneous. Broken lines indicate results based on 25 or fewer prototrophs. The left and right-hand dotted intervals on the map correspond to recombination frequencies of approximately 14 and 2.8 prototrophs, respectively, per 10^5 progeny. The scale applies to the remainder of the map and shows prototrophs per 10^5 progeny.

2.   The hybrid DNA extends for a variable distance from the fixed

      opening points.

The outcome of these two postulates will be a gradient of hybrid DNA

frequency, falling from a relatively high value at the opening point to

zero at the limit of hybrid DNA extent.   This gradient will be shown by

the frequency of postmeiotic segregation and of conversion, if these are

a consequence of hybrid DNA formation.   That genes show maxima of

conversion frequency at their ends implies that the fixed opening points

are at or beyond the gene ends, rather than within genes.

RECOMBINATION IN NEIGHBOURING GENES

      Several studies have been made of recombination in two closely

linked loci, notably by Calef (5) and Putrament (36) with ad-9 and paba-1

in Aspergillus nidulans, and by Murray (32) with me-7 and me-9 in Neuro-

spora crassa.   Two techniques have mainly been used: selecting for

recombinants in one gene and then scoring them for an unselected marker

in the other gene;   or selecting for recombination between two mutants,

one situated in each gene.   These and other investigations have shown that

D class progeny (Table 1) and linked conversion (see Fig. 3a for Murray's

data) can occur with mutants in closely linked loci.   The inference from

these findings is that hybrid DNA can extend from one gene to another and

that the excision and resynthesis associated with the correction of mis-

pairing can also extend across the boundary between neighbouring loci.

      Murray (32) found that, although hybrid DNA could extend from me-9

to the neighbouring me-7 locus, it could also sometimes originate between

these genes.   The evidence for this conclusion is shown in Fig. 3.   The

me-7 locus showed a typical asymmetrical polarity pattern, conversion

being most frequent with mutants near the ends of the gene and more

frequent at one end - in this instance, the right-hand end on the map

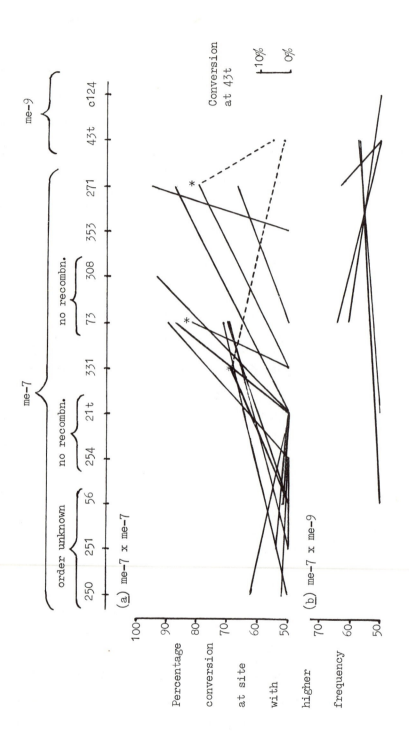

Fig. 3

Data of Murray (1969, 1970) for wild-type selection in Neurospora crassa from pairwise crosses between methionine-requiring mutants with flanking markers. The lines show as abscissa the mutants crossed and as ordinate the percentage conversion at the site with the higher frequency. The mutants are arbitrarily shown equally spaced on the map. The 22 lines are based on a total of 4194 wild-type progeny with parental flanking markers. All the crosses showed some progeny recombinant in all three marked intervals, indicative of conversion to wild-type at both the methionine mutant sites. For the me-7 x me-9 crosses this implies that hybrid DNA sometimes extended from one gene to the other.

* The crosses 56 x 331, 331 x 73 and 331 x 271 also involved 43t as an unselected marker. The right-hand end of the broken lines indicates the frequency with which conversion at the right-hand site in me-7 was associated with conversion at 43t, that is, excision in the correction of mispairing extending from one gene to the other.

(Fig. 3a) — than at the other (31). Prototroph selection from crosses between me-7 and me-9 mutants, on the other hand, revealed a much lower conversion frequency in me-9 than in the neighbouring (right-hand) part of me-7 (Fig. 3b). Evidently there is an opening point for recombination between the loci and most of the hybrid DNA entering me-7 from the direction of me-9 originates between the loci. In view of the evidence given above that hybrid DNA can extend from me-9 to me-7, it is apparent that hybrid DNA, in extending from one gene to another, can cross a site which in another meiotic cell is an initiation point for hybrid DNA. It is also evident that the excision and resynthesis associated with the correction of mispairing can cross an opening point. The significance of this discovery is discussed further below.

THE INITIAL EVENTS IN RECOMBINATION

The idea that recombination in eukaryotes is initiated from pre-determined points spaced along the chromosome and lying outside the genes is supported by the discoveries made by Catcheside and associates of dominant repressors of recombination in specific regions of the chromosomes of Neurospora crassa. Similar findings have been made by Simchen and Stamberg (38) with Schizophyllum commune. The Neurospora results are outlined in Fig. 4, which shows maps of parts of three of the seven linkage groups. The results may be summarised as follows.

1. The rec genes are dominant repressors of recombination in one or more specific regions of the genome (22, 39, 8, 2, 9).

2. The regions in which recombination is affected by a particular repressor are not linked, or at least are not closely linked, either to one another or to the repressor locus.

3. The repressors affect the polarity of recombination as if an opening point located to one side of the gene were kept shut (49, 47, 2).

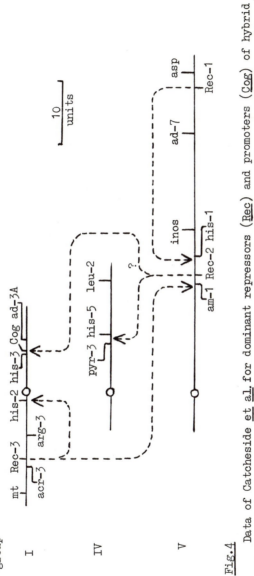

<u>Fig.4</u>

Data of Catcheside et al. for dominant repressors (<u>Rec</u>) and promoters (<u>Cog</u>) of hybrid DNA formation from specific opening-points in the chromosomes of Neurospora crassa. <u>Rec-1</u> represses proximal entry of hybrid DNA into <u>his-1</u>, but does not affect 9 other loci tested. <u>Cog</u> is a dominant promoter of recombination in its neighbourhood, including <u>his-3</u> and the region between <u>his-3</u> and <u>ad-3A</u>. <u>Rec-2</u> represses recombination between <u>pyr-3</u> and <u>his-5</u>, and either <u>Rec-2</u> or another Rec gene is epistatic to <u>Cog</u> and represses recombination between <u>his-3</u> and <u>ad-3A</u>, and distal entry of hybrid DNA into <u>his-3</u>. <u>Rec-3</u> represses recombination in <u>his-2</u> and proximal entry of hybrid DNA into <u>am-1</u>, but does not affect 8 other loci tested.

Smyth (40) found that the proximal polarity of am-1 in N. crassa was much weaker when recombination within this gene was reduced by rec-3. Nevertheless, the am-1 polarity was still predominantly proximal. The opening point affected by rec-3 is evidently on the proximal side of am-1 with respect to the centromere, but is only partially closed by rec-3, or, alternatively, there are other initiation points proximal to am-1 which contribute hybrid DNA to it. In view of the ability of hybrid DNA to overshoot opening points, these additional initiation points might be on either side of the one affected by rec-3.

Angel, Austin and Catcheside (2) found a dominant promoter of recombination which they have called cog. It affects recombination only in its immediate neighbourhood: the his-3 gene and the region between his-3 and ad-3A (Fig. 4). The rec gene which represses recombination in his-3 is epistatic to cog. When recombination in his-3 was repressed (Rec$^+$), the polarity pattern in all the crosses studied except two was proximal, as expected if an opening point distal to his-3 were kept shut (Fig. 5c). When recombination in his-3 was derepressed (rec$^-$) but still at a low level (cog$^-$), the polarity pattern, though still proximal, was much weaker (Fig. 5b). This is consistent with increased distal entry of hybrid DNA compared with Rec$^+$. When recombination in his-3 was both de-repressed (rec$^-$) and promoted (Cog$^+$), much more distal entry of hybrid DNA was evident (Fig. 5a), as expected if cog, which is situated a short distance distal to his-3, is at the opening point affected by the rec gene.

To account for the observed interactions of the cog and rec mutants on his-3 recombination, Angel et al. (2) suggested that cog is the site of action of a specific endonuclease necessary for the initiation of recombination there. They argued that the repressor molecule produced by the Rec$^+$ gene might act directly on cog, or more likely acted indirectly

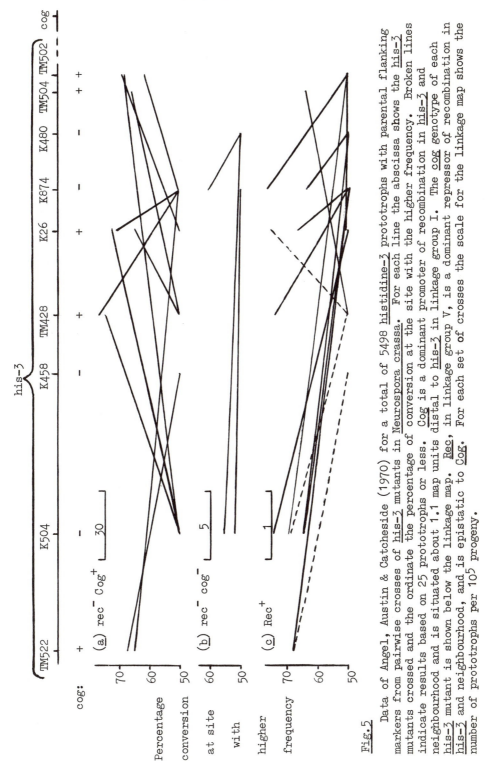

Fig. 5

Data of Angel, Austin & Catcheside (1970) for a total of 5498 histidine-3 prototrophs with parental flanking markers from pairwise crosses of his-3 mutants in Neurospora crassa. For each line the abscissa shows the his-3 mutants crossed and the ordinate the percentage of conversion at the site with the higher frequency. Broken lines indicate results based on 25 prototrophs or less. Cog is a dominant promoter of recombination in his-3 and neighbourhood and is situated about 1.1 map units distal to his-3 in linkage group I. The cog genotype of each his-3 mutant is shown below the linkage map. Rec, in linkage group V, is a dominant repressor of recombination in his-3 and neighbourhood, and is epistatic to Cog. For each set of crosses the scale for the linkage map shows the number of prototrophs per 10⁵ progeny.

via an operator site alongside the structural gene for the endonuclease. A third possibility was that the repressor acted on an operator beside cog, blocking the action rather than the production of the endonuclease.

It is possible, however, that recombination in eukaryotes is not initiated by endonuclease action but by a failure to complete the previous replication. A relationship between recombination and the preceding replication is suggested by the following.

1. Eukaryotic recombination, whether meiotic or mitotic, and whether associated with parental or recombinant flanking markers, invariably takes place after the chromosomes have replicated but involves only one chromatid from each parent.

2. Lawrence (26) found that chiasma frequency in Lilium longiflorum and Tradescantia bracteata and recombination frequency in Chlamydomonas reinhardii were affected by non-lethal doses of ionizing radiation, but that the effect was produced only in cells which were passing through one or other of two short stages: one premeiotic and the other at pachytene. Hastings (16) and Davies and Lawrence (10) found that inhibitors of DNA synthesis also affected recombination in C. reinhardii, and moreover only at the same two discrete stages. Thus, some DNA synthesis relevant to recombination seems to take place at these times.

3. Ito, Hotta and Stern (21) found that a small amount of DNA synthesis occurred at zygotene in Lilium longiflorum and Trillium erectum pollen mother cells. This DNA was richer in guanine and cytosine than the nucleus as a whole, and was thought to consist of short segments interspersed throughout the length of the chromosomes, because the addition at zygotene of an inhibitor of DNA synthesis caused chromosome fragmentation. Moreover, Stern and Hotta (42)

found that pollen mother cells of L. longiflorum reverted to mitosis
when transferred to culture medium shortly before meiosis was due to
begin, and that, contrary to a normal mitosis, this was preceded by
the synthesis of GC-rich DNA. Thus, it seems as if the synthesis of
certain GC-rich segments of the chromosomes, perhaps at the replicon
junctions, is delayed before meiosis until zygotene. Hotta and
Stern (19) have recently obtained further evidence that the zygotene
synthesis, which amounts to about 0.3% of the total DNA, represents
delayed replication.

4. Keyl (23) and Pelling (35) presented evidence that the replicon in
    eukaryotes corresponds to the chromomere, and Callan (6) suggested
    that the chromomere consists of a series of slave copies of a gene
    in tandem array with a master copy. In combination, these hypotheses
    would mean that the replicon corresponds to the gene, in the sense of
    a single functional unit. Opening points for recombination at replicon
    ends would therefore give polarity patterns indicative of opening at
    or beyond gene ends, such as are observed.

If eukaryotic recombination were triggered by open replicon junctions,
the dominant repressors of recombination in specific regions might act
to block hybrid DNA formation. According to the Whitehouse-Hastings model
(52) some DNA synthesis takes place in each chromatid before hybrid DNA
can be formed. Thus, a possible mode of action of the repressor molecules
would be to combine with a specific operator site near the replicon end
and so prevent DNA synthesis, which would otherwise be initiated at a
promoter site alongside (Fig. 6).

This hypothesis provides an explanation for a remarkable feature of
the his-3 data (2) in Neurospora crassa. When his-3 was derepressed for
recombination (rec⁻) and heterozygous at the cog site, conversion was
most frequent at the his-3 allele linked to Cog^+ (Fig.5a). If cog is a

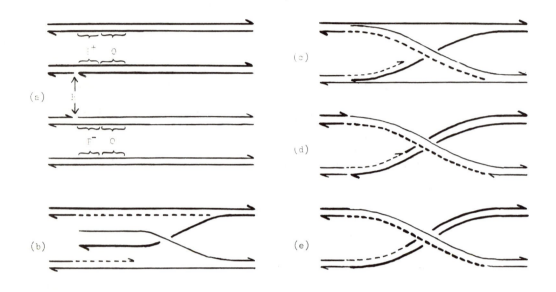

<u>Fig. 6</u>.   Recombination according to the Whitehouse-Hastings model,
         with the initiation-points unsealed replicon ends, and with
         recombination-promoter and recombination-operator segments
         alongside the opening-point. (<u>a</u>) The four chromatids are
         shown as DNA duplexes, and the parentage is indicated by the
         thickness of the lines. One strand of one molecule from each
         parent has not been joined at the end of a replicon. O, recombination
         operator; P, recombination promoter, the plus and minus signs
         indicating heterozygosity for promoter segments differing in
         efficiency; R, open replicon junction. (<u>b</u>) Dissociation of strands
         in the inner two chromatids and synthesis of new strands (shown
         as broken lines) with unbroken chains as templates. Synthesis
         less extensive in the P⁻ chromatid. (<u>c</u>) Dissociation of new chains
         from their templates and cross annealing. (<u>d</u>) Breakdown of
         template chains by nucleases. (<u>e</u>) Breaks sealed.

promoter site for a DNA polymerase, the more efficient allele (+) might
be expected to promote more extensive synthesis than $\underline{cog}^-$ (Fig. 6). This
would mean that hybrid DNA was more extensive in the $\underline{Cog}^+$ chromatid than
$\underline{cog}^-$, and so more often reached $\underline{his-3}$.

Heterozygosity for polymerase promoters might explain, at least in
part, why hybrid DNA seems so often to be of unequal extent in the two
chromatids. This inequality is exemplified by the data of Kitani and
Olive (24) for asci of $\underline{Sordaria\ fimicola}$ from crosses between wild-type
and spore colour mutants at the $\underline{g}$ locus. Those asci with 5:3 or 3:5 ratios
of wild-type : mutant spores, and with parental combinations of flanking
markers, are classified in Table 2. The excess of tetrads with three
parental chromatids over those with two would be accounted for if DNA
synthesis, and hence hybrid DNA extent, were greater in one chromatid
than the other. The alternative explanation, that hybrid DNA is of equal
extent, but that correction of mispairing favours restoration of the
parental genotype, is favoured by Kitani and Olive (25). This possibility
is contradicted, however, by the finding by Leblon and Rossignol (27) that,
at least in $\underline{Ascobolus\ immersus}$, the correction pattern depends on the
mutagen used to obtain the mutant.

An interesting feature of Kitani and Olive's data (Table 2) is that
mutants g_1 and h_3 show a marked excess of three-parental over two-parental
asci in the 3:5 class and a rather slight excess in the 5:3 class, while
the converse is true for the other mutants. This difference between the
5:3 and the 3:5 asci is significant for g_1 and h_3. Such a difference is
expected if there is heterozygosity for a polymerase promoter, because
hybrid DNA will occur at the mutant site for spore colour more often in one
chromatid than the other.

Emerson and Yu-Sun (12) found that the conversion pattern and
frequency when spore colour mutant $\underline{w6}^c$ in $\underline{A.\ immersus}$ was crossed with

Table 2

Data of Kitani & Olive (1967) for 5 : 3 and 3 : 5 asci at the g locus in Sordaria fimicola with parental arrangements of flanking markers. Parents: $\dfrac{M + N}{m \; g \; n}$, where M/m and N/n are the morphological mutants mat and corona, respectively.

Mutants at g locus		Aberrant segregation + : g			
		5 : 3		3 : 5	
		Ascus genotype and number of parental chromatids			
Sequence on map	Allele symbol	M + N M +/g n 3 m g n	M + N M +/g N 2 m + n m g n	M + N M +/g N 3 m g n m g n	M + N M g N 2 m +/g n m g n
1	g₁	29 (58%)	21 (42%)	13 (86.7%)	2 (13.3%)
2	h₂ h₂ₐ	13 ⎤ 24 (85.7%) 11 ⎦	3 ⎤ 4 (14.3%) 1 ⎦	12 ⎤ 19 (65.5%) 7 ⎦	7 ⎤ 10 (34.5%) 3 ⎦
3	h₄ h₄ᵦ	3 ⎤ 9 (81.8%) 6 ⎦	0 ⎤ 2 (18.2%) 2 ⎦	16 ⎤ 22 (73.3%) 6 ⎦	5 ⎤ 8 (26.7%) 3 ⎦
4	h₃	9 (60%)	6 (40%)	37 (90.2%)	4 (9.8%)

wild-type were influenced by a closely linked controlling factor. Their
data have features in common with those for the cog promoter site in
Neurospora.

One of the difficulties of the specific endonuclease hypothesis (2)
for recombination initiation is the comparative rarity of recombination
even when the opening points are derepressed. On the open replicon
junction hypothesis, however, the replicon ends left unjoined might be a
randomly distributed sample, if for instance there were insufficient
molecules of an essential component for joining the ends. The variability
in the position of crossovers from one cell to another would then be
accounted for. It is necessary to assume that the recombination
repressor molecules do not become attached, or no longer remain attached,
to their specific operator sites after a replicon junction is closed,
otherwise hybrid DNA from one opening point could not overshoot another.

Pees (34) has suggested that the sharp reversal of polarity near
mutant 12.21 in the middle of the lys-51 gene of Aspergillus nidulans
(Fig. 2) indicates a block to specific pairing, with the implication that
hybrid DNA cannot extend beyond this point. This hypothesis is untenable,
however, because all the 63 pairwise crosses of lys-51 mutants gave some
class D prototrophs and 28 of these crosses involved mutants on opposite
sides of 12.21. As indicated in Table 1, class D progeny require hybrid
DNA to extend to both sites. There is no evidence for any recombinational
discontinuity within lys-51.

Howell and Stern (20) have detected in Lilium pollen mother cells an
endonuclease that produces single strand breaks with 5'-hydroxyl and
3'-phosphoryl termini in the DNA. Enzymes for repairing the DNA were
also found: a phosphatase, which could dephosphorylate the 3' ends; a
polynucleotide kinase, which could phosphorylate 5' termini; and a
polynucleotide ligase, which could join adjacent 3'-hydroxyl and

5'-phosphoryl ends. The endonuclease, kinase and ligase activities reached
maxima in late zygotene or early pachytene, and their synthesis is believed
to be directly related to recombination. The existence of an endonuclease
does not conflict with the hypothesis that recombination is triggered by
open replicon junctions, because an endonuclease would still be required
to cut the second strands in each molecule (between stages c and d in Fig.
6) and to cut one strand in the correction of mispairing. A third likely
requirement for an endonuclease is discussed below.

LACK OF CROSSING-OVER IN LAMPBRUSH LOOPS

Callan's master and slaves hypothesis, which has recently received
support from molecular studies (46), has an interesting implication for
chromosome organization in relation to recombination. Callan (6) pointed
out that no recombination takes place in the slaves. The evidence for
this is threefold:

1. Chiasmata are not observed in the loops of lampbrush chromosomes.
2. The chromomeres are too densely folded at zygotene and pachytene for
    homologous pairing at the molecular level. In keeping with this, the
    greater part of the material of the chromosomes lies outside the
    synaptonemal complex.
3. By analogy with the Bar eye duplication in Drosophila melanogaster,
    if pairing occurred between slaves in homologous chromosomes, it
    would no doubt often be with the homologues relatively displaced,
    for example, slave no. 1 in one chromosome with no. 2 in the other.
    As a result of crossing-over, the number of slaves would become
    variable, and the strict geometric variation in the DNA content of
    homologous chromomeres found by Keyl (23) in the subspecies of
    Chironomus thummi would be lost.

In view of this evidence, Callan suggested that recombination was confined to the master copies.

As indicated above, there is good evidence that hybrid DNA can extend from one gene to another. In order to reconcile such overshooting with tandem arrays of slaves free of recombination, I suggested (50) that the chromomeres were temporarily detached from the chromosome axis by a crossover mechanism comparable to that proposed by Campbell (7) for the integration and excision of phage genomes from the bacterial chromosome. Such reciprocal exchange has now been confirmed for phage lambda in Escherichia coli and some others, and according to Signer's hypothesis (37) an enzyme makes staggered cuts in the two specific regions of DNA. All that might then be needed to complete the exchange would be cross-annealing of the complementary strands followed by ligase sealing. In Fig. 7 this hypothesis is applied to chromomere excision. The slaves could be re-inserted in the same way. Recent results for phage lambda (11) indicate little or no homology of nucleotide sequence between the phage and bacterial attachment sites. Evidently the recombination mechanism involves an enzyme system which recognizes a particular nucleotide sequence in the phage DNA and a different nucleotide sequence in the DNA of the host, and then cuts the chains and unites the broken ends phage to bacterium, and vice versa. The sites of staggered cleavage, spanning homologous regions, are believed to be less than 20 and probably less than 12 nucleotides apart, with the specificity of the exchange carried largely or entirely by the enzyme.

Another possibility for chromomere excision would be crossovers comparable to those between homologues (50). If some turnover of DNA were involved in these crossovers, this might account for the greater part of the DNA synthesis at pachytene in Lilium observed by Hotta and Stern (19). Unlike the DNA synthesized at zygotene, the pachytene synthesis apparently replaces pre-existing DNA and is believed to be of the nature

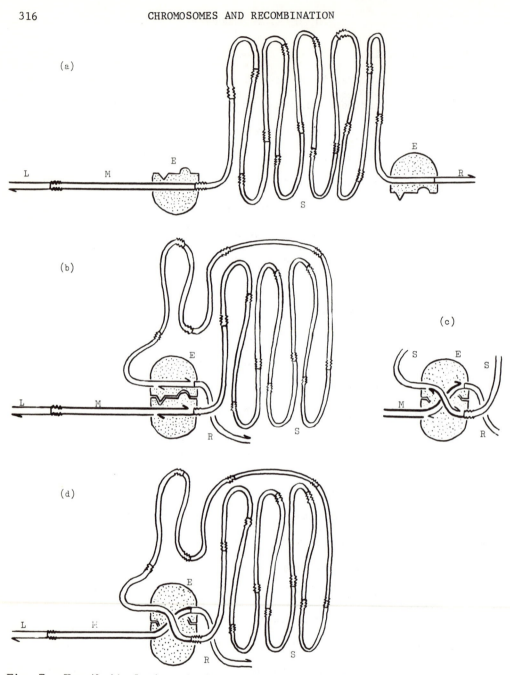

Fig. 7. Hypothetical steps in the excision of a tandem array of slave
copies of a gene. M, master copy; S, slave copies; L, R,
next genes to left and right, respectively; E, enzyme aggregate.
The nucleotide chains of the master copy are shown with thick
lines. (a) Before excision, with enzyme aggregates attached at
corresponding positions in the master copy and the final slave.
(b) Association of the two enzyme aggregates, and cutting of the
nucleotide chains by an endonuclease. (c) Exchange of pairing
partners between the points of breakage. (d) Sealing of breaks
by a ligase.

of repair synthesis, since it is not inhibited by hydroxyurea. The amount

of DNA synthesized is about 0.1% of the total DNA content of the pollen

mother cell nucleus. This is equivalent to 110 fg or about 10^8 nucleotide

pairs. From the lack of effect on the density gradient of shearing the

DNA synthesized at pachytene, when carrying a heavy label, Howell and

Stern (20) inferred that the pachytene synthesis occurred in at least 6.6

x 10^5 separate places in the nucleus. This figure is understandable if it

represents twice the number of chromomeres in the four chromatids of all

the bivalents, in addition to twice the number of places where

recombination is expected to occur between homologues. Howell and Stern

quote 36 as the mean chiasma frequency of the Lilium nucleus. Fungal

data suggest that in each meiotic cell recombination is to be expected

associated with a parental arrangement of neighbouring parts of the

chromosome, and hence not detectable cytologically, with a frequency in the

range 0.5 to 4.0 times the crossover frequency (see below). Thus, at most,

about 180 recombination events between homologues are expected per Lilium

pollen mother cell.

If chromomere excision takes place at pachytene, knowledge of its

mechanism could be important for an understanding of the integration and

release of tumour viruses, since this might occur in a similar way. The

recent discoveries of RNA- and DNA-directed DNA polymerases and of a DNA

endonuclease, exonuclease and ligase in virions of tumour viruses (3, 45,

43, 29, 30) point to the likelihood that their genomes are inserted in

and released from a chromosome of the host by a recombination mechanism.

Temporarily to detach the chromomeres at meiosis may seem a wasteful

process, but is perhaps understandable if slave copies of genes evolved

relatively late, after the eukaryote recombination system, with its hybrid

DNA overshooting one gene into another, had already become established.

There is evidently a selective advantage in having hybrid DNA entering a

Fig.8   Diagram to illustrate the crossover cancellation hypothesis for the origin of hybrid DNA associated with parental flanking markers (PFM). The hypothesis explains why crossovers, that is, hybrid DNA associated with recombinant flanking markers (RFM), show interference with one another, while PFMs show no interference with RFMs. The lines represent the nucleotide chains of a bivalent at the pachytene stage of meiosis, broken lines distinguishing those of one parental genotype. Each chromatid is represented as a DNA duplex.

(a) Before hybrid DNA formation.

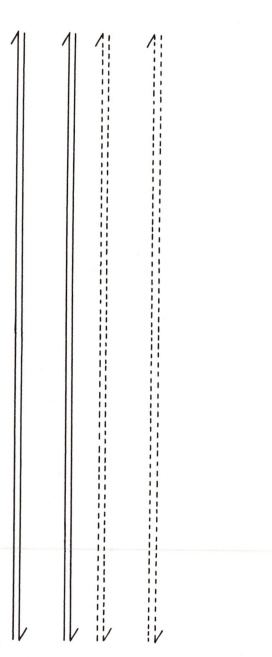

(b) Several RFMs arise, distributed at random as regards position and chromatids.

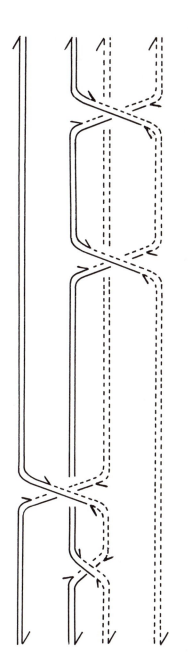

(c) Some of the RFMs are replaced by PFMs, such that the remaining RFMs are spaced out. The crossover cancellation would be brought about by an enzyme system moving along the synaptonemal complex, but what enzymes would be required would depend on the molecular structure of an RFM at stage (b).

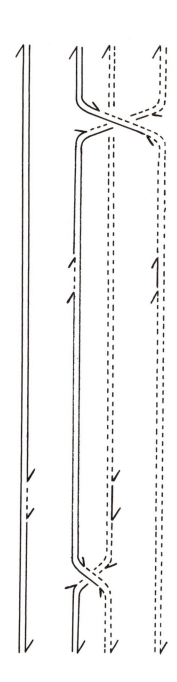

gene from either end.

RECOMBINATION WITH PARENTAL FLANKING MARKERS

Recombination associated with a parental combination of flanking markers seems to be an invariable feature of eukaryote genetic exchange, whether meiotic or mitotic. It is found regularly in conjunction with aberrant asci and in prototroph selection experiments. The relative frequencies of recombination associated with parental and with recombinant flanking markers range from about 30% parental in several genes in Aspergillus nidulans to about 80% parental in me-7 in Neurospora crassa (31). Whereas crossovers, that is, recombination associated with a recombinant arrangement of flanking markers (RFM) usually show positive interference with one another, recombination with parental flanking markers (PFM) shows no interference with crossovers (41, 14).

Crossovers clearly have selective value in promoting exchange of chromosome segments. The general occurrence of PFMs suggests that they are also favoured by natural selection, but the reason for this is not obvious. A possible explanation for the occurrence of PFMs, and for their different interference pattern from RFMs, is provided by the cross-over cancellation hypothesis (51), which has two postulates.

1.   The initial events in recombination are all RFMs, and these show no interference, either as regards their position or the pair of chromatids involved (Fig. 8b). Though randomly distributed at a superficial level of analysis, the crossovers would be initiated at fixed opening points. As already mentioned, this pattern might be related to a random distribution of open replicon junctions.

2.   At a later stage in pachytene, an enzyme aggregate moves along the synaptonemal complex and converts some of the RFMs into PFMs, such that the surviving RFMs are spaced out (Fig. 8c). The PFMs would

retain hybrid DNA segments. The molecular events required to produce a PFM from an RFM would depend on the precise structure of the initial exchange, which might be better described as a potential RFM.

The assumption underlying this hypothesis is that crossover cancellation provides a means of controlling the frequency and distribution of exchanges.

It is well known that chromosomal inversions, particularly when heterochromatin is moved (44), lead to an increase in the frequency of crossing-over in the other chromosomes. Controlling elements in the chromosome carrying the inversion might be adversely affected by the new euchromatin-heterochromatin boundaries, giving rise to less frequent cancelling of crossovers. This would explain the otherwise puzzling discovery made by Lucchesi (28) that inversions increase the frequency of X-ray induced exchanges. The hypothesis of reduced cancelling of crossovers to explain these interchromosomal effects is supported by preliminary results obtained with an inversion in linkage group III of Sordaria brevicollis (1). Crossing-over between buff and yellow spore colour loci in linkage group II was appreciably increased in the presence of the heterozygous inversion, and within the buff locus PFMs were much reduced relative to RFMs.

CONCLUSIONS

From genetical studies the basic mechanism of recombination in eukaryotes, and its implications for chromosome structure and organization now seem to be well-established: a chromosome axis of one double-stranded DNA molecule; fixed opening points from which hybrid DNA of variable extent is initiated; overshooting of hybrid DNA from one gene to another, including crossing another opening point; and correction of mispairing in hybrid DNA, involving extensive excision which may also cross gene boundaries and opening points.

Much more speculative are the other hypotheses discussed in this article: recombination initiated from open replicon junctions and controlled by promoter and operator sites; hybrid DNA of unequal extent in the two recombining chromatids; chromomere detachment by reciprocal exchange; and crossover cancellation.

An ascomycete with closely linked spore colour mutants and flanking markers, in conjunction with mutants and structural changes affecting recombination of the colour mutants, offers the best hope for testing a number of these ideas.

REFERENCES

1. A.F. Ahmad, D.J. Bond and H.L.K. Whitehouse, manuscript in preparation.

2. T. Angel, B. Austin and D.G. Catcheside, Austral. J. Biol. Sci. 23, 1229 (1970).

3. D. Baltimore, Nature, Lond. 226, 1209 (1970).

4. J.G. Brewen and W.J. Peacock, Proc. Nat. Acad. Sci. U.S.A. 62, 389 (1969).

5. E. Calef, Heredity, Lond. 11, 265 (1957).

6. H.G. Callan, J. Cell Sci. 2, 1 (1967).

7. A.M. Campbell, Advanc. Genet. 11, 101 (1962).

8. D.G. Catcheside, Austral. J. Biol. Sci. 19, 1039 (1966).

9. D.G. Catcheside and B. Austin, Austral. J. Biol. Sci. 24, 107 (1971).

10. D.R. Davies and C.W. Lawrence, Mutation Res. 4, 147 (1967).

11. R.W. Davis and J.S. Parkinson, J. Molec. Biol. 56, 403 (1971).

12. S. Emerson and C.C.C. Yu-Sun, Genetics 55, 39 (1967).

13. J.R.S. Fincham and R. Holliday, Molec. Gen. Genet. 109, 309 (1970).

14. S. Fogel and D.D. Hurst, Genetics 57, 455 (1967).

15. S. Fogel and R.K. Mortimer, Proc. Nat. Acad. Sci. U.S.A. 62, 96 (1969).

16. P.J. Hastings, Ph.D. thesis, Univ. Cambridge (1965).

17. R. Holliday, Genet. Res. 5, 282 (1964).

18. R. Holliday, Symp. Soc. Gen. Microbiol. 20, 359 (1970).

19. Y. Hotta and H. Stern, J. Molec. Biol. 55, 337 (1971).

20. S.H. Howell and H. Stern, J. Molec. Biol. 55, 357 (1971).

21. M. Ito, Y. Hotta and H. Stern, Devl Biol. 16, 54 (1967).

22. A.P. Jessop and D.G. Catcheside, Heredity, Lond. 20, 237 (1965).

23. H.G. Keyl, Chromosoma 17, 139 (1965).

24. Y. Kitani and L.S. Olive, Genetics 57, 767 (1967).

25. Y. Kitani and L.S. Olive, Genetics 62, 23 (1969).

26. C.W. Lawrence, Heredity, Lond. 16, 83 (1961); Rad. Bot. 1, 92 (1961); Nature, Lond. 206, 789 (1965); Genet. Res. 9, 123 (1967); Heredity, Lond. 23, 143 (1968).

27. G. Leblon and J.L. Rossignol, Colloque Soc. Franç. Génét., Toulouse (in press).

28. J.C. Lucchesi, Genetics 54, 1013 (1966).

29. S. Mizutani, D. Boettiger and H.M. Temin, Nature, Lond. 228, 424 (1970).

30. S. Mizutani, H.M. Temin, M. Kodama and R.T. Wells, Nature New Biol. 230, 232 (1971).

31. N.E. Murray, Genetics 61, 67 (1969).

32. N.E. Murray, Genet. Res. 15, 109 (1970).

33. A.B. Oppenheim and M. Riley, J. Molec. Biol. 28, 503 (1967).

34. E. Pees, Genetica 38, 275 (1968).

35. C. Pelling, Proc. Roy. Soc. Lond. B 164, 279 (1966).

36. A. Putrament, Molec. Gen. Genet. 100, 321 (1967).

37. E.R. Signer, Ann. Rev. Microbiol. 22, 451 (1968).

38. G. Simchen and J. Stamberg, Nature, Lond. 222, 329 (1969).

39. B.R. Smith, Heredity, Lond. 21, 481 (1966).

40. D.R. Smyth, Austral. J. Biol. Sci. 24, 97 (1971).

41. D.R. Stadler, Proc. Nat. Acad. Sci. U.S.A. 45, 1625 (1959).

42. H. Stern and Y. Hotta.   In The Control of Nuclear Activity
    (ed. L. Goldstein, Prentice-Hall, Englewood Cliffs, N.J., 1967),
    pp. 47-76.

43. S. Spiegelman, A. Burny, M.R. Das, J. Keydar, J. Schlom,
    M. Trávníček and K. Watson, Nature, Lond. 227, 1029 and 228, 430 (1970).

44. D.T. Suzuki, Genetics 48, 1605 (1963).

45. H.M. Temin and S. Mizutani, Nature, Lond. 226, 1211 (1970).

46. C.A. Thomas, B.A. Hamkalo, D.N. Misra and C.S. Lee, J. Molec. Biol.
    51, 621 (1970).

47. P.L. Thomas and D.G. Catcheside, Can. J. Genet. Cytol. 11, 558 (1969).

48. J. Tomizawa, J. Cell Physiol. 70, suppl. 1, 201 (1967).

49. H.L.K. Whitehouse, Nature, Lond. 211, 708 (1966).

50. H.L.K. Whitehouse, J. Cell Sci. 2, 9 (1967).

51. H.L.K. Whitehouse, Nature, Lond. 215, 1352 (1967).

52. H.L.K. Whitehouse and P.J. Hastings, Genet. Res. 6, 27 (1965).

DISCUSSION

SMITHIES: Would you care to comment on the ability of Stahl's
model to account for the recombination data you have described?

WHITEHOUSE: Stahl's model fails to account for Leblon and
Rossignol's data on the specificity of conversion pattern in relation
to the mutagen used to obtain the mutants. It also fails to explain
map expansion, and it does not provide a satisfactory explanation for
the data on polarity in recombination. Holliday and I (Molec. Gen.
Genet. 107, 85-93 (1970)) have discussed these and other features of
Stahl's hypothesis which seem to us to make it unacceptable.

CAMPBELL: As I remember, in Fogel's results with yeast, there was
extensive conversion, but he always found complete equality of recip-

rocal ascus types, such as 6:2 vs. 2:6. In the results you have shown, this is frequently not the case. Is there some discrepancy between the two sets of results?

WHITEHOUSE: The difference between yeast, on the one hand, and Ascobolus and Sorderia, on the other, in the pattern of conversion is interesting. It may be that the enzyme system postulated to bring about correction of mispairing in hybrid DNA differs from one species to another, and that in yeast it is not influenced by the particular combination of mismatched bases in the hybrid DNA, but merely excises a segment of either strand chosen at random.

THE EVOLUTION OF CELLULAR TAPE READING PROCESSES AND MACROMOLECULAR COMPLEXITY

by

Carl R. Woese
Department of Microbiology
University of Illinois
Urbana, Illinois

INTRODUCTION

I feel, somehow, that this Symposium signals a major turning point in the course of biology. It demonstrates increasing interest by a broadening spectrum of biologists in evolution. Its being directed to the evolution of genetic systems, rather than confined as is customary to evolution in genetic systems, indicates the groping for new and significant directions in the study of evolution. But my feeling is generated most of all by something deeper than this -- the firm belief that the kinds of problems soon to be encountered in the study of evolution will ultimately force a drastic revision in what we now take to be the fundamental concepts of biology.

Modern biology tends to adopt an attitude toward evolution that historically should prove odd if not actually amusing. Evolution classically has been the foundation upon which the science of biology rests. This view still receives obeisance. Yet in some areas of biology, particularly molecular biology, evolutionary considerations are of secondary importance at best. The situation here resembles the peculiar and peripheral way the classical physicist sometimes invoked the Creator. A Creator was required to provide the atoms, set them in their appointed places, generate the rules, and start the clock of the universe going. Then He bowed out of the picture. Evolution fills a similar function in the molecular concept of E. coli.

My point can also be seen in another way. If one were to say that a complete understanding of an enzyme's function requires a knowledge of its structure, any biologist would probably agree. However, were one to say that we need to understand an enzyme's evolution in order to comprehend fully its structure, not only would it be hard to find agreement, but the statement would not even be considered sensible in some quarters.

In view of this flippant treatment accorded evolution, it will be necessary to survey certain of the basic assumptions modern biology makes about the nature of the universe in general and the nature of biology in particular, for herein lies the source of the problem.

Molecular biology is clearly reductionistic. It has demonstrated to its own satisfaction that all biology (with the possible exception of the uncharted reaches of the mind) can in principle be reduced to (derived from) the tenets of physics as now understood.[1] Thus, biology becomes a mere derivative branch, albeit complicated, of the fundamental science, physics. Furthermore, in that the conceptual structure of biology is considered virtually complete, all the biological problems that now remain unsolved (interesting or not) become mere workings out of the details inherent in this structure.[1]

If this view of biology were correct, evolution would indeed be a mixed bag of peculiarities, a basically unrelated collection of "historical accidents," an unordered wandering through an immense evolutionary phase space. However, such a reductionistic view of biology is, fortunately, not the only way to look at the problem, and in my opinion is highly unlikely to prove a productive way in the final analysis. The philosophical weaknesses of this general position, i. e. , "scientific

materialism, " have been discussed by Whitehead. [2] And a number of

biologists have challenged the correctness of taking a purely reduction-

ist view of biology, on largely intuitive grounds.   I think many of us

here tonight would agree that its failure to cope with evolutionary con-

cepts--to make evolution the focal point of the conceptual structure of

biology, or even to make of evolution a connected whole--is sufficient

reason to question the soundness of the present conceptual structure of

biology.

Perhaps the area of most immediate concern for the evolutionist

here, where he is first likely to encounter the weaknesses or distortion

in the conventional wisdom, is in the particular area I have chosen to

discuss, the evolution of those "tape reading" processes that together

constitute the essence of the "Gene. " The basis for our present con-

cepts of nucleic acid replication, transcription, and translation lies in

"templating, " a notion introduced into biology about 30 years ago by

Pauling and Delbruck. [3] Templating has two aspects, "recognition" and

"alignment. " "Recognition" is seen in terms of "complementary fit"--

a lock and its key, a hand in a glove. Alignment is the juxtaposition of

two units in a way that promotes polymerization.

From the evolutionist's point of view the most important char-

acteristic of templating is its basically static nature. It is pure spatial

geometry; its essence is devoid of time ordered process.

Templating is admirably suited to describing nucleic acid repli-

cation and transcription. Base pairing and stacking of monomer units

provide precisely the type of "recognition" and "alignment" described,

and account for nearly all that is now known about the gene per se. [4]

In a sense it is unfortunate that templating is such a simple and

beautiful model for the nucleic acid replication process, for by its implicit omission of evolution from the description, evolution comes to have have no essential role to play in gene replication. How can such a process today differ in its basics from the nucleotide interactions that might be found in a primitive ocean, a purely chemical milieu? (Of course one can object that polymerases are "evolved" entities, but even that does not have a convincing ring, for the main role they play, the moving of the replication point, might also be some property inherent in base bairing, etc.) Thus, gene replication and transcription appear to be (and are treated as) processes that are not concerned with evolution in any fundamental way.

Translation is another matter, however. This process differs in many respects from the other tape reading process, and the evolutionist must here examine the conventional wisdom closely before accepting it as a part of his concept of cellular evolution. The reason is that the conventional wisdom has never explicitly and clearly faced the question of whether the nature of translation is in essence an evolutionary problem--which would make translation fundamentally unlike the other tape reading processes.

When the nature of gene replication was revealed in the Watson-Crick structure, attempts were made to model translation similarly-- as anyone familiar with Gamow's work on the genetic code knows. [5] Crick was quick to point out the unlikelihood that nucleic acid <u>could</u> "template" amino acids in protein synthesis. [6] At this juncture one then had two alternatives: (1) to discard templating as the proper way to look at translation, or (2) to introduce some modification of templating that would permit its retention as a valid model for translation. The latter alternative was chosen, and the famous Adaptor Hypothesis was the result. [7] Each amino acid became outfitted with an "adaptor" such that the resulting unit <u>could</u> now be templated--the adaptor itself

being recognized through base pairing to the template--which event then
somehow aligned the amino acid as well. From that day to this the
transfer RNA molecule has been seen as an adaptor.

(I note parenthetically that this ostensibly simple modification--
the adaptor--opened the door ever so slightly to evolution, for now the
linkage between amino acid and codon was determined by the properties
of an evolved entity, the activating enzyme, making at least the codon
assignments subject to evolution. However, the subsequent history of
the field has made it clear that the evolution of translation never became
a matter for serious concern; experimental approaches to the problem
have by their nature been predicated on the assumption that it is possible
in principle to "understand" the process without giving thought to its
evolution--and none has been!)

What do we loose by viewing translation in templating-adaptor
terms? Firstly, tRNA clearly becomes a static entity. (See for example, the definition of "adaptor" in Webster's dictionary.) The
adapted-amino acid is the concrete unit, the building block, that is
processed by some protein synthesizing machine. This, I feel, accounts
for the fact that tRNA is very often spoken of in a passive tense, while
the ribosome is referred to in an active tense. This also, I feel,
accounts for our attempts to determine the structure (the single functional form) for tRNA, or the structure for the anticodon loop, etc.
Furthermore, I would claim this view of translation is responsible for the
nature of "alignment" never being seriously questioned. Alignment is
just some simple placing of the incoming amino acid near the peptide.
That it might actually be some fairly intricate process, worthy of study
in its own right, is seldom if ever suggested.

It could be that its most serious flaw is the way in which the
template-adaptor view conceives codon recognition. Is codon recognition

just this matter of correct "fit" between codon and anticodon (as we

think nucleotide recognition in gene replication may be)? Or perhaps

does the real biological specificity here derive from something more

than a mere static fitting together? This question receives emphasis

from the observation that incorrect tRNAs can bind to message RNAs

on the ribosome, but they nevertheless, are not decoded. The best

example here perhaps, is the aspartyl-tRNA (codon GAU or GAC)

binding perfectly well to a poly (GAA) (repeating sequence) mRNA but

not effecting asp incorporation in an in vitro system.[8]

Since template-adaptor thinking is so much a part of modern

biology, it is difficult to envision transfer RNA as anything but an

adaptor. The static adaptor--more properly, the adapted amino acid--

clearly connotes an entity having a single (functional) state. Thus, its

reasonable alternative is a machine--a device that passes through a

cycle of states. An adaptor is considered only in terms of the sup-

posedly static recognition and alignment functions of translation--

forcing us to invoke the ribosome (and its "factors") as being what makes

it all go.[1] If we are willing to consider tRNA, not necessarily as an

adaptor, but as a machine, then hitherto unimagined roles for tRNAs

in mRNA movement and in "non-static" kinds of "alignment" processes

can be imagined.

THE METAPHYSICS OF EVOLUTION

In the past, discussion of the evolution of translation has re-

volved about the special selective advantages various features of the

process have. Thus, discussion would center upon matters such as why

the codon is a triplet, why there are 20 amino acids encoded, and why

it is this particular twenty; on why the codon assignments are universal,

why wobble pairing and degeneracy exist, whether codon reassignment is the only possible mechanism for evolving the order to the codon assignments, and so on. [9] One could argue the pros and cons of whether codon assignments were evolved in order to minimize the deleterious effects of mutations, whether certain amino acids were more primitive in the code than others, and so on. [10, 11, 12] These types of discussion never get very far, but their real fault lies in the fact that they focus attention on a mixed bag of unrelated evolutionary arguments, with the result that we tend to miss the forest for the trees.

What is important at this point is not to fall into this type of reductionist trap, not to make of evolution this mixed bag of more or less unrelated "historical accidents. " We must try to view the evolution of translation as some integrated whole, and to see this evolution even as part of a larger connected evolution of the cell, and so on.   The fragmented reductionist view of evolution that is so prevalent must be countered by appropriate, holistic constructs.

Of course, to do as I suggest requires nothing short of a metaphysical purge, a rejection of the reductionist materialism that pervades biology today. The philosophical underpinnings for a new biology, a biology in which evolution is the central consideration, are clear in Whitehead's philosophy of "organism, "--stressing process as fundamental, a view in which existence and evolution begin to fuse. [13] Bohm has recently begun a restructuring of the conceptual basis of science in a Whiteheadian mode, stressing the primacy of order, attempting a proper redifinition of "randomness, " and emphasizing hierarchical nature of order. [14] His introduction of the notion of a hierarchy of time-like parameters should also appeal intuitively to the evolutionary biologist.

In terms of a Whiteheadian metaphysics, evolution seems to define itself as the problem of how order at one "level" of the universe relates to order at adjacent "levels." It is clear, particularly to the biologist, that the material Universe seems to constitute a hierarchy of (more or less) discrete "size levels." The biologist routinely works on the molecular vs. the cellular vs. the organismal vs. the societal levels, and recognizes the atomic vs. the subatomic levels of physics that underlie biology. If we define one such level as a "microuniverse" and the next higher (clearly distinguishable) level as a "Macrouniverse," the evolutionist seeks to define the nature of the "elements" in the microuniverse from which "macroentities" in the macrouniverse are constructed. He seeks the Laws that govern this concrescence. He is concerned with the extent to which (and assumptions under which) properties of the macroentities (macroproperties) are or are not derivable from properties of various aggregates of elements in the microuniverse (i.e., from statistical properties of defined classes of "related" microstates of the microuniverse). In particular, he is concerned with the extent to which such Laws of concrescence might be invariant--i.e., independent of what level one chooses to define as the microuniverse. (I think one can readily see that were such invariant Laws of concrescence to exist, biology would be an entirely different kind of science than it is generally taken to be!)

EVOLUTION OF MACROMOLECULAR COMPLEXITY

Certain recent developments in molecular biology and in cybernetics promise to yield considerable insight into specifics of the macoorder-microorder problem. I should like now to discuss these as they pertain to macromolecular concrescence, the evolution of complexity (macroorder) on the molecular and molecular-aggregate level.

Consider first what general characteristics might be essential to elements in a microuniverse if they are to be used in building

macroorder. An element has to "feel" as well as be felt by its environment; it has to "respond to" its environment, as well as to have an
effect upon it. This more or less means the element must be capable
of existing in two or more "reasonably stable" states, but states that
can be attained through perturbations whose energies are in the
"thermal" range (perhaps near its upper reaches) for that microuniverse.

Over and above their multiple states, the elements must then
have the capacity to "interconnect" with one another, to become coupled.
In other words, states of one element must be capable of influencing the
stability of states of another element. In this way not only stability of
the state of an element (i. e. , its sensitivity to perturbation) can be
effected, but the nature of the interaction of the coupled whole with the
environment can become more complicated, more "specific" (than
would be the case for the uncoupled parts).

I am particularly intrigued by recent investigations into the
properties of networks of bistable (and multistable) elements that are
interconnected in various random ways--e. g. , random choice of which
element connects to which others, or random choice of the input-output
logic for an element. I refer here to the work pioneered (I believe) by
Ashby and coworkers and extended by Kauffman. [15, 16] What amazes
everyone about these randomly constructed networks is first of all that
they tend to behave as "entities, " i. e. , they settle down into cyclic behavior patterns. What next impresses one are their "biological"
characteristics: (1) the "entities" (cycles) can sometimes be quite
complex, (2) they have homeostatic character--in that when perturbed,
they tend to return to their original form, and (3) cycles built using
certain logics show properties that are invariant with, are not a function of, which particular elements are connected to which others.

It is generally agreed, I think, that these types of networks are a decent first approximation model for a number of biological systems. Allosteric proteins and the on-off nature of gene repression immediately suggest bistable (or multistable) elements. Attempts have been made to model regulation and particularly development in metazoans in random network terms.[16]

The question I would raise tonight in the context of evolution of the primitive cell, is whether similar networks are valid models for macromolecular structure, function, and their evolutions. On the macromolecular level what would correspond to "bistable" elements? How are such elements interconnected (i. e., coupled)? How is such a network constructed? And in this sense "construction" means evolution. What properties will our macromolecular networks have?

I will now try to summarize the evidence that proteins and in particular the translation apparatus are entities of this general nature. However, I will call the networks in these cases "ensemble machines" for reasons that will become apparent. Ensemble machine is a concept in which not only macromolecular structure and function, but their evolutions as well, are properties of coupled bistable (multistable) elements.

At the present state of our knowledge, we seem to have an indication of what an "element" is in at least one case (as we shall see). And, I would claim that the general rules for constructing (evolving) ensemble machines are becoming evident. Furthermore, the properties of these networks, these ensemble machines, give some insight into how macroorder might arise from a microuniverse.

First, consider the construction rules (which turn out to be quite different than those used by Ashby or Kauffman). These, I claim,

have been at least partially recognized as special cases in evolution for
several years now. All I now suggest is that we recognize them for what
they are--examples of a general rule--probably the mechanism, the
only mechanism, used in nature to evolve macromolecular complexity.

Let me, initially, merely describe the construction rules, the
mechanism for evolutionary assembly--and attempt to rationalize their
various features later. The mechanism, called a "co-dimerization
cycle" works as follows (see Figure 1): A peptide A encoded by gene a
evolves to become a dimer A•A. Next the underlying gene, a, duplicates, permitting the evolution of a dimer, A'•A*, of related, not
identical halves. At some subsequent point in time the two related
genes, a' and a*, become joined in such a way that the resulting protein
retains the essential properties of its ancestral quaternary structure,
although it is now in reality merely a tertiary structure, A'-A*. (This
covalently linked "pseudo-dimer" is called a "co-dimer.") As evolution
proceeds the two "halves" of the co-dimer become increasingly unlike
one another (but in complementary, interacting ways); so a point is
reached where we would want to define the co-dimer as a new protein,
B. B is about twice the size of the starting protein A, and I would
claim twice as "intricate." The cycle now begins over again using B as
the starting protein, which then dimerizes, etc., eventually yielding
protein C, about twice the size of B, and so on. The exact order of
all steps in the cycle is not critical, thus they need not always occur
in the exact same sequence in every instance. This cycle is pictured
as starting with very small proteins, say of the order of decapeptides,
and continuing to work until proteins of the modern variety emerge. [17]
Dus et al., for example, claim to see a repeated unit of length about
one dozen amino acids in a bacterial cytochrome C--which in terms of
the co-dimerization mode of genesis would mean three rounds of its

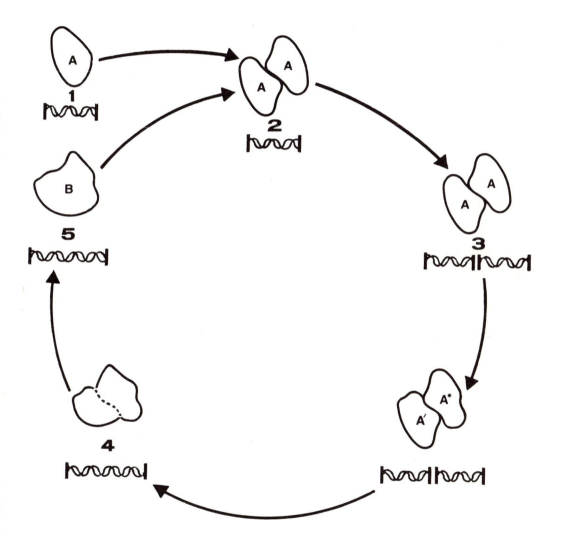

Figure 1.   The Co-dimerization cycle for evolution of macro-
             molecular complexity.

cycle, i.e., $2^3 \times 12$ is about the number of amino acids in cytochrome C.[18] As we shall see the same sort of cyclic genesis is envisioned as generating the ribosome--except that it is the RNA that undergoes co-dimerization. The ribosomal proteins consequently increase in number (and so ultimately in number of kinds).

I will not attempt to defend this as the universal mechanism for evolving macromolecular complexity. The experimentalist must have the final say here in any case. However, I do want to amplify and try to rationalize certain of the cycle's features. For one, the cycle demands that dimerization of a protein (sometimes) leads to selective advantage-- although it is clear that the dimer is no more complex than the corresponding monomer in its information content (defined by the number of amino acids in a polypeptide chain). However, there appears to be another quality to (biological) information, perhaps best called "intricacy" (or the more conventional term "meaning"). And I would claim a dimer can in principle evolve to a state of greater intricacy than any monomer of the same information content.[17]

Then there is the assumption that to have evolutionary significance the dimer must comprise identical (or related) subunits. This is difficult to rationalize at present, for the capacity for unrelated genes to become linked and so to produce proteins with unrelated "halves" clearly exists. Perhaps the probability of forming quaternary structure from everted tertiary structure (forming intermolecular "tertiary" structure from what was intramolecular tertiary structure) might prevent anything but related peptides from forming meaningful association. In any case development of a proper theory of coupling between identical (vs. unrelated) networks of the type that constitute macromolecules is needed before one could hope to make definitive pronouncements here. (If coupled oscillator systems are properly analogous, dimerization of identical (related) subunits might be rationalized in terms of resonance phenomena.)

The gene joining step, which reduces a quaternary structure to a tertiary structure is seen as advantageous on the grounds that this covalent linkage is ultimately the most versatile evolutionarily--i. e. , this manner of linking the two halves of the molecule places the fewest constraints on the subsequent evolution of the two halves of the molecule. It is not unreasonable that this covalent linkage step in the cycle be delayed for one or more rounds--and hemoglobin can be viewed as an example of just this, i. e. , as a dimer of a noncovalently linked subunit $(\alpha \cdot \beta)$. [17]

If one starts a co-dimerization cycle with a bistable element, and couples this (dimerizes it) to a second element in such a way that two of the four possible states of the resulting pair become favored (stabilized), then the resulting entity itself is to a first approximation again a bistable entity. One can then couple two of these larger entities, via a second co-dimerization cycle, in such a way that two (or perhaps two subsets) of their total set of microstates are further stabilized, and so on. The coupling between two elements or two groups of elements envisioned here is not of an all or none type, it is merely one that tends to stabilize to some degree the "preferred" microstates, i. e. , certain defined classes of microstates. [19] In this sense it differs from the coupling involved in the Ashby-Kauffman networks. It differs further from the latter type of coupling in that it is not taken to be always between individual bistable elements. In the present instance coupling is of a "hierarchical" nature, that is, the influence of the state of one unit (be it element or group of elements) on its dimeric partner is to a first approximation initially a coupling

between the two units each as a whole; it is not predominantly between individual elements within the units.[19]

Built along such a hierarchical coupling scheme the molecule
ultimately evolved is composed of about 2^n bistable elements (where n
is, of course, the number of co-dimerization cycles) and is generally
functionally bistable (or quadristable, etc.)--in the sense that it
undergoes transitions between two different functionally significant
macrostates (two classes of microstates).

The nature of the coupling and the general construction principles
for these ensemble machines are such that when the machines are large
enough--contain a sufficient number of elements--then they are virtually
immune to random perturbations in their microuniverse. This is <u>not</u>
because the structure has become so rigid that thermal fluctuations do
not affect it, do not ever cause state transitions in individual elements.
Rather, it is because the probability of thermal fluctuation causing
state transitions in exactly the right combination of individual elements
that is required to effect a macrostate transition, a transition of the
machine as a whole, <u>this</u> probability becomes vanishingly small as the
number of individual elements in the ensemble increases.[19]

Thus in evolutionary terms we have started with a macromolecule
(granted a small one) whose functioning (state transitions) was more or
less at the mercy of random perturturbations in its microuniverse. The
macromolecule has evolved, grown to become an ensemble machine,
and in so doing its function has become increasingly immune to the
vicissitudes of its microuniverse. Let me stress that the way this is
accomplished is not analogous to the usual energy barrier model--
something so high that nothing is energetic enough to hurdle it. The
situation is more like a <u>maze</u> too difficult to pass through it. Thus,
I like to look at these ensemble machines as Biological Mazes.

No amount of energy _per se_ can get you through a maze; you need _information_. *

Therefore, only information, molecules with specific structure are likely to effect macrostate transitions in ensemble machines; they do so by "knowing" the correct combination of elements whose states require changing to bring about a macrostate transition.[19]

The ensemble machines I am describing seem isomorphic with certain kinds of electronic circuitry. Although it is difficult to see in mechanical terms, the electrical analogs have certain properties that may prove extremely interesting in ensemble machine theory. To this point, I have discussed only how ensemble machines might be made immune to noise. However, as anyone who has built radio sets from childhood on up knows, the specificity of response to a signal (the discrimination) can be made more and more precise by building a bigger set (i. e., more elements coupled in the proper way). Could it be then that increased capacity to discriminate between similar structures (i. e., increased biological specificity) is also an inherent property of the evolving ensemble machine? In other words, the range of structures that will trigger macrostate transitions can narrow as the ensemble machine becomes more complex.

EVOLUTION OF THE TRANSLATION APPARATUS

It is clearly my contention that the translation apparatus is an ensemble machine. And until we are willing to conceptualize it in this sort of evolutionary way, we will be hard put to understand it. Therefore, I should now like to propose a specific model for how a translation process can be designed along the ensemble machine principles of evolution.

* Parenthetically, I find a "maze" is heuristically a more satisfying way to conceive of that phenomenon usually called quantum mechanical tunneling (i. e., "tunneling" through an energy "barrier").

Given charged tRNAs, the central event in translation is then the tRNA-codon-ribosome interaction, which as everyone knows, involves a "recognition" of the codon, the formation of a peptide bond, and the moving of the message RNA relative to the mechanism, by one codon unit. The focal point for at least a part of this process lies in the so-called "anticodon arm" of the transfer RNA molecule, a structure that always has a double stranded "stalk" of five base pairs, capped by a central "loop" of seven nucleotides, the middle three of which are the anticodon proper.

To this point I have spoken only in the abstract of bistable elements. Now we are about to meet one face to face! In 1967 Fuller and Hodgson, using molecular models, showed that the anticodon arm could be put into either of two stable configurations.[20] The first of these, called the FH or "+" form, is shown in Figure 2a. As can be seen the 3'-5' chain of the double stranded stalk can form a base stacked single stranded helical extension that goes into the so-called loop region for a length of five additional bases. This leaves two bases in the loop to stretch from the top of this single stranded helical extension back down to the other chain of the double stranded stalk (i. e. , its 5'-3' chain). Note that the anticodon proper comprises the topmost three bases in the single stranded extension, and there is room in such a structure for the anticodon to base pair with the codon.*

In Figure 2b the alternative, hf or "-" form for the anticodon arm is shown.[20] It is just like the FH, or "+" form, except that the single stranded helical extension of five bases originates from the 5'-3' chain of the stalk, not its 3'-5' chain. Because the "-" and "+" forms are symmetrical, the anticodon proper is again found at the top of the

---

* This form is called "+" because the direction of reading of the codon is the same as that defined by the process of extending the double stranded stalk.

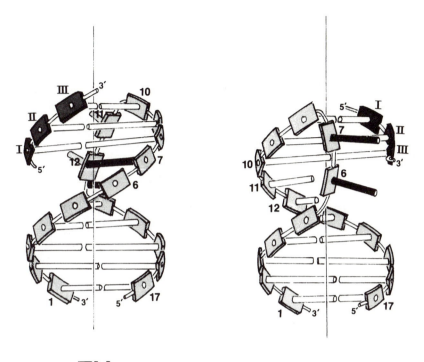

FH   **hf**

Figure 2.   The two forms for the anticodon loop of tRNA.

2a (left) the FH or "+" form
2b (right) the hf or "-" form

Numbering of bases is according to the convention of
Fuller and Hodgson.   Codons are shown in Roman
numerals.

single stranded extension, but, as you will note, inverted relative to its orientation in the FH or "+," form. (And therefore the direction of codon reading is opposite to the direction defined by the process of extending the double helical stalk--hence the designation "-".)

Since all of us in 1967 were probably under the influence of the template-adaptor concept of translation, we did not see these two forms "+" and "-" as defining a bistable element. Thinking tRNA to be a static entity, Fuller and Hodgson, I think like anyone would have done, asked the question "Which one of these two forms of the anticodon loop is the correct one?" For reason I needn't go into, they decided the "+" (FH) was the correct form.[20] Recently Uhlenbeck, working in Doty's laboratory and later on his own, has found strong experimental support for the idea that the tRNA molecule can exist in either the "+" or the "-" form, although under his conditions the "-" (hf) form is the energetically favored one.[21] (Uhlenbeck, personal communication.)

This particular bistable element, the anticodon arm, is of particular interest in that one can utilize its state transitions to design a very simple primitive mechanism that not only does the codon recognition but is the basis for the mechanical movement in translation--i. e. , the pulling of the message RNA--as well.

Figure 3 shows what I have called the translation complex. Note that when the incoming tRNA is in the "-" (hf) form but the previous tRNA is in the "+" (FH) form, then the two anticodon arms and their respective codons can form one enormous double helical type of structure equivalent to twenty base pairs in length.[22] It is hard to believe a structure such as the translation complex has no biological significance! Since a tRNA in the "-" (hf) state can form this structure with the preceding tRNA, while a tRNA in the "+" (FH) state must form the

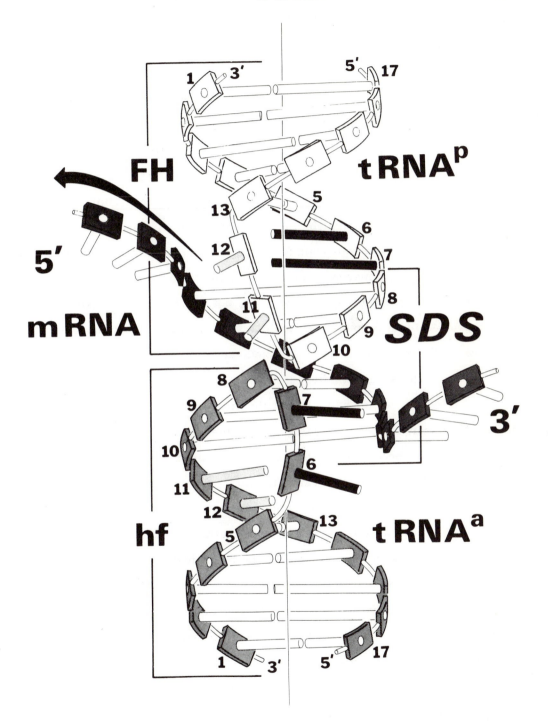

Figure 3.   The translation complex - see text for details.

structure with the succeeding tRNA, it follows that a tRNA should undergo a state transition, "-" → "+", during the process of translation.[22]

Figures 4a and 4b show how these state transitions can be used to drive the message RNA through the translation apparatus. Figure 4a again shows the translation complex at some arbitrary point in a translation cycle, say when codons 3 and 4 are being read, by the "peptidyl" and the "amino acyl" tRNAs respectively. Note that codon number 5 is necessarily excluded from the axis of the translation complex at this point. In Figure 4b, the most recent tRNA (reading the 4th codon) has undergone a transition from the "-" (hf) form to the "+" (FH) form. A necessary consequence of this is that codon 3 can no longer share a common helical axis with codon 4, but now codon 5 can exist in a common axis with codon 4. As soon as $tRNA_5$, appears to read codon 5, a translation complex is reestablished, and the cycle begins again, this time with the state transition in $tRNA_5$, and so on.[22]

I think the reader can see that the mechanism for movement of the mRNA described here is analogous to a mechanical ratchet. For reasons that will become clear below, the mechanism should perhaps be called a "double reciprocating ratchet," when it applies to translation today.

Now, if we were to construct a translation machine in the evolutionary sense, I think something like a ratcheting mechanism is what we would want to use as a starting point. The mechanism itself is very simple, which it would have to be in order to arise under primitive earth conditions. The mechanism operates in terms of the anticodon arms. Therefore, the type of primitive tRNA we would have to invoke could be simpler than today's Medusa-like version. The archetype tRNA need be only about as complex as the anticodon arm of tRNA today--i.e., a single self-complementary arm structure. The existence of self-complementary nucleic acids in a primitive environment can be readily

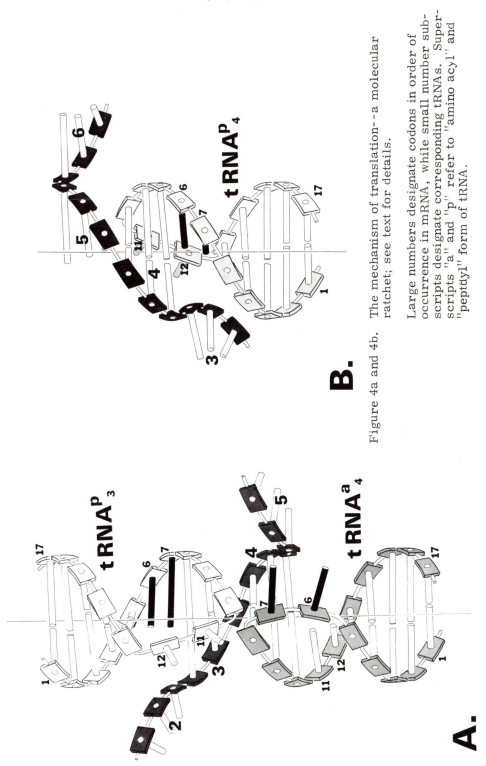

Figure 4a and 4b.   The mechanism of translation--a molecular
ratchet; see text for details.

Large numbers designate codons in order of
occurrence in mRNA, while small number sub-
scripts designate corresponding tRNAs. Super-
scripts "a" and "p" refer to "amino acyl" and
"peptdyl" form of tRNA.

rationalized--e.g., they would be more stable than nucleic acids with random secondary structure.[23]

Of course so simple a primitive tRNA presents a problem when it comes to "recognizing" and attaching an amino acid. But the problem is really no more acute than it would be for today's versions of tRNA were we to take away the activating enzyme (a structure that needed some kind of preexisting translation mechanism in order to begin its own evolution). Without attempting to justify the assumption, I will postulate that the primitive tRNA carried the amino acid or the growing peptide, as the case may be, within its "loop" region--say on the anticodon-adjacent adenine residue, the one that is so highly substituted in modern tRNAs. (I would also claim that these loops had properties that allowed them to "recognize" the amino acid in the first place.)[20] With this mode of attachment the workings of the ratchet mechanism would permit the α-amino group of the amino acid on one tRNA to attack the proper carbonyl carbon of the peptide, held on the other tRNA in the translation complex--a point that can be demonstrated with the use of molecular models.

The main drawback to so simple a machine is its imprecision. The mechanism is driven by the random perturbations in its environment. But these by their nature will make the machine function in an inexact, somewhat unpredictable way. For example, the process of amino acid recognition would be inexact at best for so simple a device. Perhaps an amino acid could be recognized only to the extent of being one of a general type of amino acid, which would lead to "group" codon assignments.[24] There is no reason to suspect codon recognition to be exact either (given wobble pairing as a possibility). And ratchets do move backward sometimes, which in this case could mean termination of peptide output. Then there is reading frame maintenance, and so on.

What sort of proteins would such a machine make? At best I would guess, they have to be very short, and they would be what I have called "statistical proteins"; a "statistical protein" being defined as a group of proteins none of which tend to have exactly the same primary structure, but all of which are approximate translations (by today's standards) of one given nucleotide sequence.[24, 25] Needless to say, such proteins would be very limited in their enzymatic capacities. But at so early a stage in evolution, something is better than nothing; any kind of catalytic capacity, however crude and nonspecific, might be advantageous.

The evolutionary route from there to here is, of course, along ensemble machine lines. The primitive ratchet mechanism, involving two bistable elements at a time might be stabilized in certain ways by coupling additional bistable elements into the system, as discussed above. Perhaps the new bistable elements could evolve as an extra arm on the archetype tRNAs. And I would not be at all surprised if this were ultimately found to be the reason behind the evolution of several of the tRNA arms. (As you know, the so-called common arm of tRNA can in principle exist in the same "-" (hf) and "+" (FH) forms that are found in the anticodon arm. And the common arm is self-complementary as well; i.e., the ψCGA sequence pairs with itself.[22])

As the translation ensemble machine evolved there would be no need for all new units added to be covalently linked to the tRNAs; in fact there could conceivably come a point in evolution beyond which covalent linkage would be disadvantageous. Therefore, a physically linked structure might ultimately emerge, and this, of course, would be a two-fold symmetric structure--one half associating with each of the tRNAs in the translation complex. The ribosome would then increase in size and complexity by the above co-dimerization type of evolution, until it reached its present, perfected form.

I should like to stress several points about such an evolutionary model: (1) although ribosome and tRNA are physically linked only, they are to be considered to form a <u>tightly</u> coupled mechanism. Thus, the original simple "-" → "+" state transition in primitive tRNA that pulled the message RNA tape, evolves to become a part (the central part) of an all-or-none macrostate transition in the ribosome-tRNA complex as a whole. (2) The ribosomal RNAs, like their tRNA counterpart, contain many "arm" structures (and so potential bistable elements). Therefore, I feel that nucleic acid arm structures are the main class of bistable elements from which the ribosome ensemble machine is built. (3) As proteins evolved they of course, came to assume more and more important roles in the evolving ribosome, though they may have had little or no role to begin with, and the essence of translation today still rests, I feel, in the properties of the RNA components. The so-called "T" and "G" factors are an interesting case in point. I have said above that in the early stages of its evolution an ensemble machine can be driven by the random perturbations occurring in its microuniverse. However, as the machine becomes more and more perfected (evolved), it becomes progressively immune to being driven by random processes and requires information in order to pass from one macrostate to another. On the present model I would conclude that the T and G factors probably evolved to be the information, the signals the machine came to need in order to pass from one macrostate to the next. (4) I think a very useful way to look at ribosome function is in terms of an analogy: if we consider the ribosome-tRNA complex to be an enzyme, the tRNA would be analogous to the enzyme's "active site," the ribosome to the "rest of the enzyme." In this sense then, the ribosome has <u>no</u> function in its own right, but it is the bulk, the complexity of the enzyme, that makes for precise, accurate functioning of the site. (5) Finally, in that the translation complex has a two-fold (rotational) symmetry and the ribosome is pictured as reflecting this symmetry in its evolution, the ratchet model suggests a type of "site model" for the ribosome quite unlike the conventional A-site-P-site model.[22] This latter is what might be called an asymmetric serial-passage machine--it has two sites (of different kinds); the codons enter one and pass to the other, from which they then exit. What the ratchet suggests is a machine having two functionally equivalent sites; and a codon passes through only one of these--i.e., the "-" (hf) → "+" (FH) transition does not involve a translocation, but rather occurs in situ. Thus the even numbered codons would pass through one site, the odd numbered codons through the other site. In the above terms, this would be a symmetric parallel-passage machine.

I could discuss evolution of the other tape reading processes in the cell, but this would add nothing to the generality and so to any evolutionary value of what has already been said. Just let me end by saying that nucleic acid replication may not be all a matter of base pairs. At least in the molecular mechanism of movement of the polymerases, there may still remain an interesting evolutionary problem.

References

1. Stent, G. Proc. Am. Acad. Arts and Sci. 99 909 (1970).

2. Whitehead, A. N. "Science and the Modern World." The Mac-Millan Company, 1925.

3. Pauling, L., and Delbruck, M. Science 92 77 (1940).

4. Watson, J. D., and Crick, F. H. C. Cold Spr. Harb. Symp. Quart. Biol. 18 123 (1953).

5. Gamow, G., Rich, A., and Ycas, M., Adv. Biol. Med. Phys. 4 23 (1956).

6. Crick, F. H. C. "A Letter to the RNA Tie Club" (unpublished) quoted by M. Hoagland in "The Nucleic Acids" III. Chargoff, E. and Davidson, J. (eds.). Academic Press, N. Y. 1960.

7. Crick, F. H. C. Symp. Soc. Exp. Biol. 12 138 (1958).

8. Ghosh, H. P., Soll, D., and Khorana, H. G. J. Mol. Biol. 25 275 (1967).

9. Woese, C. J. Mol. Biol. 43 235 (1969).

10. Sonneborn, T. M. In "Evolving Genes and Proteins" Pg. 377, V. Bryson and H. Vogel (eds.). Academic Press, New York, 1965.

11. Goldberg, A. L. and Wittes, R. E. Science 153 420 (1966).

12. Crick, F. H. C., J. Mol. Biol. 38 367 (1968).

13. Whitehead, A. N. "Process and Reality" The Macmillan Company (1929).

14. Bohm, D. In "Towards a Theoretical Biology 2" pg. 18, 41, C. H. Waddington (ed.), Edinburgh University Press (1969).

15. Walker, C. C., and Ashby, W. R. Kybernetics 3 100 (1965).

16. Kauffman, S. A. J. Theoret. Biol. 22 437 (1969).

17. Woese, C. R. J. Theoret. Biol. 32 000 (1971).

18.   Dus, K., Sletten, K., and Kamen, M.   Jour. Biol. Chem. 243 5507 (1968).

19.   Woese, C. R.   Manuscript in preparation.

20.   Fuller, W., and Hodgson, A.   Nature 215 817 (1967).

21.   Uhlenbeck, O. C., Baller, J., and Doty, P.   Nature 225 508 (1970).

22.   Woese, C.   Nature 226 817 (1970).

23.   Woese, C.   Bioscience 20 471 (1970).

24.   Woese, C.   Proc. Nat. Acad. Sci. U.S. 54 1546 (1965).

25.   Woese, C.   "The Genetic Code: The Molecular Basis for Genetic Expression." Harper and Row, New York (1967).

DISCUSSION

CAMPBELL: You expressed doubts early in your talk about the importance of historical accidents in evolution. Would you clarify the basis of those doubts?   In the systems you have described, it seems that once a certain level of complexity has been reached, it is increasingly hard to go back and try other possibilities, so that the system is essentially committed to one pathway that was initially chosen by chance.

WOESE: Your question goes to the heart of the micro-order-macroorder relationship, and I find it difficult to answer in any convincing way.   Let me state the underlying metaphysics as I see it, and see if that helps.   Any "thing" be it an atom, a cell, or a star, can be defined by, it _is_ a certain class of "related" microstates of the "elements" in its microuniverse.   At any given "moment" in _its_ time

the macroentity is represented by one particular microstate
in this class; at another "moment" it is represented by
another member of this class, etc. Which one of the microstates _is_ the macroentity at one given moment is a matter of
historical accident, and the class of microstates may be so
immense that only the bearest fraction of these are realized
during the existence of the entity. The "macroproperties"
of the macroentity, its "characteristics", then are statistical properties, time-averaged properties over the class of
microstates. (They are averaged over all macroentities "of
this type" as well.) I will not discuss here the problem of
"detecting" macroproperties, which is critical to their
definition. Just as a macroentity "renews itself" (switches
from one microstate to another within the same class) from
moment to moment - until its ultimate destruction - a macroentity has to undergo a process of "coming into existence",
of "evolution", to begin with - again be it an atom, a cell,
or a star. This "becoming" or evolving, has to be governed
by the same sorts of Laws of Nature as are its continuance,
its self-renewal - i.e., the former cannot be processes
whose outcomes are much less certain than that of the latter,
given defined starting conditions. Now when this metaphysics
is applied to biological evolution the following sorts of
assertions emerge: The fact that the _E. coli_ ribosome and
the _B. subtilis_ ribosome have the particular primary struc-

tures they do is a matter of historical accidents. However, the difference between an E. coli ribosome and a B. subtilis ribosome must be a difference in microstate. The macroproperties of these ribosomes are the same. And that such a ribosome would evolve (given the primitive earth as it was) is a foregone conclusion. That an organism with the macroproperties of E. coli would evolve is a foregone conclusion. That an animal with the macroproperties of a sheep would emerge is a foregone conclusion. (However, note that in terms of its "macroproperties", when properly defined, a sheep may not differ from a cow, etc.) I guess the best way to respond to your question then, is this: As long as the evolutionist looks at properties that are idiosyncracies of individual microstates of a macroentity (which seems mostly what he has been doing up to now), then evolution will necessarily be seen as historical accident, as a random walk in an immense phase space, or as an exercise in miracles. However, when the evolutionist comes to the point that he can define what are the true macroentities (and so their macroproperties) in biology, then he will find Laws of Evolution; and particular macroentities will emerge from evolutionary processes (under given conditions) just as certainly as particular subatomic particles emerge from the decay of given atomic nuclei. In your question you speak of the evolutionary "system....committed to one

pathway" (my emphasis). Your phrasing strongly suggests
that you will accept an evolution that is in some sense not
historical accident every step of the way, but only at
certain crucial turning points (the pathway in your words
"was initially chosen by chance"). For me even these turning points are accidental only in a microorder sense. One
should still be able to define "macroorder" properties for
them. Frankly, I am not yet totally convinced my metaphysics is correct (so have rarely discussed it publicly).
Many things about it bother me. However, I am convinced
that the conventional metaphysics for evolution is grossly
incorrect. I am also convinced that it is in the interest
of science for groups of scientists to work from different
metaphysical bases. And therefore, I am convinced that I
am fulfilling my role as a scientist in following the above,.
essentially Whiteheadian, metaphysics in attacking the
problem of evolution.

    LEE: In your model, perhaps with a slight modification,
can mRNA go backward? (I know it doesn't go backward in
reality.)

    WOESE: I see no reason why this mechanism cannot run
backwards in principle, and so produce 3'-5' translation of
mRNA. Which direction evolved probably has to do with
whether a peptidylated archetype tRNA is more stable in
the "+" (hf) vs the "−" (FH) form, which could conceivably

turn on where the peptide (or amino acid) is attached to primitive tRNA, etc.

STRICKBERGER: Do the evolutionary concepts you have offered lead to irreversibility? (At which points?) Would the establishment of "imperturbable" molecular conditions enable a loss of complexity and more primitive structures to recur?

WOESE: I am a believer in the "time's arrow" view of evolution, and I don't see where anything I've said runs contrary to the notion. You can see irreversibility in the workings of the co-dimerization cycle, for example. Once the two halves of the original true dimer have become joined and have evolved, in a "complementary" way, to form a larger and more intricate protein, it is unlikely that you could take the two "halves" of this new protein apart and expect them to function as did their ancestor peptide. As evolution proceeds, the two "halves" become tightly coupled, their natures change, and function of one depends upon the presence of the other. You really can't reverse this evolution; the cycle won't work backwards. I'm not sure what you mean by "establishment of 'imperturbable' molecular conditions", so don't know how to reply. It could mean merely lowering of temperature. That I don't think would lead to reemergence of primitive structures - simpler structures perhaps, but not necessarily the primitive ones. You could argue that up to a <u>point</u> lowering perturbation leads to emergence of increasing complexity (compare various points in the solar system). Perturbations are both the blessing and the curse. They are essential for creation and maintenance of complexity, yet they tend to destroy complexity.

KOSOWER: (1) Why do you insist on only two states?
(2) In your terms, at a certain level of complexity of macro-states, it would no longer be possible to go (evolve) in all directions.

WOESE: I don't insist on bistable elements, merely upon multistable elements. Occam's Razor then suggests we first attempt to approximate a system using the simplest of these, bistable elements unless we have reason to do otherwise. The work of Kauffman suggests that one generally does not gain versatility or complexity by using elements with more than two states. I agree with your second point, a very general one. I somehow feel, however, that we wouldn't agree completely if we discussed it in detail. In any case, your point is akin to one of Dr. Boyer's, and I'll pursue it further there.

BOYER: (1) If ensemble theory leads ultimately to imperturbability of the elements how then does evolution of, say, proteins continue to proceed? (2) Does this then force the attention of evolution wholly upon regulatory elements? (3) Is this necessarily so? Might not evolution be harmonic and in certain times might it not return to the primitive.

WOESE: The imperturbability is to random fluctuations
in the protein's environment. In any case, the elements
themselves are not imperturbable, the state of the protein
as a whole, its macrostate, is what is imperturbable. But
this does not mean that the specific nature (properties) of
the macrostate cannot be changed by mutationally altering the
character of the elements. So in one sense, protein evolution should still be possible. However, I do feel that in
another sense the nature of protein, its _general_ nature, has
finished evolving, perhaps as far back in time as the emergence of the first metazoans. (It's sort of like the situation with cars these days. They have changed (evolved)
over the past decade, but they aren't really any different
than they were ten years ago.) I agree with your second
point in that I too (if I read you correctly) feel the
thrust of evolution is no longer centered on designing types
of macromolecules, but rather on designing more complex
architecture using macromolecular building blocks (or even
larger ones). I guess you can say this means emphasis on
regulation. What I'm not sure you want to say is "regulatory elements" rather than "regulatory systems". The way
I'd like to see it said is that at one stage in evolution
Nature designs (evolves to "perfection") networks of coupled
elements (ensemble machines, etc.). The coupling here is
sort of analogous to regulation within the network. When
design is near "perfection" for one class of networks, Nature

then uses these networks as units, as elements, for beginning
to design (evolve) the next higher level of networks, and so
on. I would say that evolution is now working on design
several levels above the level that I have called "ensemble
machines". Your final point is interesting and probably
evolutionarily important. The neoteny that some evolution-
ists claim is responsible for the evolution of Homo sapiens -
and so perhaps for his brain - is akin to what you suggest.
"Time's arrow" does argue against what you suggest happening
too often or regressing too far, I would guess. Certainly
I consider it a moot point, for the present.

SMITHIES: Can you devise any tests of your translation
mechanism based on the odd and even numbered codons being
translated on different subunits of the ribosome?

WOESE: A number of tests can be devised for the aspect
of the model to which you refer - i.e., its having two
functionally equivalent sites that process codons alternately.
Although having the sites on operationally definable (separable) subunits makes testing the model easier, there are
approaches that work even when this is not the case. Also
the type of test you devise turns on whether or not initiation always begins in a particular one of the sites. Let me
give one example of a genetic approach to the problem of the
odd vs. the even codons (what we call a "parity" experiment).
One constructs mutant strains in which the "parity" of a
codon is changed. This can be done by adding genetically

three bases to a cistron (preferably in some insensitive
place, and far proximal to the test codon). Then one measures some "higher order" property of this codon - e.g. its
level of suppression in any of a variety of ways (assuming
the codon to be nonsense or missense). Functionally equivalent sites are highly unlikely to be absolutely <u>identical</u>.
A clear demonstration that suppression level generally
varies as a function of a codon's "parity" (whether it is
odd vs. even) would be hard to interpret in any terms but
two parallel sites processing codons alternately. Though
less elegant in design, straightforward biochemical experiments involving "site tagging" approaches (that either <u>do</u>
or do <u>not</u> yield results that are a function of codon
"parity") may prove most convincing in proving or disproving
this "macro-" aspect of the ratchet model.

DANCIS: What happened before there was tRNA? Was
there an association between amino acids and nucleic acids
that led to polypeptides?

WOESE: There is no reason to rule out other modes of
nucleic acid catalyzed protein synthesis. One conceivable
mechanism of this type, called "positional templating"
involves small peptides (or even monomeric forms) of basic
amino acids becoming aligned in the major groove of a
nucleic acid double helix (analogous to polylysine-DNA
complexes) as a prelude to their condensation into larger

peptides.   (Woese, _Bioscience_ 20, 471 (1970)).   Of course

this is not "translation" of the nucleic acid.

SPOFFORD: I'd like to know more about the orientation

of the amino acids that are at the other end of the anti-

codon loops - or still farther in the more evolved tRNA's.

I'd tried visualizing how they would both come out near each

other when the plane of the one anti-codon loop is inverted

relative to the other.

WOESE: If there is one thing that attempts to model

tRNA tertiary structure have taught us, it is that there are

surprisingly few steric restrictions on the relative posi-

tioning of tRNA arms.   The translation complex (Figure 3 in

the body of my text) defines an axis of (two-fold rotational)

symmetry.   It should be possible for the ends of the CCA

arms of the two tRNA's that form the translation complex to

meet in the vicinity of this axis (which is, of course,

perpendicular to the helix axis of the complex).   One of the

models for tRNA tertiary structure, proposed by Melcher - a

model in which all of the arms are sort of stacked, sandwich

fashion, and pointing in about the same direction - should

help to convince you of this point (Melcher, G., _FEBS Lett._

3, 185 (1969)).

MELTON: I very much like your model as a molecular

mechanism.   However, I don't see wherein it is metaphysi-

cally different from previous models.   First, the attractive

feature whereby A and P are not physical sites but rather
alternate functional states of two equivalent sites is a
feature of earlier proposals such as the roller model.
Second, in the structural transition you postulate tRNA is
still passive. It requires an external driving force such
as GTP and "translocation" factors and is not due to any
inherently driven pulling or writhing action of the tRNA.

WOESE: Let me make the following points in reaction to
your comments: (1) The ratchet mechanism is the only
molecular model for translation so far proposed. Thus on
the molecular level, the other models you mention cannot be
compared to it. (2) On a grosser level - where one uses
terminology, such as "sites", that is not now capable of
definition in molecular terms - the ratchet model still
differs radically from the conventionally accepted models,
such as the A-site-P-site model or Spirin's machine, both
of which fall into the class of models I have called asymmetric serial-passage machines; the ratchet mechanism can
be interpreted to suggest a symmetric, parallel-passage
type of machine. (3) On the level of a very general
description of the mechanism - which I take you to imply
by your "roller" comment - I have to admit the ratchet falls
into a class that has been described before. I am reminded
of Crick's wonderfully colorful classification of all
possible models for ribosome movement into one of three
categories - rollers, sliders, or wrigglers (at the 1963

Cold Spring Harbor Symposium). I personally think a "wriggler" would fit the ratchet model best - but that category does cover a multitude of sins. (4) To my knowledge, all attempts to conceptualize the mechanics of translation to date have either implicitly or explicitly assumed the mechanical movement of the process somehow to be an essential property of the ribosome. This the ratchet model flatly contradicts. The essence of movement is taken to reside in particular tRNA interactions - which the ribosome may come to reinforce, make more accurate, etc. over the course of evolution, just as the body of the enzyme "reinforces", etc. the properties of the enzyme's active site. Here is where one sees the difference in metaphysics manifesting itself: A reductionist, particle-geometry-oriented, molecular biology that spawned the template-adaptor concept of translation, readily came to accept tRNA as a _static_ entity (as I said above), and so could not see tRNA as a machine. I would further contend that this same metaphysics, which denies the central importance of evolution in nature, also cannot see what the real "problem of the ribosome" is; and this is largely responsible for the _lack_ of significant progress that has occurred over the past decade in our understanding of the basic nature of translation. The "problem of the ribosome" is precisely the "problem of the _evolution_ of the ribosome". (5) I really cannot see your point about tRNA being passive now, just because energy and information have to be supplied to trigger the macrostate transition. No machine has the inherent capacity to move itself (short of a perpetual motion machine)! They all in principle have to be "plugged in" and some have to be "turned on" as well. tRNA today granted, is an element in a tightly coupled network of elements. But just because one couples together a bunch of small machines to make a larger, more complex machine doesn't alter the basic nature of the small machines. The movement that pulls the mRNA tape in translation was and still _is_ an inherent property of a simple tRNA machine.

SO MUCH "JUNK" DNA IN OUR GENOME

Susumu Ohno
City of Hope National Medical Center

The mammalian genome (haploid chromosome complement) contains roughly 3.0×10^{-9} mg of DNA which represents about 3.0×10^9 base pairs. This is at least 750 times the genome size of *E. coli*. If we take the simplistic assumption that the number of genes contained is proportional to the genome size, we would have to conclude that 3 million or so genes are contained in our genome. The falseness of such an assumption becomes clear when we realize that the genome of lowly lungfish and salamanders can be 36 times greater than our own (Ohno and Atkin, 1966).

In fact, there seems to be a strict upper limit for the number of gene loci which we can afford to keep in our genome. Consequently, only a fraction of our DNA appears to function as genes. The observations on a number of structural gene loci of man, mice and other organisms revealed that each locus has a 10^{-5} per generation probability of sustaining a deleterious mutation. It then follows that the moment we acquire 10^5 gene loci, the overall deleterious mutation rate per generation becomes 1.0 which appears to represent an

unbearably heavy genetic load. Taking into consideration
the fact that deleterious mutations can be dominant or
recessive, the total number of gene loci of man has been
estimated to be about 3 X 10^4 (Muller, 1967; Crow and Kimura,
1970). Even if an allowance is made for the existence in
multiplicates of certain genes, it is still concluded that,
at the most, only 6% of our DNA base sequences is utilized
as genes (Kimura and Ohta, 1971). Aside from conventional
structural genes and regulatory genes, this 6% should
include the *promotor* region and *operator* region which are
situated adjacent to each structural gene, for these regions
can certainly sustain deleterious mutations. More than 90%
degeneracy contained within our genome should be kept in
mind when we consider evolutional changes in genome sizes.
What is the reason behind this degeneracy?

Certain untranscribable and/or untranslatable DNA base
sequences appear to be useful in a negative way (the
importance of doing nothing). If functional genes customarily
occupied the region around the centromere, evolutional
changes of chromosome complements would not have occurred as
often as has been observed. Because the centromeric heterochromatin which represents a long tandem repeat of a short
untranscribable sequence (Jones, 1970; Southern, 1970) can
be lost or duplicated without deleterious consequences,
speciation more often than not has been accompanied by
chromosomal changes. The same can be said of those DNA base
sequences which are used as partitions between the genes.
It may be of selective advantage to space adjacent genes far
enough apart by inserting a stretch of untranscribable and/or

untranslatable DNA base sequence as a partition. In this way, the deleterious effect of *nonsense* or *frame-shift* mutations can be confined to a single locus, instead of allowing it to spread to other genes. Indeed, Miller and Beatty (1969) have shown long partitioning sequences between genes for 18S and 28S ribosomal RNA of the nucleolar organizing region. Furthermore, the recent recovery of longer than usual human hemoglobin α- and β-chains (Milner *et al.*, 1971; Flatz *et al.*, 1971) can be interpreted to mean that the hemoglobin α- as well as β-chain gene is followed by a partitioning sequence and that a mutation can bring about a partial translation of a normally silent partitioning sequence.

Inasmuch as the only requirement to be qualified as partitioning sequences is to be untranscribable and/or untranslatable, it is not likely that these sequences came into being as a result of positive selection. Our view is that they are the remains of nature's experiments which failed. The earth is strewn with fossil remains of extinct species; is it a wonder that our genome too is filled with the remains of extinct genes?

So long as a particular function is assigned to a single gene locus in the genome, those mutations which affect the *active* site of a gene product would produce deleterious consequences and, therefore, be eliminated by natural selection. Thus, natural selection permits only trivial changes, while forbidding changes in the basic character of a gene. Indeed, the amino acid sequences of fibrinopeptide A and B in which only C-terminal arginine appears to represent a real *active* site evolved 1500 times faster than histone IV where the *active* site appears to consist of an entire molecule (Kimura and Ohta, 1971).

It then follows that the creation of a new gene with hitherto nonexistent function is possible only if a gene becomes sheltered from relentless pressure of natural selection. This shelter has apparently been provided either by polyploidization or by tandem duplication. Redundant copies of genes thus produced are now free to accumulate formerly forbidden mutations and thereby to acquire new functions (Ohno, 1970).

The chance of acquiring a new function by unrestricted accumulation of mutations, however, should be as small as that of an isolated population emerging triumphant as a new species. Degeneracy is the more likely fate. The creation of every new gene must have been accompanied by many other redundant copies joining the ranks of silent DNA base sequences, and these silent DNA base sequences may now be serving the useful but negative function of spacing those which have succeeded. Triumphs as well as failures of nature's past experiments appear to be contained in our genome.

References

1. S. Ohno and N.B. Atkin, Chromosoma 18, 455 (1966).

2. H.J. Muller, in Heritage from Mendel (University of Wisconsin Press, Madison, 1967) pp. 419-447.

3. F. Crow and M. Kimura, An Introduction to Population Genetics Theory (Harper & Row, New York, 1970).

4. M. Kimura and T. Ohta, Nature 229, 467 (1971).

5. K.W. Jones, _Nature_ 225, 912 (1970).

6. E.M. Southern, _Nature_ 227, 794 (1970).

7. O.L. Miller, Jr. and B.R. Beatty, _Science_ 164, 955 (1969).

8. P.F. Milner, J.B. Clegg and D.J. Weatherall, _Lancet_
    I, 729 (1971).

9. G. Flatz, J.L. Kinderlerer, J.V. Kilmartin and H. Helmann,
    Lancet I, 732 (1971).

10. S. Ohno, _Evolution by Gene Duplication_ (Springer-Verlag,
    Heidelberg, 1970).

DISCUSSION

BOYER: The calculation of the permissible number of
structurally active loci (a la King and Jukes) from lethal
mutation rates depends on how well the value of 10^{-5} muta-
tions/locus/generation represents the whole of the genome.
We only measure what we see. Immutable or nearly immutable
loci are not examined. We don't yet know the real proportions. It thus seems to me that the permissible number of
structural loci is - as yet - a somewhat suspect way to
arrive at figures of 1% structural utility to 99% junk.

OHNO: Although deleterious mutations affecting such
gene products as actin, myosin and microtubule protein have
not been detected, I believe that this is simply because
these mutations are early embryo lethals. The estimate has
been made on inbred strains that the mouse genome contains
10^4 gene loci which mutate at the rate of 10^{-5}/locus/gener-
ation to recessive lethals. Furthermore, if immutable gene
loci really exist, by their very definition, they would not
have contributed to evolution.

DNA OF FUNGI

Roger Storck

Department of Biology, Rice University, Houston, Texas 77001

FUNGAL SYSTEMS

Fungi are non-photosynthetic eukaryotes. The dominant feature of
their organization is the filament or hypha. During growth filaments
elongate, branch and hyphal tips fuse to form a three-dimensional mycelium.
Biosynthetic activities are localized in these tips and a cytoplasmic
streaming toward them results in the vacuolization of the older parts of
the filaments. Depending on the type of fungi hyphae are either septate
or not. In both cases however, cytoplasm and cytoplasmic organelles such
as nuclei and mitochondria migrate toward tips since septa have a central
perforation. A growing mycelium can best be viewed as a coenocytic system
in which a nucleated cytoplasm moves into a system of tubes. In some
fungi, filaments are replaced either temporarily or permanently by isolated
cells. Asexually, fungi reproduce by filament fragmentation, budding,
scissiparity or spore production. Sexual reproduction which ultimately
is expressed in spores, entails karyogamy which is in higher fungi preceded
by a dikaryotic stage which may last through several cell divisions. In
most fungi, diploidy exists during a very short period of the life cycle.
Plasmogamy often leads to an heterokaryotic condition in which nuclei with
different genotypes are mixed in the same system of tubes. Diploidization
and meiosis in most cases take place in the sexual organs at determined

times in the life cycle. In some organisms, a "parasexual cycle"[1] which

entails the following sequence of events: heterokaryosis, nuclear fusion,

recombination and segregation, occurs at low frequency, in undifferentiated

and unspecified parts of hyphae. The male and female sex organs which are

differentiated forms of filaments may either be similar in size and shape

(isogamentagia) or distinct. In the latter case, the contribution of one

mating partner may be limited to a haploid nucleus. Such a situation

permits the demonstration of cytoplasmic inheritance. In fungi, size and

shape of the sex organs should not be confused with sex competibility or

mating types. The homothallic type is characterized by a mycelium which

is self fertile whereas the heterothallic one is self sterile and requires

in order to reach sexual fulfillment a fusion with a thallus of the oppo-

site mating type.

FUNGAL TAXONOMY

The number of existing fungal species has been estimated to be as

large as one hundred thousand[2]. It is therefore not surprising that

several systems of classifications have been formulated. One of these[3]

which has the advantages of being simple and understandable has been

adopted for the present study. It is presented in Table 1 in an abridged

form. Many species of Oömycetes are aquatic. The others are obligate

terrestrial parasites. There is increasing evidence that some orders are

diploid. Zygomycetes are strictly terrestrial. Hyphae are usually non-

septate. Asexual reproduction is by means of conidia or sporangiospores

(spores contained in a sporangium). Gametangial copulation leads to the

formation of zygospores. The characteristic feature of Ascomycetes is

the ascospores. These occur in an ascus often in definite numbers. The

Hemiascomycetidae include yeast and yeast-like organisms. Unlike the

Euascomycetidae they do not produce asci containing fruiting bodies or

ascocarps. Filaments are septate and asexual reproduction is by means of

TABLE 1

FUNGAL GROUPS

LOWER FUNGI <

HYPHOCHYTRIDIOMYCETES
Zoospores with one
anterior flagellum

CHYTRIDIOMYCETES
Zoospores with one
posterior flagellum

OÖMYCETES
Zoospores biflagellated
Oöspores

ZYGOMYCETES
Non-motile spores
Zygospores

HIGHER FUNGI <

ASCOMYCETES
Ascospores
  Hemiascomycetidae
  Ascocarp absent
  Euascomycetidae
  Ascocarp present

DEUTEROMYCETES
(IMPERFECT FUNGI)
No sexual reproduction

BASIDIOMYCETES
Basidia, basidiospores
  Heterobasidiomycetidae
  Basidia septate or divided
  Homobasidiomycetidae
  Basidia simple, unicellular

conidiospores (conidia). Organisms such as Neurospora, morels and truffles belong to this sub-class of Ascomycetes. The Deuteromycetes or imperfect fungi lack sexual reproduction and this makes their taxonomy difficult. Most organisms produce asexual spores similar to those characteristic for Ascomycetes, others have a yeast or yeast-like morphology. Asci formation has been discovered in many species which were originally thought to be imperfect fungi. In this class organisms were also found which have a sexual mode of reproduction typical for Basidiomycetes. The major characteristic of this class is the basidium; a spore bearing structure. Often, the basidium bears four basidiospores. Hyphae are as

in Ascomycetes and Deuteromycetes usually septate. A typical hyphal

structure called "clamp connection", insures the separation into two

daughter cells of sister nuclei arising from conjugate division of a

dikaryon. Jelly fungi, rusts, smuts and several groups of yeast-like

fungi are included in the Heterobasidiomycetidae while mushrooms, puff-

balls and stinkhorns are placed in the sub-class: Homobasidiomycetidae.

BASE COMPOSITION OF FUNGAL DNA

The base composition of DNA from several hundred fungal species has

been determined. As is true for other organisms, buoyant density and

melting temperature measurements were used in the majority of cases. In

consequence it must be assumed that the calculated guanine plus cytosine

contents in moles percent (% GC) are correct and that fungal DNA does not

contain significant amounts of abnormal bases and sugars. The results of

these studies have been reviewed recently[4]. The reader is referred to

that review and also to more recent publications [5,6,7,8,9,10,11] for

detailed information since only the major features which have emerged

from these studies will be discussed here. It is of relevance at this

point to remind the reader that the total number of species surveyed is

only a small sample of the total fungal "population" and also that the

sampling of some fungal groups is often influenced by availability of

isolates and their ability to grow under standard laboratory conditions.

As shown in Table 2, the compositional diversity of fungal DNA is

similar to that found in the other groups of microorganisms, prokaryotic

as well as eukaryotic. More data are needed in order to find out if the

differences between these ranges are significant or not.

In fungi as in other groups of microbes, organisms described under

the same specific epithet display as a rule a minimum heterogeneity in

their base composition. For example, a study[4], performed under the same

TABLE 2

RANGES OF % GC IN MICROBES

Taxonomic group	Range of % GC
Bacteria [a]	25-75
Blue-green algae [b]	35-71
Algae [c]	37-68
Protozoa [c]	22-68
Fungi	27-70

a, Hill[12]; b, Edelman et al.[13]; c, Mandel[14].

set of conditions, of 322 species of filamentous fungi totaling 492

isolates reveals that 80% of these species have a range which is less

than 5 when expressed in % GC and that this minimal range value is found

for only 30% and 15% of the genera and family respectively (Figure 1).

For six species out of seven for which at least six isolates were analyzed

the standard deviation is, as shown in Table 3, less than 2% in each case

TABLE 3

STATISTICAL INDICES OF INTRASPECIFIC
FREQUENCY DISTRIBUTION OF GC CONTENT VALUES

Species	N	$\bar{X}$	R	S
Actinomucor elegans	12	41.3	40.5-43	0.72
Mycotypha microspora	7	45	42---47.5	1.69
Radiomyces embreei	8	46.1	44---50	1.67
Rhizopus oligosporus	12	40	38.5-41	0.76
Syncephalastrum racemosum	13	50.3	48.5-52.5	1.26
Thamnidium anomalum	12	45.2	44.5-46	0.51
T. elegans	8	51	37---60.5	7.52
Penicillium atramentosum	26	49.8	49---51	0.54
Syncephalastrum racemosum	26	48.3	47---51.5	0.91

For the first seven species N is the number of species analyzed and for
the two last species N is the number of % GC determinations performed
on the same DNA sample.
$\bar{X}$ = arithmetic average, R = range, S = standard deviation.

and in some cases it is equal to the standard deviation for 26 determina-

tions of % GC on the same sample. It might thus be possible in principle

to use this statistical index for the quantitative definition of a species.

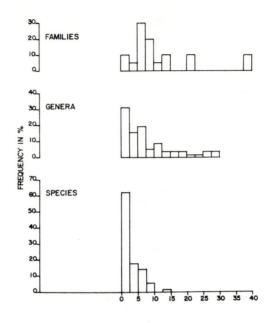

Figure 1. Frequency distribution (expressed in percent) of range size in % GC within species, genera, and families.

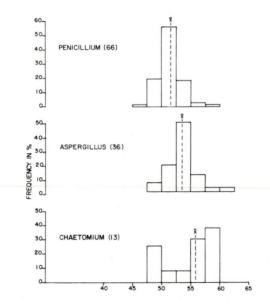

Figure 2. Frequency distribution (expressed in percent) of the GC content values in the genera Penicillium, Aspergillus, and Chaetomium. Numbers in parentheses indicate the number of species analyzed in each genus. $\overline{X}$ = average.

It is independent of the actual average GC % characteristic for a species and it has been proven in several instances that standard deviations values significantly higher than 2% were indicative of errors. Tailing values in distributions were due either to an original misidentification or a chance contamination. More rewarding are those cases in which it could be shown that the species assignment was faulty. Generalization of the concept expressed above cannot yet be made, since there are exceptions such as _Thamnidium_ _elegans_ (Table 3) for which no good explanation is yet in sight. In the first study[4] of this species a total of eight isolates were analyzed. Five yielded % GC values clustered between 53 and 55.5 and the others were 37, 40.5 and 60.5% respectively. Analysis of these extreme values have been found to be correct after analysis of new transfers from the original culture collection[15]. Furthermore, starch gel electrophoresis of enzymes representing several different genetic loci does not reveal differences between these _T. elegans_ isolates[15]. It will be interesting to find out as more research is done what will be the solution to this problem. It is hoped that additional cases of intraspecific heterogeneity in DNA base composition will prompt more investigations of biochemical properties since they might turn out to be more profitable than GC content determinations _per se_.

Frequency distributions of % GC for genera and families display as expected greater heterogeneity than for species. As shown in Figures 2 and 3 some of these distributions are skewed and indicate bimodality. It will be shown later that in the case of _Cryptococcus_, _Rhodotorula_ and _Sporobolomyces_, this bimodality is real and has resulted in a new classification of some of these organisms.[6]

The frequency distributions of GC content values for some of the classes which have been extensively investigated are presented in Figure 4. It should be kept in mind that the shape of these histograms is

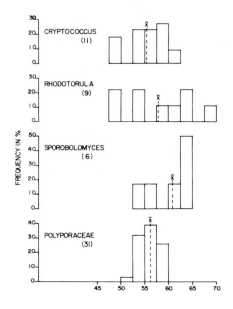

Figure 3. Frequency distribution (expressed in percent) of the GC content values in the genera Cryptococcus, Rhodotorula, and Sporobolomyces, and in the family Polyporaceae. Numbers in parentheses indicate the number of species analyzed in each taxonomic group. X̄ = average.

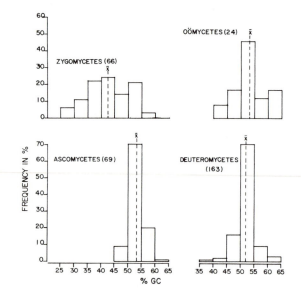

Figure 4. Frequency distribution (expressed in percent) of GC content values. Numbers in parentheses indicate the number of species analyzed in each taxonomic group. X̄ = average.

probably influenced more by the unequal sampling of orders and families

than by the total size of the sample. It should also be pointed out that

the census of species indicates great differences in class size. Whereas

there are probably less than 1,000 species of Oömycetes and Zygomycetes,

as many as 15,000 species of Ascomycetes have been described. With the

data at hand, it appears from a comparison of the histograms of Figure 4

that Zygomycetes are more heterogenous than the other classes, and that

Deuteromycetes and Ascomycetes are very similar. It must be specified

that the analysis of the Zygomycetes was limited to the Mucorales; the

largest order of this class, and that the data for Ascomycetes are for

Euascomycetidae exclusively. Additional information about % GC distribu-

tion in classes and sub-classes can be found in Table 4. Inspection of

TABLE 4

STATISTICAL INDICES OF THE FREQUENCY DISTRIBUTION OF
GC CONTENT VALUES FOR FUNGAL CLASSES AND SUBCLASSES

Class and Subclass	N_s	N_i	$\overline{X}$	R	S
Oömycetes	24	27	53	40.5–62	5.31
Zygomycetes	66	155	42.6	27.5–59	7.62
Hemiascomycetidae [a]	73	112	40.2	29---48.5	2.72
Euascomycetidae	69	90	53.4	48.5–60	2.47
Deuteromycetes	163	220	52.1	35.5–64.5	3.27
Asporogenous yeasts [b]	91	162	48.2	31.0–70.0	6.60
Homobasidiomycetidae	42	62	55.0	44---59.5	3.59

N_s and N_i = analyzed number of species and isolates respectively.
$\overline{X}$, R and S as defined in Table 3.
a. Exclusively ascosporogenous yeasts.
b. Imperfect yeast forms.

this table indicates that each taxonomic group can be characterized by

statistical indices. All the ranges overlap. In terms of average, there

is a dichotomy: 3 groups are below 50% GC and the others are above.

Before discussing the possible phylogenetic significance of this finding,

an evaluation of the taxonomic merits of GC content values for each of

these fungal groups will be presented.

Only two families of Oömycetes, each belonging to a different order, were analyzed. (This class includes four orders). The only finding worth mentioning is that for some genera of the Saprolegniacae the groupings of species on the basis of % GC are not an expression of traditional classification but correlate well in some cases with recent suggested changes for this classification which are based on the comparative morphology of the oöspore. As already indicated, the only Zygomycetes analyzed were Mucorales. All but three families were surveyed. As shown in Table 3, all the species for which more than six isolates were analyzed yielded with one exception not much more compositional diversity than was found when repeated GC determinations were made on the same sample. This small intraspecific diversity could be indicative of evolutionary stability as opposed to species with a large compositional diversity. No correlation was found between the average % GC of families and Benjamin's concept[16] of relationship of these families. In one family, namely the Cunning-hamellaceae, ranges of 27.5 to 32.5 and 43 to 47.5 were found for the genera Cunninghamella and Mycotypha respectively. This finding suggests, in agreement with studies on spore structure[17], that Mycotypha be removed from this family.

Out of these 90 isolates of Euascomycetidae only one had a value of 60% GC and only six a % GC above 57.5. Thus, the dispersion of GC content value in this group is relatively small as can be seen in Table 4. Not enough families were included in these studies and meaningful taxonomic conclusions about the class as a whole are not possible. In most of the families, the range in % GC is narrow. In one genus (Chaetomium) for which 17 isolates were analyzed, and for which a range of 48.5 to 60% was found, no correlation could be detected between groupings by GC content values and various strikingly distinct morphological characters. With but four exceptions, all the species of Deuteromycetes surveyed belong to the order <u>Moniliales</u> (there are three additional orders in this class). The various families are characterized, as is the case of <u>Asco-</u> <u>mycetes</u>, by narrow % GC distribution. The overall GC content average is close to that for <u>Ascomycetes</u>. There is no difference in the % GC of families which have few morphological differences. This appears also to be the case for those genera such as <u>Fusarium</u> in which there is a great morphological variation exhibited. The survey of the <u>Deuteromycetes</u> included 36 species of <u>Aspergillus</u> and 66 of <u>Penicillium</u>. These two genera have been studied in great detail and various form-genera have been established on the basis of morphology and physiology. In no case was it possible to find a correlation between the % GC and these groupings. Clearly, the taxonomic resolving power of the base composition of DNA is small or nihil for taxa which have such a great homogeneity. The survey of <u>Homobasidiomycetes</u> has so far been mostly limited to the <u>Hymenochaeta-</u> <u>ceae</u> and the <u>Polyporaceae</u>[11]. The other organisms which have been analyzed belong to the families: <u>Agaricaceae</u>, <u>Lycoperdaceae</u> and <u>Schizophyllaceae</u>[4]. As shown in Table 4, the average % GC value for <u>Homobasidiomycetidae</u> is higher than for the other classes. Only two values below 50% GC were found (both 44% GC). After their exclusion the range narrows considerably and instead of being 44-59.5 it becomes 50 to 59.5. With regard specifically to the <u>Polyporaceae</u>, it should be mentioned that GC content shows some promise of taxonomic value at the species, generic, family and higher levels but that firm conclusions will not be possible before more organisms are studied.

Yeast and yeast-like fungi pose a special problem for the taxonomist since most of the morphological and life-cycle features used for filamentous fungi are not applicable in their case. As a result, zymologists have relied heavily on biochemical criteria. These, with morphological and cultural traits, have been used with various degrees of success. Although

yeasts have long been considered to be true fungi, there have been in
many cases difficulties in the assignment to existing taxonomic groups.
The studies of DNA base composition have helped to improve that situation.
As seen in Table 4, yeasts have been placed in two groups. The first
contains those organisms which form ascospores and are part of the sub-
class Hemiascomycetidae. Their GC content is much less than that of the
Euascomycetidae and the ranges of % GC of these two subclasses do not
overlap. The group of yeasts listed in Table 4 as "Asporogenous" includes
imperfect forms of ascosporogenous yeasts and perfect and imperfect forms
of Heterobasidiomycetidae. It was found that ascogenous yeasts and their
imperfect counterparts apparently always have a % GC lower than 50%,
whereas those related to the Basidiomycetes lie in the range from 50-70%.
It might be added that for the genera Cryptococcus, Rhodotorula and
Sporobolomyces, which belong to this group of yeasts, an additional
intrageneric bimodality was uncovered[18]. This is illustrated in Table 5.

TABLE 5

STATISTICAL INDEXES OF FREQUENCY DISTRIBUTION OF GC CONTENT OF DNA
FROM CRYPTOCOCCUS, RHODOTORULA, AND SPOROBOLOMYCES[a]

Taxonomic group	N_{ss}	$\overline{X}$	R	S	"t"value
All genera	32	58.1	49.4-70.0	5.48	33.90
%GC ≤ 56.0	16	53.2	49.0-56.0	1.89	
%GC ≥ 58.0	16	63.0	58.0-70.0	2.99	
Cryptococcus					
All Strains	11	55.7	49.0-65.5	4.86	5.55
%GC ≤ 55.0	6	51.8	49.0-55.0	1.84	
%GC ≥ 58.0	5	60.4	58.0-65.5	2.78	
Rhodotorula					
All strains	6	57.9	52.5-70.0	6.56	6.33
%GC ≤ 54.5	4	53.5	52.5-54.5	0.25	
%GC ≥ 63.5	2	66.8	63.5-70.0		
Sporobolomyces					
All strains	15	59.9	51.5-65.0	4.71	35.20
%GC ≤ 56.0	6	54.4	51.5-56.0	1.51	
%GC ≥ 60.5	9	63.6	60.5-65.0	1.42	

[a] P for all genera and for Cryptococcus, Rhodotorula, and Sporobolomyces
was <<0.01
N_{ss} = number of stains analyzed
$\overline{X}$, R and S as defined in Table 3.

A "t" test for the consistency of the means indicated that this bimodality
is significant. The biological significance of this dichotomy is yet
unknown.

PHYLOGENY

Two theories have dominated thoughts on fungal evolution. According
to Gäumann[19] fungi with the exception of the Oömycetes arose from ancestral
flagellates and developed monophyletically, whereas the Oömycetes originated independently from the heterosiphonaceous algae. As shown in
Figure 5, Zygomycetes were, following this scheme, the direct precursors
of Ascomycetes which in turn preceded the Basidiomycetes. According to
the second theory whose main proponent was Bessey[20,21],Ascomycetes came
from ancestral red algae and eventually gave rise to the Basidiomycetes
whereas the Phycomycetes (lower fungi) originated from heterocont unicellular algae and evolved along three pathways. Both these theories are
based on comparative morphology. Both theories have been the object of
numerous discussions and the reader is referred to three of the most
recent ones[22,23,24] for further information. In recent years, these
theories have been subjected to investigations based on biochemical
criteria. The first of these investigations dealt with the distribution
among 26 fungal species of the two biosynthetic pathways of lysine[25,26].
One pathway which is via diaminopimelic acid (DAP) was found only in
species representative of the classes; Hyphochitridiomycetes and Oömycetes.
The other which is routed through aminoadipic acid (AAA) is present in all
other fungal classes (Table 6). These results clearly favor Gäumann's
evolutionary scheme. The second study dealt with the sedimentation
patterns of enzymes involved in the biosynthesis of tryptophan[27]. A
survey of 22 species showed that four different patterns could be
recognized in fungi. The assignment of these patterns to fungal groups

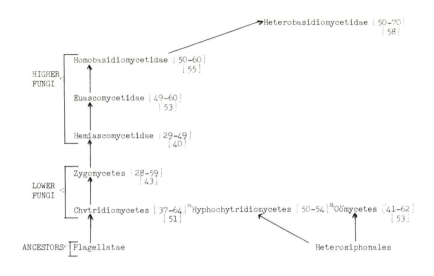

Figure 5.   Evolutionary scheme of fungi.   Adapted from Gäuman[19]. The brackets
indicate range and average of GC content.   Values for Deuteromycetes are
being included in higher fungi classes according to affinity.
a, Storck and Alexopoulos (Unpublished results)
For the Chytridiomycetes and the Hyphochytridiomycetes; 8 and 2 species
were surveyed respectively.

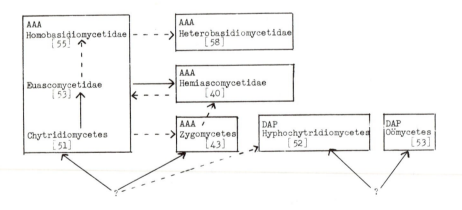

Figure 6.   Evolutionary scheme based on data of Tables 6, 7 and 8.   Each block
corresponds to a unique combination of cell wall composition type and sedimentation pattern of enzymes for tryptophan biosynthesis.   The numbers in brackets
correspond to the average % GC of DNA.   The abbreviations AAA and DAP in the
left upper corner of each block indicate the type of lysine pathway (Table 6).
The dotted arrows indicate evolutive relationships according to Gäumann
(Figure 5).

TABLE 6

LYSINE BIOSYNTHETIC PATHWAYS

Fungal group	Type of pathway
Hyphochytridiomycetes	DAP
Chytridiomycetes	AAA
Oömycetes	DAP
Zygomycetes	AAA
Hemiascomycetidae	AAA
Euascomycetidae	AAA
Heterobasidiomycetidae	AAA
Homobasidiomycetidae	AAA

DAP, diaminopimelic acid; AAA, aminoadipic acid. From Vogel[25,26].

is shown in Table 7. It links Chytridiomycetes to the Euascomycetidae and

the Homobasidiomycetidae. Zygomycetes and Heterobasidiomycetidae share in

common another pattern. Hemiascomycetidae and Oömycetes find themselves in

TABLE 7

SEDIMENTATION PATTERNS OF TRYPTOPHAN BIOSYNTHETIC ENZYMES

Fungal group	Sedimentation pattern type
Hyphochytridiomycetes	
Chytridiomycetes	I
Oömycetes	IV
Zygomycetes	III
Hemiascomycetidae	II
Euascomycetidae	I
Heterobasidiomycetidae	III
Homobasidiomycetidae	I

From Hütter and DeMoss[27].

an isolated position, each one having a different pattern. As a result of

these findings the authors have suggested to abandon Zygomycetes and

Hemiascomycetidae as intermediates between the Chytridiomycetes and the

Euascomycetidae. The only features that this theme shares in common with

Gäumann's scheme are the connections between Euascomycetidae and Homo-

basidiomycetidae and the independent origin of the Oömycetes. It might

be mentioned in passing that in agreement with the results of the DNA

DNA OF FUNGI

base composition discussed earlier, the imperfect yeast forms such as
Cryptococcus laurentii and _Rhodotorula glutinis_ have the same enzyme
pattern as _Sporobolomyces salmonicolor_, _Tremella mesenterica_ and _Ustilago
maydis_ which are _Heterobasidiomycetidae_. The third type of investigation
is focused on the kind of polysaccharides found in the cell wall of fungi.
The taxonomic and phylogenetic implications of this type of studies have
been recently reviewed[28,29]. Bartnicki-Garcia, the author of these two
reviews, has based his approach to phylogeny on the type of pair of polysaccharides which appears to be the principal component of "vegetative
walls". Six such pairs have been found in the fungi which have been the
object of the present study. Their distribution among fungal classes
which is seen in Table 8 also suggests to break away from Gäumann's
scheme by side-lining the _Zygomycetes_ and _Hemiascomycetidae_ from the main
line going to the _Euascomycetidae_ and the _Homobasidiomycetidae_.

TABLE 8

CELL WALL COMPOSITION

Fungal group	Polysaccharides pair
Hyphochytridiomycetes	Cellulose-Chitin (III)
Chytridiomycetes	Chitin-β-Glucan (V)
Oömycetes	Cellulose-β-Glucan (II)
Zygomycetes	Chitin-Chitosan (IV)
Hemiascomycetidae	Mannan-β-Glucan (VI)
Euascomycetidae	Chitin-B-Glucan (V)
Heterobasidiomycetidae	Chitin-Mannan (VII)
Homobasidiomycetidae	Chitin-β-Glucan (V)

From Bartnicki-Garcia [28,29]
The roman numerals are those used by·this author for designa-
     tion of cell wall category.

When the % GC values were introduced into Gäumann's scheme, as was
done in Figure 5, it appeared that evolution was marked by a progressive
increase in GC content of DNA from _Zygomycetes_ to _Basidiomycetes_. Of
significance was the fact that the average % GC for the _Zygomycetes_ is

closer to that of the Hemiascomycetidae than to that of the Euascomycetidae.
This could be taken to indicate that the yeasts are primitive rather than
reduced forms. The fact that the average % GC value for the Oömycetes is
53% could be accepted on the basis that these organisms are not on the
main evolutionary line. The weakness of this suggestion lies in the fact
that the average value of % GC in Chytridiomycetes, admittedly based on a
limited sample (8 species), is equal to 51. Perhaps more important is
that this evolutionary increase in % GC appears to be irreconcilable with
the studies of other biochemical criteria which has been presented earlier.
An attempt has been made to offer an evolutionary scheme compatible with
most of the available morphological and biochemical data. This scheme
is presented in Figure 6. It consists of six blocks. The four blocks on
the left belong to AAA lysine pathway and those on the right belong to
the DAP pathway. The total number of blocks corresponds to the total
number of observed combinations of a polysaccharide pair (Table 8) and a
tryptophan enzyme pattern (Table 7). The largest block corresponds to the
monophyletic series suggested by these biochemical studies and contains
three classes with average % GC varying between 51 and 55. This scheme,
in agreement with that of Gäumann, suggests that Heterobasidiomycetidae
evolved from the Homobasidiomycetidae. With regard to the yeasts and
yeast-like cells which are placed in this group, the degeneracy theory
used by some would be favored in this scheme. It also favors the view
that Hemiascomycetidae are derived from the Euascomycetidae and that the
true yeasts have been derived from mycelioid forms by reduction. The
Zygomycetes are presented as an isolated group which might have evolved
from an "unknown" which is also the precursor of the Chytridiomycetes.
The DAP groups namely Oömycetes and Hyphochytridiomycetes are isolated
from the AAA groups but have both an average % GC slightly higher than
50%. If the scheme presented in Figure 6 eliminates the need for an

increase of DNA GC content as a function of evolutionary time, it does

not offer a mechanism to explain changes in cell wall composition and

tryptophan enzymes in the cases of the Heterobasidiomycetidae and the

Hemiascomycetidae nor one accounting for a decreased DNA GC content in

the latter sub-class. An evolutionary scheme which will not be discussed

here has been offered which eliminates the need for changes in cell wall

composition and enzyme patterns[29]. Unidirectional modification of DNA

base composition accompanying evolution has been observed in bacteria

for the Actinomycetes. These organisms have % GC ranging from about 63 to

75%[12] and are considered as the end of an evolutionary line in view of the

occurrence of a mycelial stage bearing spores. A theory, based on ultra-

violet irradiation as a selective factor, has been presented to explain

this evolutive increase in % GC of DNA[30], but it cannot be applied to

fungi. There is at present no other theory or experimental fact which

suggests an explanation for the changes in fungal DNA base composition

which appear to have taken place in fungi. Although no consideration is

given in this presentation to mitochondrial DNA, it might be mentioned

that a significant decrease in the buoyant density of this type of DNA

in Saccharomyces has been obtained as a result of cytoplasmic mutation[31].

GENOME SIZE

Comparison of % GC of DNA is as a rule of little value, as the

evidence presented here indicates, for resolving genomic differences

between organisms which have a high degree of genetic relatedness. The

prediction, based on similarity of base composition, that there is phylo-

genetic relatedness may be false. This was demonstrated in a study of

several Saccharomyces species by measurements of DNA/DNA and DNA/RNA

hybridization[5]. There is no doubt that such type of studies which are

now being pursued in several laboratories will considerably improve our understanding of taxonomic and phylogenetic relationships among fungi.

The minuteness of nuclei of most fungi has been responsible for the
slow progress of our understanding of nuclear division in somatic structures. In consequence, the inventory of chromosome formulas is limited.
The results of a critical survey of the literature done by this author[32],
are presented in Table 9 in a condensed form. It shows that chromosome
numbers of fungi are relatively small and that their differences do not
appear to have a phylogenetic significance. It might be pointed out that

TABLE 9

HAPLOID CHROMOSOME NUMBERS (n)

Fungal group	N	$\overline{X}$	R
Hyphochytridiomycetes	–	–	–
Chytridiomycetes	3	19	14-28
Oömycetes	1	3	–
Zygomycetes	2	15	14-16
Hemiascomycetidae	3	9.4	2-17
Euascomycetidae	35	6.4	4-18
Heterobasidiomycetidae	–	–	–
Homobasidiomycetidae	17	9.6	3-16

N, Number of studies; R, range of haploid chromosome
    numbers; $\overline{X}$, average.
From Biology Data Book (in press)

polyploidy in addition to poor cytological resolution could be responsible
for the highest counts. The most reliable numbers are those which result
from a combination of cytological and genetical methods. Such are those
for example of Aspergillus nidulans (n=8) and Neurospora crassa (n=7).
This last number is also found not only in other Neurospora species but
also in related genera such as Gelasinospora and Sordaria.

Measurements of renaturation kinetics of denatured DNA have demonstrated that such rates can be used for the determination of the number
of different nucleotide sequences present in DNA. Studies of DNA from a
variety of organisms have demonstrated the number of different nucleotide
sequences is greater in the higher forms of life and in the more primitive

ones. These studies have in addition proven that while such sequences
accounted for the totality of the haploid DNA content (genome size) of
viruses and bacteria they did not in the case of eukaryotes. Indeed, the
nuclear genomes of these organisms contain a large fraction (5-10% or
more) of "repeated nucleotide sequences." (The reader is referred to a
recent and thorough review[33] of these studies for detailed information).
This type of investigation provides us with the best available tool for
the molecular study of phylogenetic relationships.

In the case of fungi, very little information is yet available about
genome size and genome complexity. The spore content of DNA has been
measured for several organisms[34]. The values obtained do not in most
cases give direct information on the genome size since some spores are
multinucleated and the number of nuclei per spore is unknown. In the
few cases where this number was known, amounts of DNA varied from 0.048
to 0.088 pgm per haploid nucleus, which correspond to about 3×10^{10} and
5×10^{10} daltons respectively. A few measurements of reassociation
kinetics yield as shown in Table 10 comparable values. It should be

TABLE 10

GENOME SIZE OF FUNGI

Fungal group	Genome size (in daltons)[a]
ZYGOMYCETES	
Mucor bacilliformis	2.0×10^{10} [b]
M. racemosus	7.0×10^{9} [c]
HEMIASCOMYCETIDAE	
Saccharomyces cerevisiae	9.2×10^{9} [d]
EUASCOMYCETIDAE	
Neurospora crassa	2.2×10^{10} [e]
Talaromyces vermiculatus	3.0×10^{10} [b]
HOMOBASIDIOMYCETIDAE	
Coprinus lagopus	3.0×10^{10} [f]

a, calculated from $C_0t_{0.5}$ values; b, R. Seidler
(personal communications); c, R. Storck (unpublished);
d, Bicknell and Douglas[5]; e, Dutta et al[36]; f,
Penn and Dutta[37].

indicated that several of these stem from studies which are of a
preliminary nature. These values suggest that the size of the fungal
genome is approximately 5 to 10 times larger than that of bacteria[35].
There is yet no evidence available suggesting that fungal nuclear DNA
contains as is the case for all other eukaryotes a significant fraction
of "repeated DNA." If none is found, there will be one more reason for
saying that fungi are remarkable organisms.

Acknowledgments: I express my gratitude to S. K. Dutta, C. Shaw,
R. Seidler and D. Stout for communication of their unpublished results.

REFERENCES

1. Pontecorvo, G., _Ann_. _Rev_. _Microbiol_. _10_, 393 (1956)

2. Bisby, G. R. and G. C. Ainsworth, _Brit_. _Mycol_. _Soc_. _Trans_. _26_, 16 (1943).

3. Alexopoulos, C. J., _Introductory Mycology_, John Wiley and Sons,
    New York and London, 1962.

4. Storck, R. and C. J. Alexopoulos, _Bacteriol_. _Rev_. _34_, 126 (1970).

5. Bicknell, J. N. and H. C. Douglas, _J_. _Bacteriol_. _101_, 505 (1970).

6. Meyer, S. A. and H. J. Phaff, _Spectrum_ _1_, 1 (1970) University of
    Georgia.

7. Wickerham, L. J. _Spectrum_ _1_, 31 (1970) University of Georgia.

8. Nakase, T. and K. Komagata, _J_. _Gen_. _Appl_. _Microbiol_. _16_, 511 (1970).

9. Nakase, T. and K. Komagata, _J_. _Gen_. _Appl_. _Microbiol_. _17_, 43 (1971).

10. Nakase, T. and K. Komagata, _J_. _Gen_. _Appl_. _Microbiol_. 17, 77 (1971).

11. Storck, R., M. K. Nobles and C. J. Alexopoulos, _Mycologia_ LXIII, 38
    (1971).

12. Hill, L. R., _J_. _Gen_. _Microbiol_. _44_, 419 (1966).

13. Edelman, M., D. Swinton, J. A. Schiff, H. T. Epstein and B. Zeldin,
    Bacteriol. _Rev_. _31_, 315 (1967).

14. Mandel, M. _In_ M. Florkin, B. T. Scheer, and G. W. Kidder (ed.),
    Chemical Zoology, vol I, 541, Academic Press Inc., New York, 1967.

15. Stout, D. and C. Shaw, (Personal communication).

16. Benjamin, R. K., _Aliso_ _4_, 321 (1959).

17. Young, T. W. K., *J*. *Gen*. *Microbiol*. *55*, 243 (1969).

18. Storck, R., C. J. Alexopoulos and H. J. Phaff, *J*. *Bacteriol*. *98*, 1069 (1969).

19. Gäumann, E., Die Pilze, *Birkhäuser*, Basel, 1964.

20. Bessey, E. A., *Mycologia* *34*, 355 (1942).

21. Bessey, E. A., *Morphology* *and* *Taxonomy* *of* *Fungi*, Hafner Publishing Co., New York, 1950.

22. Martin, G. W. *In* G. C. Ainsworth and A. S. Sussman (ed.) "The Fungi" vol. 3, 635, Academic Press, New York, 1968.

23. Raper, J. R. *In* G. C. Ainsworth and A. S. Sussman (ed.) "The Fungi" vol. 3, 677, Academic Press, New York, 1968.

24. Savile, D. B. O. *In* G. C. Ainsworth and A. S. Sussman (ed.) "The Fungi" vol. 3, 649, Academic Press, New York, 1968.

25. Vogel, H. J. *In* V. Bryson and H. J. Vogel (ed.) Evolving Genes and Proteins, P. 25, Academic Press, New York, 1965.

26. Vogel, H. J., J. S. Thompson and G. D. Shockman, Symposia of the Society for General Microbiology *XX*, 107 (1970).

27. Hütter, R. and J. A. DeMoss, *J*. *Bacteriol*. 94, 1896 (1967).

28. Bartnicki-Garcia, S., *Ann*. *Rev*. *Microbiol*. *22*, 87 (1968).

29. Bartnicki-Garcia, S., *In* J. B. Harborne, Phytochemical Phylogeny, p. 81, Academic Press, New York, 1970.

30. Singer, C. E. and B. N. Ames, *Science* *170*, 822 (1970).

31. Garnevali, F., G. Morpurgo and G. Tecce, *Science* *163*, 1331 (1969).

32. *In* Biology Data Book (*in* *press*). Published by the Federation of American Societies for Experimental Biology.

33. Kohne, D. E., *Quart*. *Rev*. *Biophys*. *3*, 327 (1970).

34. Storck, R. *In* J. B. G. Kwapinski (ed.) Molecular Microbiology, (*in* *press*). Van Nostrand Reinhold Co., New York.

35. Laird, C. D., *Chromosoma* (Berl). 32, 378 (1971).

36. Dutta, S. K. and D. E. Kohne, Proceedings XII International Botanical Congress, Seattle, Washington.

37. Penn, S. and S. K. Dutta, *Genetics* *64*, s50 (1970).

DISCUSSION

JACOB: Would you explain how you compared the sedimentation patterns of the enzymes isolated from the different organisms? Could these patterns reflect the variation in isolation and therefore not a very good supplemental method for classification.

STORCK: As indicated in the legend of Figure 9, these determinations were performed by R. Utter and G. A. DeMoss. I refer you to their paper for detailed information (J. Bacteriol. 94, 1896 (1967)). I would like to add that they purified these enzymes and, as a result, the objection you raised is ruled out.

STEIN: Are there correlations between GC ratios (and other biochemical data) and the degree to which sexuality has been developed in these groups?

STORCK: If you limit the comparison to Zygomycetes versus the higher fungi the answer appears to be yes.

DNA IN HIGHER PLANTS

H. REES

Department of Agricultural Botany, U.C.W., Aberystwyth

INTRODUCTION

The amount of chromosomal DNA in the nuclei of different species of higher plants varies enormously. There is almost a hundred times as much DNA in the nuclei of Lilium longiflorum than of Linum usitatissimum. The variation is far greater than we find in most other Orders of plants and animals, particularly in higher vertebrates such as birds and mammals (fig. 1). A question that springs to mind is why the higher plants should be so tolerant of quantitative DNA change, and why other groups are so intolerant. A more fruitful way of framing this question perhaps is to ask what special benefits follow as consequences of this DNA variation? To a large extent we should expect the consequences to depend upon the nature, the quality, of the DNA which is gained or lost. Quality, in turn, will depend on the mechanism whereby the DNA change is achieved. In broad and general terms there are, of course, three ways by which the DNA amount is varied,

1. By polyploidy and aneuploidy.
2. The amplification, or deletion, of segments within chromosomes
3. The evolution of supernumerary, B chromosomes.

All three contribute substantially to the DNA variation in Angiosperms. According to Stebbins (1963) 35 per cent of higher plant species are polyploid or of polyploid ancestry. B chromosomes, especially common in the Graminae, occur in at least 15 per cent of Angiosperm species (Darlington, 1956; Battaglia, 1964). Recent surveys provide abundant evidence, also, for the amplification (or deletion) of DNA within the chromosomes of higher plants.

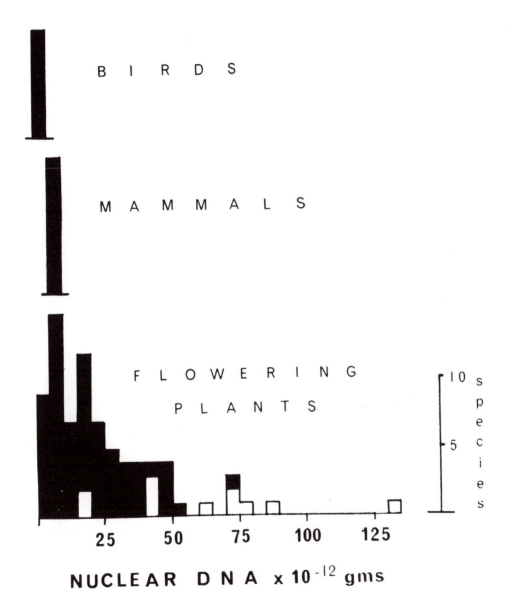

NUCLEAR D N A x 10^{-12} gms

Fig. 1.   The distributions of nuclear DNA amounts in species of Birds, Mammals and Higher Plants.   All DNA values are 2C (4C in tetraploids).   Unshaded blocks represent polyploid species.   Data derived from Mirsky and Ris (1951), Vendrely (1955), Rendel (1955), Baetcke, Sparrow, Nauman and Schwemmer (1967), Sunderland and McLeish (1961), Martin (1966), Evans, Rees, Snell and Sun (in press), Van't Hof (1965), Paroda and Rees (1971).

It will be convenient to consider, first, the structural aspects of DNA change and subsequently, in the light of these considerations, some of the consequences of the DNA variation to the Genetic Systems of Higher Plant populations.

THE STRUCTURAL BASIS

Polyploidy and Aneuploidy

The structural causes of polyploidy and of aneuploidy are familiar enough and there is no need to enlarge upon them here.   The nature of the DNA gained is also clear, at least in so far as it copies precisely the DNA of complete chromosomes or of chromosome complements.

Amplification and Deletion

In fig. 2 are the chromosome complements of four species within the genus Lathyrus.   All are diploids with the same number of chromosomes ($2n = 14$) although, as the figure shows, the size of chromosome varies considerably. Those of Lathyrus hirsutus are about three times larger by volume than the chromosomes of L. angulatus.   Chromosome size is closely correlated with nuclear DNA content (fig. 3) so that the nuclei of L. hirsutus have about three times as much DNA as the nuclei of L. angulatus.   The Lathyrus survey is typical of many which established that an extensive variation in chromosomal DNA accompanied the divergence and evolution of diploid species within genera of higher plants.   What was, and still is, surprising is the magnitude of DNA differences between closely related species.

As for the origin of the DNA change there are two contending possibilities (fig. 4),

a. A differential polynemy i.e. a variation in the lateral multiplicity of DNA strands within chromosomes.

b. A lengthwise amplification or deletion of chromosome segments.

There are attractive features about the concept of a differential polynemy.   All loci, all genes, are implicated in the structural change and, consequently, the ratio of genes relative to one another is maintained.   No

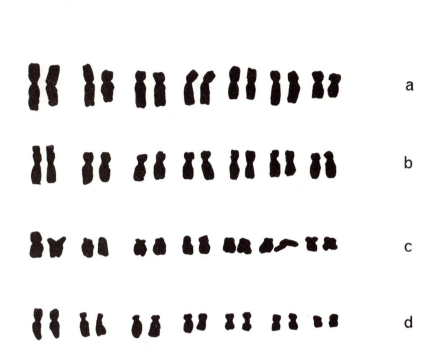

Fig. 2.   Mitotic metaphase chromosomes in Lathyrus. a L. hirsutus, b L. tingitanus, c L. articulatus, d L. angulatus.

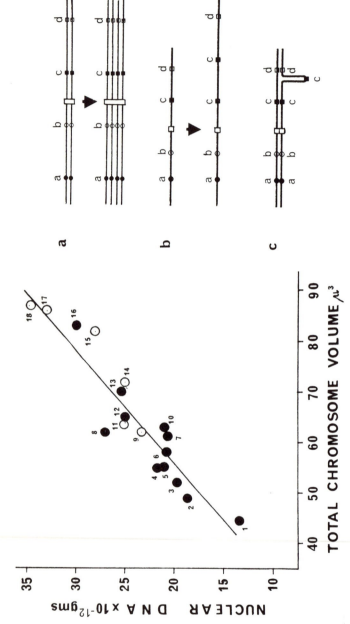

Fig. 3.   2C nuclear DNA amount plotted against chromosome volume (at mitotic metaphase in Lathyrus species. 1, Lathyrus angulatus; 2, L.articulatus; 3, L.nissolia; 4, L.cicera; 5, L.annuus; 6, L.aphaca; 7, L.sphaericus; 8, L.tingitanus; 9, L.maritimus; 10, L.ochrus ; 11, L.maritimus; 12, L.odoratus; 13, L.sativus; 14, L.niger; 15, L.tuberosus; 16, L.hirsutus; 17, L.latifolius; 18, L.sylvestris. Open circles, outbreeders; closed circles, inbreeders.   After Rees and Hazarika (1969).

Fig. 4.   Alternative mechanisms of variation in chromosomal DNA. a, A differential polynemy, i.e. a variation in chromosome "strandedness". b, a segmental repetition or amplification. c, configuration expected at pachytene in an F1 hybrid between species with different nuclear DNA amounts and assuming the difference in DNA to have arisen by segmental amplification (or deletion).

problem of genetic "imbalance" is posed by the DNA gain or loss.   This is
in sharp contrast to the second alternative, namely a segmental amplification
or deletion, where the dosage of some genes is altered whereas, for others, it
remains unchanged.   For this reason the distinction between the two alternatives
bears profoundly upon the nature of the DNA gained or lost and, hence, upon
its properties.   Coupled with the fact that there is still controversy as to
whether a or b accounts for the DNA variation it is necessary to consider,
albeit briefly, the evidence upon which the evidence for and against the two
alternative hypotheses are based.

a.   A Differential Polynemy

The evidence falls into four categories

i.   Discontinuity of DNA values.

In the sub-genus Vicia of the genus
Vicia the chromosomal DNA values make up a markedly discontinuous series,
a series moreover which closely approximates a 1:2:4 progression (Martin and
Shanks, 1966; Martin, 1968).   Martin concludes that the discontinuous
progression reflects the differential multiplicity of DNA strands in the nuclei
of the different species.   Rothfels, Sexsmith, Heimberger and Krause (1966)
propose a similar explanation for a (less convincing) sequence in Anemone (see
also Rothfels and Heimberger on the Droseracae, 1968).

More recent and more extensive work in Vicia by Yean (1971) shows,
however, that discontinuity is not the rule for all species even within the
genus Vicia.   In other genera, such as Allium (Jones and Rees, 1968) and
Lathyrus (Rees and Hazarika, 1969), there is no suggestion of discontinuity.

ii.   Chromosome Size.

Rothfels et al (1966) point out that the relative
lengths of chromosomes within complements of related Anemone species are much
the same even when the absolute lengths, and the nuclear DNA amount, varies
by a factor of two or more.   The same is true for many other genera, Lathyrus
(fig. 2) and Allium, for example.   On the face of it the gain or loss of DNA
is distributed uniformly throughout the complement, precisely what would be
expected with increase or decrease in strandedness within chromosomes throughout
the complement.

Fig. 5 a shows the mitotic complements of Allium cepa and A. fistulosum,
the chromosomes ranked in order of size.   A. cepa has 27 per cent more nuclear

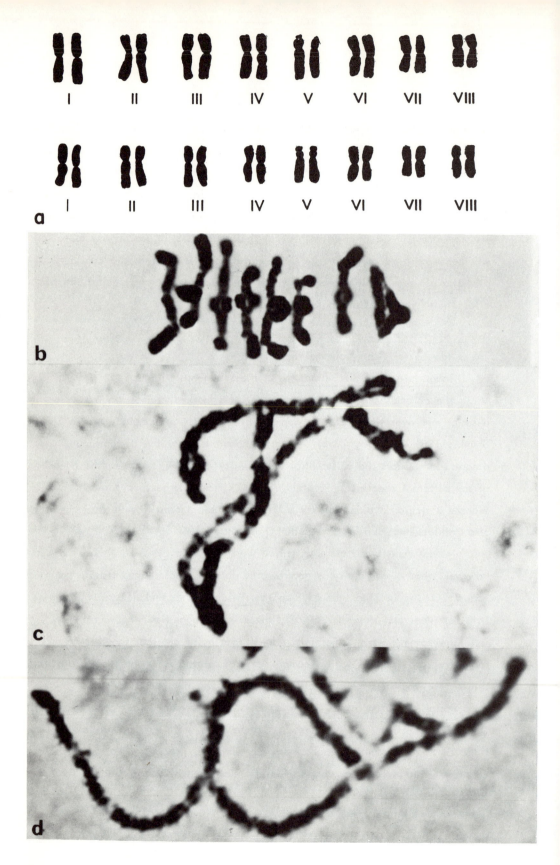

Fig. 5.　a.　The mitotic metaphase chromosomes of <u>Allium</u> <u>cepa</u> (top) and <u>Allium</u> <u>fistulosum</u> (bottom) ranked in order of size.　<u>b</u>.　First metaphase of meiosis in the F$_1$ hybrid <u>A. cepa</u> x <u>A. fistulosum</u>.　Note asymmetry of bivalents.　<u>c</u> and <u>d</u>, Extensive loops at pachytene in the F$_1$ hybrid (see. fig. 4<u>c</u>).

DNA than A. fistulosum.   The A. cepa chromosomes, as expected, are about
a third larger in total volume than those of A. fistulosum.   It will be observed
also that, much as Rothfels reported in Anemone, the relative lengths are
remarkably similar within the two complements.   A reasonable inference would be
that each A. cepa chromosome has a third more DNA than its A. fistulosum
homologue, an inference in keeping with a difference in the multiplicity of
chromosome strandedness between nuclei.   Ranking the chromosomes in order
of size, however, is no guarantee of ranking in order of homology.   Fig. 5b
shows the first metaphase of meiosis in the hybrid A. cepa x A. fistulosum.
All bivalents display asymmetry, as we would expect.   It is also abundantly
clear, however, that some bivalents are much more asymmetrical than others.
The difference in length between chromosomes within bivalents varies from
about 70 per cent to 10 per cent.   Clearly gain or loss of DNA was not
uniformly distributed between chromosomes within complements and, consequently,
it is not interpretable on the grounds of a differential polynemy.   Equally clearly,
conclusions based on comparisons between mitotic metaphases alone, as in Anemone
and Lathyrus, are less than convincing.

iii.   DNA Diminution in Polyploids.   There are reports of tetraploids with DNA
amounts less than double those of their reputed diploid ancestors.   Dowrick and El
Bayoumi (1969) describe such cases in Chrysanthemum.   In the scale insects it is
reported that a tetraploid species has less DNA than its reputed ancestral diploids
(Hughes-Schrader and Schrader, 1956; Hughes-Schrader, 1957).   Such observations
have been used to support the case for a differential polynemy (see Lewis and
John 1963).   It is argued that the chromosomes of the tetraploids have fewer
strands than the chromosomes of the diploids from which they are derived.   The
arguement is weak on two counts.   First, there is no certainty that the tetraploids
are direct descendants of the diploids with which they are compared.   Second, even
if they were, the DNA diminution might just as well reflect a lengthwise deletion
as a reduced multiplicity of DNA strandedness.

iv.   Electron Microscopy.   Wolfe and Martin (1968) claim that the number of
fibrils in the chromosomes of Vicia faba is greater than in V. Sativa chromosomes
which contain less DNA.   Electron micrographs of chromosome fine structure are
notoriously ambiguous.   Indeed, as we all know, it is by no means established
that chromosomes are made up of more than one DNA strand, let alone whether

the number of strands varies between species (see Whitehouse, 1968).

To my mind not one of the four kinds of evidence carries conviction.
The case for a differential polynemy, however vigourously propounded (see
Martin, 1968), is not proven.

b.   Lengthwise Amplification or Deletion.

The case for segmental amplification or repetition is, in contrast to the above,
well established.   The evidence, from three kinds of investigation, is indeed
compelling.

i.   Polytene chromosomes in Chironomus.   For the most convincing evidence
we have to turn to the work on the chironomids, Chironomus thummi thummi,
Ch. thummi piger and their F1 hybrid.   Keyl (1947, 1962; see also Keyl and
Strenzke, 1956) has shown that the DNA in certain bands of loci of Ch. thummi
thummi chromosomes is increased by a factor of 2, 4, 8, or even 12 in comparison
with corresponding loci in Ch. thummi piger.   The case for a localised amplification
is virtually unassailable.   The only possible alternative is to say the least unlikely,
viz. a localised variation in the multiplicity of DNA strands within one and the
same chromosome.

ii.   Pachytene analysis in Allium.   It will be recalled that "homologous" chromosomes
in Allium cepa and A. fistulosum differ in length and, hence in DNA content, by
10 to 70 per cent.   At pachytene in the hybrid A. cepa x A. fistulosum unpaired
loops and overlaps range from 10 to 70 per cent of the total length of the segments
paired (Jones and Rees 1968).   The presence, and dimensions, of the unpaired
segments argue for DNA change by lengthwise repetition or deletion (see fig. 5).
Comparable configurations have been described in a hybrid between Lolium perenne
and L. temulentum   (Rees and Jones, 1967).   Martin (1968) quotes work by
Tobgy (1943) on a Crepis hybrid in which the chromosome size and presumed DNA
differences between parents were not reflected in loops and overlaps at pachytene.
With due respect I would suggest that the hybrid should be re-examined.

It is worth noting that the loops observed at pachytene in both Lolium and
Allium were few and long, showing that the capacity for segmental gain or loss
is restricted to few segments within chromosomes.

iii.   DNA base sequences.   The capacity for extensive, localised repetition of
base sequences within DNA's is now well established by direct biochemical analysis

in numerous organisms.   Higher plants are no exceptions.   Britten and Kohne
(1968) have shown, for example, that as much as 80 per cent of the wheat
complement is made up of highly repeated base sequences.   That differences in
the amount of DNA between species arise through repetition is also suggested by
the disparity in base ratios between species differing in nuclear DNA amount.
For example in Allium cepa and A. fistulosum the GC fractions are 35.6 and
37.9 respectively (Kirk, Rees and Evans, 1970).   The difference is highly
significant.   Yean (1971) has shown a substantial disparity in base composition,
also, between Vicia species with different amounts of DNA.

In the light of this evidence the structural basis of quantitative DNA
change within chromosomes appears indisputable.   It is achieved, exclusively,
through lengthwise repetition or deletion of localised chromosome segments.   The
precise mechanism, in molecular terms, remains conjectural (see Keyl, 1965).   As
for the quality of the DNA gained or lost we must assume that it varies.   In
genetic terms it may embody informative base sequences which may, or may not
be transcribed or, alternatively, uninformative or nonsense sequences.   In
cytological, and perhaps complementary, terms it may comprise euchromatin, facultative
or constitutive heterochromatin.

B Chromosomes

There is compelling evidence that DNA loss by segmental deletion, as well
as gain by repetition, has accompanied the evolution of many species within genera
of higher plants (Stebbins, 1966; Ress and Hazarika, 1969).   It follows that a
fraction, sometimes a substantial fraction, of the DNA within chromosomes of many
species is dispensable.   In the case of B chromosomes it is the invariable rule that
all, as distinct from part, of their DNA is dispensable, because in all populations
carrying B chromosome some individuals, often the majority, are without B's.
B's in other words are relatively inert. .   Characteristically they carry no single
genes with major effects upon the phenotype.   In so far as their DNA incorporates
sequences with specific information it would appear that such information is suppressed.
In this connection it is not perhaps surprising that B chromosomes are often, although
not invariably, heterochromatic.

B chromosomes, one must assume, have evolved from the normal, A

chromosomes of the complement.   They are generally smaller than the A's and
their DNA content consequently representative of only part of the DNA in the
A chromosomes from which they derive.   In view of the absence of homology
between B's and A chromosomes at meiosis it is probably that the composition and
organisation of their DNA have in fact diverged substantially from that of the
original A chromosome fragments.   The divergence may take the form of localised
repetition of base sequences within the B's (Gibson and Hewitt, in press).

CONSEQUENCES

From what we know of the mechanism by which the amount of DNA is
varied we should be in a position to explore and to compare the genetic
consequences of DNA gained through polyploidy and aneuploidy on the one hand
with that due to B chromosomes or to segmental amplification within chromosomes
on the other.   With polyploidy and aneuploidy all gene sequences within
complements or within chromosomes are multiplied to a modest extent.   With B
chromosomes some gene sequences from the A chromosome ancestor may persist
unchanged, others proliferated to an extensive degree (see Gibson and Hewitt, loc.cit.).
Segmental amplification within chromosomes may be of varying extent, ranging
from a modest gene duplication to a prodigious repetition (Britten and Kohne,
1968; Thomas, 1969).   With suitable markers the consequences of quantitative DNA
change are, in certain cases, readily established and distinguished.   Indeed
Nilsson-Ehle gave a lucid account of the effects of varying the dosage of alleles
controlling grain colour in polyploid wheat and oats as long ago as 1909.   There
followed numerous accounts in polyploids and aneuploids of other species.   In sharp
contrast reports and accounts of the consequences of amplification within chromosomes
are few and far between.   The most familiar, no doubt is that on the Bar locus in
Drosophila.   More recently Ritossa and Spiegelman (1965) have described the
consequences of more extensive amplification at the bobbed locus, also in Drosophila.
There are other accounts but, as already mentioned, they are few.   When we turn
to B chromosomes markers of any kind are virtually non-existent.   The one exception
is a gene or gene cluster located at the end of the long arm of the standard B
chromosome in rye which determines non-disjunction of the B chromosomes at first
pollen grain mitosis.   Its role was established by the investigation of B

chromosomes deficient for the distal segment of the long arm (Müntzing and
Lime de Faria, 1952).

In view of the scarcity of markers a formal Mendelian approach is of limited
use in assessing the genetic consequences or the adaptive significance of the DNA
gained through B chromosomes or the repetition of base sequences within chromosomes.
A crude, and on the face of it, ingenuous possibility is to ask whether evolutionary
gain or loss of DNA per se, any DNA, has predictable consequences upon growth
and development.   From earlier extensive work on polyploids there are, however,
encouraging indications that such an approach might well prove fruitful.   Thus,
it is generally true to say that polyploidy in plants causes,

1.   An increase in cell size and mass.

2.   A reduction in initial growth rate.

3.   Broader and thicker leaves.

4.   An increase in overall plant size.

These consequences are not of course invariable (Stebbins, 1963).   The
effects vary between species and between varieties within species.   Nevertheless
as generalisations they are indisputable.   Are there comparable generalisations to
be made about the DNA gained, or lost, by repetition and deletion or by the gain
or loss of B chromosomes?   Stebbins (1966) is of the opinion that there are.   All
the evidence to hand tends firmly to support his view.

Cell Size and Mass.

In fig. 6 are the fresh weights of root meristem cells from diploid species
with widely different nuclear DNA contents.   There is a consistent increase in
cell weight with increasing DNA.   Fig. 7 shows a similar increase in cell
weight due to increase in B chromosomes in rye.   In both cases the increase
in cell weight is associated with increase in cell size.   As with polyploidy a
quantitative DNA variation, whether due to B chromosomes or to segmental
amplification, imposes a direct control upon cell size and mass, upon cell
growth and development.

Mitotic Cycles.   Stebbins also suggests that a quantitative DNA variation may
be of particular significance in regulating the duration of cell division (see also
Rees and Hazarika, 1969).   Previous work by Van't Hof and Sparrow (1963) and
by Van't Hof (1965) had suggested an increasing duration of mitotic cycles with

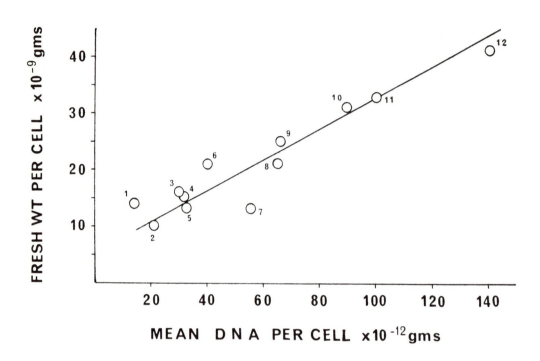

Fig. 6.   Mean fresh weight of root tip meristem cells plotted against the mean nuclear DNA per cell. 1, Agave attenuata; 2, Phalaris coerulescens; 3, Zea mays; 4, Phalaris hybrid; 5, Phalaris minor; 6, Galtonia candicans; 7, Vicia faba; 8, Narcissus pseudonarcissus; 9, Allium cepa; 10, Scilla campanulate; 11, Tulipa gesnariaria; 12, Lilium longiflorum.   Data from P.G. Martin (1966).

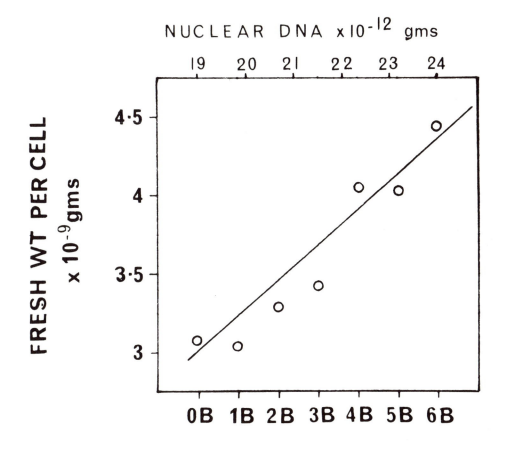

Fig. 7. Fresh weight of root meristem cells in rye plants with from 0 to 6 B chromosomes. DNA values are 2C. Data from R.N. Jones.

increasing nuclear DNA content.   Our own results are in agreement with those
of Van't Hof and Sparrow (cf. Martin, 1968;).   The results of our survey are
presented in fig. 8.   It will be seen that, overall, the duration of the mitotic
cycle increases with increasing DNA.   There are, at the same time, some
variations on the general theme,

    1.   The duration of the mitotic cycle is, on average, 4 hours longer in
          Dicotyledons than in Monocotyledons.

    2.   The rate of increase in the duration of the cycle is disproportionately
          high where the increase in DNA is caused by the addition of B
          chromosomes (see Ayoanadu and Rees, 1968 b).   The contrary
          appears to hold for polyploids.

It is not surprising that the phenotypic consequences of DNA gain vary
in relation to its origin and quality.   What is also worth emphasising however
is that DNA gain from whatever source has, to a large extent, certain common
and predictable effects.

The rate of cell division along with cell expansion together determine
the rate of growth of tissues and organs.   Since both components of cell growth
are correlated with nuclear DNA quantity we might well expect to find some
correlation between the amount of nuclear DNA and the general pattern of growth
of organs and even organisms.   M.D. Bennett has recently made surveys of seed
weights' within genera containing species with varying DNA amounts.   In many
of these eg. Allium and Vicia he finds strong correlations between the seed
weights and the nuclear DNA quantity   (personal communication).

It is appropriate at this point to make clear that I am not for one moment
suggesting that the amount of nuclear DNA determines exclusively or is even
paramount in determining the rate of cell division or rate of cell growth.   These
characters are of course subject to the control of both nuclear and extra-nuclear
genes with specific effects during growth and development.   What is being
emphasised is that alteration in the amount of DNA does impose its own
inevitable and indeed predictable influence upon the expression of such
characters whether it be the size of the cell or the rate of its division.   In
this predictive sense this particular category of "mutations" is unique.

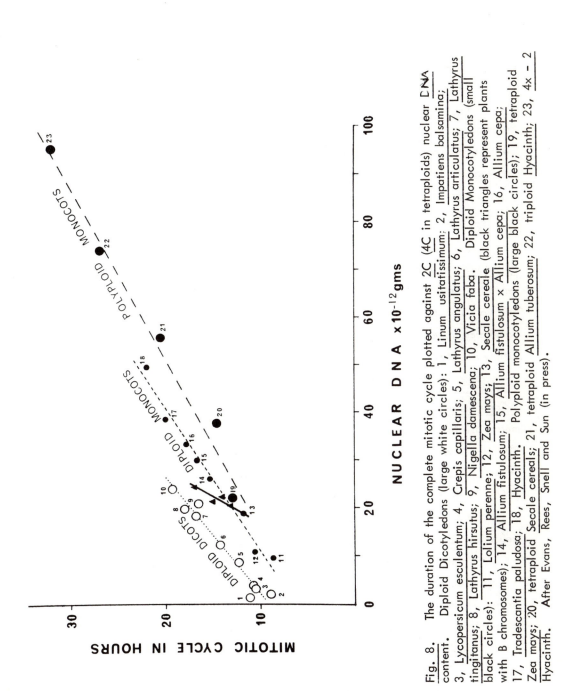

Fig. 8. The duration of the complete mitotic cycle plotted against 2C (4C in tetraploids) nuclear DNA content. Diploid Dicotyledons (large white circles): 1, Linum usitatissimum; 2, Impatiens balsamina; 3, Lycopersicum esculentum; 4, Crepis capillaris; 5, Lathyrus angulatus; 6, Lathyrus articulatus; 7, Lathyrus tingitanus; 8, Lathyrus hirsutus; 9, Nigella damescena; 10, Vicia faba. Diploid Monocotyledons (small black circles): 11, Lolium perenne; 12, Zea mays; 13, Secale cereale (black triangles represent plants with B chromosomes); 14, Allium fistulosum × Allium cepa; 15, Allium fistulosum × Allium cepa; 16, Allium cepa; 17, Tradescantia paludosa; 18, Hyacinth. Polyploid monocotyledons (large black circles); 19, tetraploid Zea mays; 20, tetraploid Secale cereals; 21, tetraploid Allium tuberosum; 22, triploid Hyacinth; 23, 4x – 2 Hyacinth. After Evans, Rees, Snell and Sun (in press).

The Genetic System

Mutation is but one of three ingredients of Genetic Systems which control the extent and flow of heritable variation in populations (Darlington, 1947; Darlington and Mather, 1949).   The others are recombination and the breeding system.   As we have seen quantitative DNA changes make their own special contributions to the sum of mutational events affecting the development and growth of the phenotype.   These quantitative DNA changes exert also a profound influence upon recombination at meiosis and upon breeding systems.

Recombination

Polyploidy.   The consequences of polyploidy are too well known to need repeating here.   Most important from the standpoint of recombination is that the addition of complete sets of chromosomes in autopolyploids increases the range of genotypes in respect both of alleles of the same gene and of combinations of different genes.   Above all, as Lewis (1967) has emphasised, polyploidy serves to increase the proportion of heterozygotes in the population.

B Chromosomes.   Fig. 9 shows how, in maize, the chiasma frequency at first metaphase of meiosis, and thereby recombination, is increased in the presence of B chromosomes (Ayonadu and Rees, 1968a; see also Hanson, 1967).   This "boosting" of recombination by B chromosomes is widespread (John and Hewitt, 1965; Jones and Rees, 1967; Barlow and Vosa, 1970) but not invariable.

In Lolium (fig 9) and Aegilops species the B's depress the chiasma frequency (Cameron and Rees, 1967; Simchen, Zarchi and Hillel, 1971).

Whether the effects of B's on chiasma formation are achieved through their influence on chromosome pairing at pachytene or upon events subsequent to pairing is not known.   That B's do, however, influence the pairing process is established by the work of my colleague G.M. Evans.   Briefly Evans's work, in Lolium, shows,

    1.   That B chromosomes suppress pairing and chiasma formation between homeologous chromosomes in the hybrid Lolium temulentum x L. perenne. (In the absence of B chromosomes pairing and chiasma formation is extensive; fig. 10).

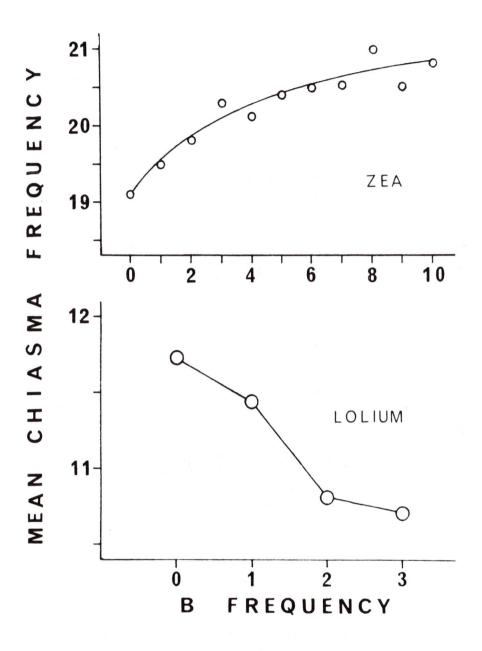

Fig. 9.   Variation in the chiasma frequencies of pollen mother cells in plants
with varying numbers of B chromosomes.   In Zea mays the chiasma frequency
is increased, in Lolium perenne decreased with increasing B frequency.
(Data on Zea from Dr. U. Ayonoadu, these on Lolium by Dr. P. Williams)

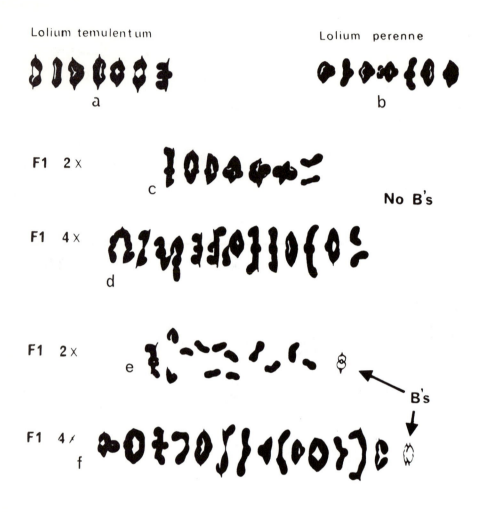

Lolium temulentum

Lolium perenne

a

b

F1 2x

c

No B's

F1 4x

d

F1 2x

e

B's

F1 4x

f

Fig. 10.   B chromosome control over chromosome pairing in Lolium hybrids.
a, Lolium temulentum. b, L. perenne (with smaller chromosomes). c,
Asymmetrical bivalents and univalents in the F1 diploid hybrid L. temulentum
x L. perenne. d, The tetraploid induced from the F1 diploid, with
quadrivalents, trivalents, bivalents and univalents. e, The F1 diploid showing
almost complete asynapsis in the presence of 2B chromosomes.   Homeologous
pairing is suppressed. f, the tetraploid hybrid with 4 B's.   Pairing restricted
to homologues; typical amphidiploid behaviour.

2.   That in tetraploids derived from the species hybrid pairing is
restricted to homologous chromosomes in the presence of B's, so that
only bivalents are formed at first metaphase.   The tetraploid , in
consequence, behaves like an amphidiploid.   Without B's, and in
sharp contrast, multivalents resulting from pairing between homeologues
as well as homologues, are common at first metaphase.   The Lolium
B's operate in much the same way as the "diploidising" locus in
polyploid wheat (Riley, 1960).   To what extent this property is shared
by B's from other sources remains to be seen.

Amplification and deletion.   One of two kinds of effect upon recombination
may be envisaged following repetition and deletion of segments within
chromosomes.   The first follows directly from change in chromosome length.
Clearly where the distance between genes is increased or decreased to any
substantial degree, following amplification or deletion, one would expect an
increase or decrease respectively in the amount of recombination between them.
There is no direct evidence for an adjustment in linkage by this means although
there is of course ample evidence establishing a correlation between chromosome
length and recombination.   The second kind of effect is one imposed indirectly
by a genotypic control associated with gain or loss of a chromosome segment.
For two excellent examples of this kind we have to turn once more to animals,
to grasshoppers.   Certain populations of Chorthippus parallelus are polymorphic
with respect to a terminal, heterochromatic segment in the S8 chromosome (John
and Hewitt, 1966).   The chiasma frequency in spermatocytes is increased in
males homozygous for the supernumerary segment (16.4) relative to the
heterozygotes (16.3) and "basic" homozygotes (14.5).   A comparable
polymorphism is described by Southern (1970) in Metrioptera brachyptera.   In
contrast to Chorthippus, however, the supernumerary segments depress rather
than boost the chiasma frequencies.   In both cases the control is unequivocal
but, as with B chromosomes, the direction of change varies between species.
The similarity of control excercised by B's and by supernumerary segments is
worth emphasising.   The similarity in control is paralleled by similarity in
organisation.   The supernumerary segments in the grasshopper, like the B
chromosomes in many species, are heterochromatic.

The Breeding System

One obvious consequence of polyploidy is the creation of a genetic barrier to gene flow between the polyploids and their diploid ancestors. Amplification within chromosomes also contributes towards genetic isolation in so far as it alters the structure and thereby the "homology" of chromosomes at meiosis.   In the Allium cepa x A. fistulosum hybrid discussed earlier the reduced homology is reflected by the high proportion of cells with univalents (over 90 per cent).   B chromosomes, as we saw in the Lolium temulentum x L. perenne hybrid, may reinforce such divergence in homology although it is not entirely clear how this particular phenomenon is of significance in an evolutionary sense.

CONCLUSION

Perhaps the two most surprising aspects of the nuclear DNA variation in higher plants are, first, the magnitude of the variation and, second, the highly predictable consequences of this variation upon cell growth and development, especially cell size and division.

It is not clear why these higher plants should be so tolerant of change. It is however tempting and not unreasonable to speculate that alterations to cell dimensions and to the rate of cell replication may be physiologically more acceptable in higher plants than in organisms such as the mammals whose organisation is so much more complex in respect both of the elaboration and variety of cell form and function.

The predictability of many of the consequences of DNA gain or loss is surprising because the DNA gained or lost, by different means, is clearly of varied composition.   The DNA fraction which varies by segmental amplification or by the addition of B chromosomes is clearly not "representative" of the whole complement as is the case with polyploidy.   Neither is it a random sample. The relative "inertness" of much of this DNA, coupled with its frequently distinctive heterochromatic nature, testify to a certain singularity in composition. How, therefore, does one explain the consequences common to so many categories of quantitative DNA variation between species and between genotypes within species?

An interpretation in formal genetic terms would be that determinants with similar, regulatory effects are widely distributed throughout all chromosome material, such that loss or gain by whatever means has certain common inevitable consequences.   The consequences, as we have seen, have more to do with rates of metabolic processes rather than with direction.   Precisely by what means remains to determined, although it is worth emphasising that "determinants" in this connection need not necessarily be thought of in terms of genes exercising direct control over specific enzyme products.   For example, extensive proliferation of base pairs, whatever the sequences, may demand a longer period of DNA synthesis during mitosis and, thereby, decrease in the duration of the mitotic cycle overall.   This, in turn, could have manifold effects upon the phenotype of the cell and, indeed, of the development of the whole organism.

Finally, it is worth reminding ourselves that not all the quantitative DNA variation between species need necessarily be interpreted on grounds of fitness or in terms of adaptive change.   Much of the DNA may indeed be strictly redundant.   The best evidence for this is the retention of redundant organs, for example the elaborate but entirely superfluous cross pollinating mechanisms in self-pollinating species of higher plants.   And it is not difficult to visualise how the redundant genes are retained.   A most likely possibility is that they are closely linked to genes of indispensable function, so that their removal by deletion awaits in each case the rare chance of two precisely located breaks within a very short chromosome segment.   There is in short an element of "inertia" (in the sense used by Mather, 1953) towards genetic change.

REFERENCES

Ayonoadu, U., and Rees, H. (1968a).   Genetica 39. 75.

Ayonoadu, U., and Rees, H. (1968b).   Exptl. Coll. Res. 52, 284.

Barlow, P.W., and Vosa, C.G. (1970).   Chromosoma, 30, 344.

Battaglia, E. (1964).   Caryologia 17, 245.

Britten, R.J., and Kohne, D.E. (1968).   Science 161, 529.

Cameron, F.M., and Rees, H. (1967).   Heredity 22, 446.

Darlington, C.D. (1947).   "The Evolution of Genetic Systems".   Cambridge

Darlington, C.D. (1956).   "Chromosome Botany". Allen and Unwin.

Darlington, C.D., and Mather, K. (1949).   "The Elements of Genetics".
      Allen and Unwin.

Dowrick, G.J., and El Bayoumi, A.S. (1969).   Genet. Res. 13, 241.

Gibson, I., and Hewitt, G.M. (In press).   Chromosomes Today Vol. 3.

Hanson, G.P. (1961).   Maize Co-operation News Letter, 36, 34.

Hughes-Schrader, S., and Schrader, F. (1956).   Chromosoma 8, 135.

Hughes-Schrader, S. (1957).   Chromosoma 8, 709.

John, B., and Hewitt, G.M. (1965).   Chromosoma 16, 548.

John, B., and Hewitt, G.M. (1966).   Chromosoma, 18, 254.

Jones, R.N.,and Rees, H. (1967).   Heredity 22, 333.

Jones, R.N., and Rees, H. (1968).   Heredity 23, 591.

Keyl, H.G. (1947).   Chromosoma 8, 739.

Keyl, H.G. (1962).   Chromosoma 13, 464.

Keyl, H.G. (1965).   Experientia 21, 191.

Keyl, H.G., and Strenzke, K. (1956).   Z. Natur forsch 11, 727.

Kirk, J.T.O., Rees, H., and Evans, G.M. (1970).   Heredity in press.

Lewis, K.R. (1967).   The Nucleus 10, 99.

Lewis, K.R., and John, B. (1963).   "Chromosome Marker", Churchill, London.

Martin, P.G. (1966).   Exptl. Cell. Res. 44, 84.

Martin, P.G. (1968).   'Replication and Recombination of Genetic Material',
      Canberra.

Martin, P.G., and Shanks, R. (1966).   Nature 211, 650.

Mather, K. (1953).   Symp. Soc. Exptal. Biol. VII, 66.

Nilsson-Ehle. (1909).   'Kreuzungsuntersuchungen an Hafer und Weizen'. Lund.

Muntzing, A., and Faria, Lima de. (1952).   Hereditas, 38.

Rees, H., and Hazarika, M.H. (1969).   Chromosomes Today 2, 158.

Rees, H., and Jones, G.H. (1967).   Heredity 22, 1.

Riley, R. (1960).   Heredity, 15, 407.

Ritossa, F.M., and Spiegelman, S. (1965).   Proc. Natl. Acad. Sci. U.S.
      53, 737.

Rothfels, K., Sexsmith, E., Heimburger, M., and Krause, M.O. (1966).
      Chromosoma 20, 54.

Rothfels, K., and Heimburger, M. (1968).   Chromosoma 25, 96.

Simchen, G., Zarchi, Y., and Hillel, J. (1971).   Chromosome 33, 63.

Southern, D. (1970).   Chromosoma 30, 154.

Stebbins, G.L. (1963).   "Variation and Evolution in Plants".   Columbia
      University Press, New York.

Stebbins, G.L. (1966).   Science 152, 1463.

Thomas, C.A. (1969).   In:- "The Neurosciences: A Study Program".
      Rockefeller University Press.

Tobgy, H.A. (1943).   Genetics 45, 67.

Van't Hof, J. (1965).   Exptl. Cell. Res. 39, 48.

Van't Hof, J., and Sparrow, A.H. (1963).   Proc. Natl. Acad. Sci. U.S.
      49, 897.

Whitehouse, H.L.K. (1968).   Scientia C111, 585.

Wolfe, S.L., and Martin, P.G. (1968).   Exptl. Cell. Res. 50, 140.

Yean, L.W. (1971) Ph.D. thesis.   Univ. Adelaide.

DISCUSSION

SPOFFORD: I was interested in the relation you showed between recombination and total DNA. It gives one kind of function for "junk" DNA - of separating the informational genes so that recombination would recur between instead of within genes. Would you care to comment on this possibility?

REES: In the graphs I showed it was the addition of B chromosomes

which contributed to extra nuclear DNA and brought about changes in recombination. I agree, however, that DNA inserted between genes could
have the effect you suggest.

DANCIS: While proximity of an unnecessary gene to a necessary one
may prevent the loss of the former, it will not prevent loss of its
function. How do you explain the maintenance of color in self-fertilizing species?

REES: Because, I suppose, no suppressors of "color" genes have
turned up.

CONGER: You showed that both cell size (volume?) and cell-cycle
time increased with increase in amount of DNA per cell. Is there a
reciprocal relationship between these, such that _total_ mass (or volume)
increases at the same rate with time, independent of amount of DNA/cell?
For example, do big cells (large amount of DNA) which are, say, twice as
big as small cells, divide only half as fast as the small ones?

REES: I don't know. It is certainly important to find out.

NUCLEAR DNA AMOUNTS IN VERTEBRATES

Konrad Bachmann, Olive B. Goin and Coleman J. Goin
Department of Biology, University of South Florida, Tampa, Florida 33620
and
Department of Zoology, University of Florida Gainesville, Florida 32601

INTRODUCTION

Twenty years ago, Mirsky and Ris[1] presented the first survey of diploid nuclear DNA amounts throughout the animal kingdom. Their sample included 54 species of vertebrates. Since then nuclear DNA values for more than 350 species have been determined (table 1). The new values confirm and extend three general observations by Mirsky and Ris about the phylogenetic relationships among vertebrate DNA amounts:

(1) There is a great variety of nuclear DNA amounts with no obvious overall trend throughout the vertebrates. In particular, there is no general increase in the nuclear DNA amount in vertebrate evolution.

(2) Within natural groups of related species, for instance in families of bony fishes, there is a remarkable uniformity of DNA amounts. The genome size in such groups is rather stable even when the related species differ greatly in size or outward appearance.

(3) Within vertebrate groups the more advanced forms tend to have lower DNA values than the more primitive species. This is seen in fishes, particularly bony fishes, in reptiles, and in the reptile-bird transition. It suggests that decreases in the nuclear DNA amount could be a fairly regular feature of evolution. In

Table 1

Diploid nuclear DNA amounts reported for various chordate groups. All
values have been converted into picograms. Where the calibration is
doubtful, values have been put in parentheses. Numbers following the
group names indicate fraction of described species for which the DNA
amount is known.

TUNICATA (2/2000)	0.2, 0.4
CEPHALOCHORDATA (1/30)	1.2
AGNATHA	
Cyclostomata (3/50)	2.7-5.5
GNATHOSTOMATA	
Chondrichthyes (6/550)	3-15
Osteichthyes	
Actinopterygii	
Chondrostei (3/25)	3.5(-8.5?)
Holostei (2/7)	2.4, 2.9
Teleostei (176/20,000)	0.9-8.8
Sarcopterygii	
Dipnoi (4/5)	100-284
Crossopterygii (1/1)	6.5
<u>Incertae Sedis</u>	
Brachiopterygii (1/10)	12
Amphibia	
Caudata (23/300)	36-200
Gymnophiona (2/158)	7, 50
Anura (98/2,200)	3-28
Reptilia	
Chelonia (5/335)	5
Squamata (10/5,700)	3-5(8?)
Crocodilia (3/21)	5-5.7
Aves (11/8,600)	1.5-3
Mammalia	
Prototheria (2/6)	(7)
Metatheria (3/242)	5-10
Eutheria (14/3,800)	5-10

addition, spectacular increases in nuclear DNA amounts must have
taken place in the evolution of lungfishes and urodele amphibians.

For most vertebrate groups not enough species have been examined
to allow a statistical evaluation of uniformity or variability of genome
size. Extensive surveys exist mainly for teleost fishes [1,2]. The modal
DNA value for this group is on the order of 1.7 picograms (1.7×10^{-12} g).
This is rather low, and small measuring errors may have a considerable
influence on individual determinations. Our own work has concentrated
on frogs and toads of the amphibian order Anura[3]. In this order diploid
genome sizes are grouped around a modal value of 8.4 pg. This higher
value allows a more precise determination of differences among closely
related species. When genome sizes of many teleost or anuran species are
arranged in a frequency diagram (fig. 1), the values in both groups
conform to a logarithmic normal distribution around a single mode[2]
(fig. 2). Such a distribution pattern of genome sizes seems to exist
for many natural taxonomic groups. Among the anurans, for instance, the
genome sizes of the species in several genera, Bufo, Eleutherodactylus,
Hyla, and Rana, have their own characteristic distributions which closely
fit a logarithmic normal distribution pattern (figs. 3 and 4). The
morphological similarity which led to the inclusion of these species in
a single genus, order, or subclass, has its counterpart in the genome.
This might well be expected qualitatively, but it is quite remarkable
to see how closely genome size alone reflects species relationships.
Two parameters can be calculated precisely from these distribution
curves: (a) the modal genome size for the group (1.7 pg for teleosts,
8.4 pg for anurans), (b) the variability of genome size within the
group, properly expressed as a ratio, for instance between largest and
smallest DNA amount or between one of these and the modal value.

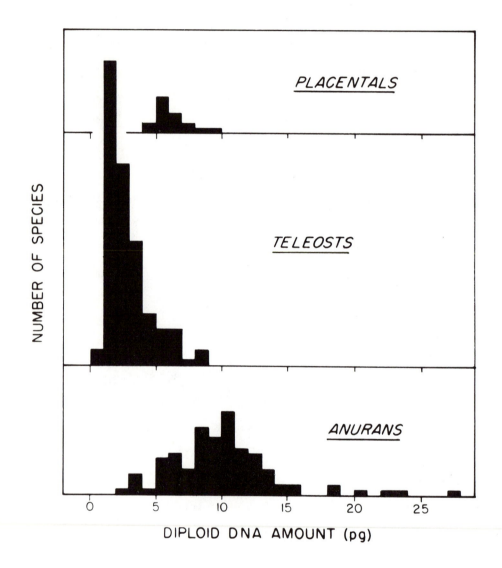

Fig. 1. Distribution of diploid nuclear DNA amounts in three vertebrate groups.

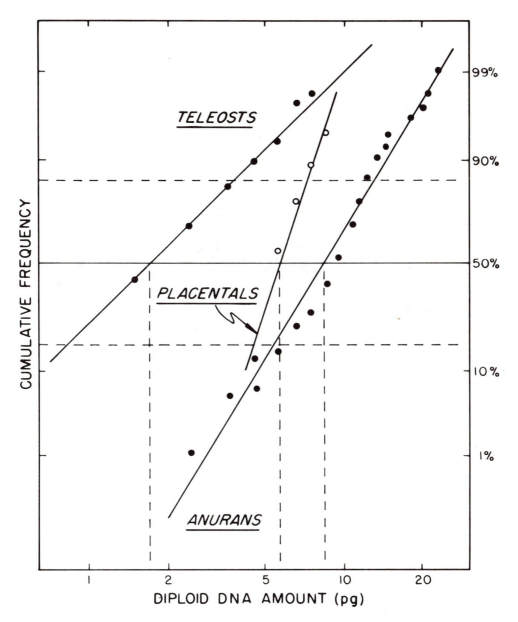

Fig. 2.   Distribution of diploid nuclear DNA amounts in three vertebrate
groups plotted logarithmically on probability paper.

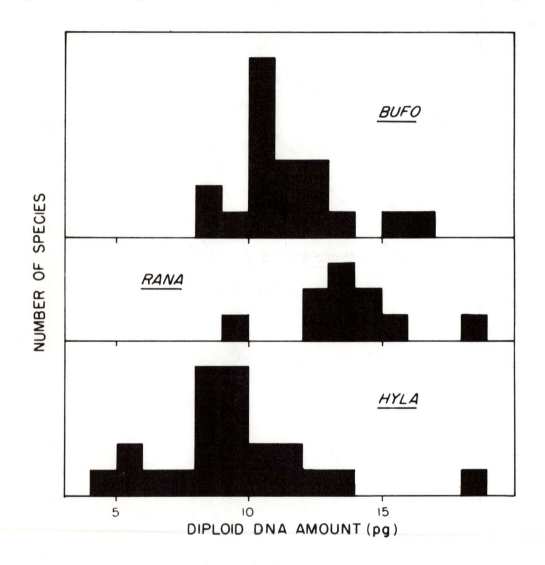

Fig. 3. Diploid DNA amounts in three anuran genera.

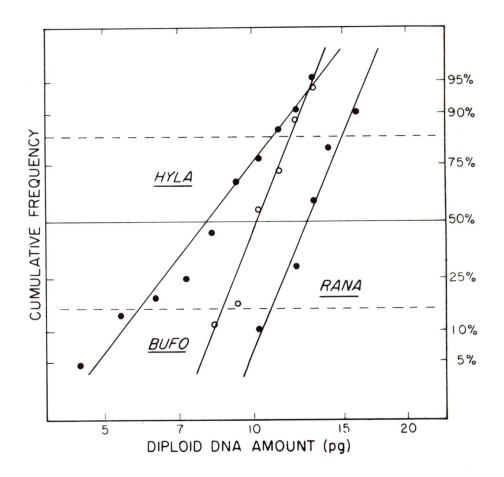

Fig. 4. Diploid DNA amounts in three anuran genera plotted logarithmically
on probability paper.

GENOME VARIABILITY

   While the biological meaning of any particular group-specific genome
size is by no means evident, the range of DNA values found in any group
reflects the degree to which the genome is modified in different species.
It should be noted that the distribution curves are very smooth, even
for the limited samples. In figures 2 and 4 ·this is expressed by the
close fit of the cumulative frequencies to a straight line. Such a
smooth distribution indicates that changes in genome size in evolution
are small, numerous, and cumulative. Major changes such as implied in
evolution by polyploidy must be considered exceptional. Species of
polyploid origin among the teleosts[4] and anurans[5] fit in the general
distribution. The anurans with the highest specific DNA amounts certainly
are not of polyploid origin. Since polyploid evolution obviously plays
a role among lower vertebrates[6] smaller continuous changes must act upon
the genome to erase the appearance of stepwise modes in the distribution
curves.

   The toad genus Bufo may serve as one specific example for the present
state of the evidence on genome variability. There are some 250 species
of toads distributed worldwide with the exception of Australia, New
Zealand, and New Guinea. The karyotypes of more than 50 species are
known. All have a haploid set of eleven chromosomes with the exception
of a group of African species which have ten chromosomes[7]. Clearly,
polyploidy plays no role in the evolution of this genus. The Bufo genome
is not as conservative as may appear from chromosome counts alone.
There is always a secondary constriction located on one of the
chromosomes, but in different species the chromosome carrying this
constriction may be the largest in the set, the smallest in the set, or
any one of several intermediate ones. Exchange of chromosome parts must
therefore play a considerable role in genome evolution in these toads.

Diploid DNA amounts have been determined in 19 species of <u>Bufo</u>. These range from 8 to 16 pg with a clear mode between 10 and 11 pg[8] (fig. 3 and 4). Genome size in this group varies by a factor of two in the absence of polyploidy. Ullerich has compared the DNA content of every single chromosome with that of its homologs in the sets of three European species with 10, 11, and 15 pg DNA per nucleus respectively[9]. The DNA differences in these species are not confined to single chromosomes; rather they seem to be due to a roughly proportionate change in all chromosomes. Such a proportionality might suggest lateral strand multiplication as the evolutionary mechanism for DNA increase. Two observations argue against strand multiplication: (1) The smooth normal distribution of the nuclear DNA values in the genus could be achieved only if numerous unit strands make up the chromosomes of every species and if multiplication is not a regular duplication of all strands. This seems unlikely. (2) Lampbrush chromosomes from the oocytes of these species have been examined in the electron microscope after treatment with trypsin and ribonuclease[10]. In all three species the same number of DNase sensitive fibers with a diameter of 20-35 $\overset{o}{A}$ have been found. The most likely explanation for the differences in genome size among the <u>Bufo</u> species is local deletion or duplication of DNA. One instance of such a duplication has been documented[11]: In <u>Bufo marinus</u> there exist populations in which the relative size of the secondary constriction and the amount of DNA that hybridizes with ribosomal RNA suggest a duplication of the nucleolar organizer.

Deletions and duplications along the axis of the chromosome seem to be the basis for most of the variety in vertebrate DNA amounts. In addition, polyploidy plays a limited role in the lower vertebrates. It is possible that polyploidy, though rare, has played a crucial role at critical points in vertebrate evolution. Clearly however, smaller genome changes take place with much higher frequency and lead to a smooth

continuous distribution of the DNA values around a modal value for the

group. The extent of this variation seems to be specific for the different

groups. As far as the limited data suggest, there is a decrease in the

range of genome sizes from teleosts through anurans to placentals, three

groups which have undergone extensive speciation at different levels of

evolutionary complexity. One of the factors determining the different

variabilities in genome sizes is undoubtedly the higher incidence of

polyploid speciation in the two lower groups when compared to the

placentals. The very narrow distribution of mammalian genome sizes has

been noticed before[6]. Of course the sample is very limited, but even

very specialized mammals (bats) have a genome only slightly smaller than

that of other mammals. This very restricted range of mammalian DNA values

is particularly puzzling in view of the profound qualitative differences

among the DNAs from related mammalian species[12]. Is there any advantage

in maintaining a certain modal genome size in certain groups?

THE MODAL DNA VALUE

What is this modal value? Is it the ancestral value of the group

from which species radiated towards higher or lower DNA amounts?

Moreschalchi, for instance, considers a value of about 5 pg ancestral

for all vertebrate groups[13]. Or does the modal DNA amount represent a

final evolutionary equilibrium toward which related species gravitate,

something like a most convenient or most efficient genome size? If that

is so, what is the ancestral DNA value? Is it at the higher end of the

range or at the lower one? Do vertebrate genomes evolve by reduction

of an initially highly redundant ancestral genome or do they evolve by

the addition of redundancy to a basic group-specific genome? What forces

direct genome evolution?

None of these questions can be answered fully. During recent years, however, some evidence has been compiled which suggests what the answers

might be. We have approached this problem by asking the following

question: Is there anything particular about species with exceptionally

high or low DNA amounts farthest away from the modal value; or is there

any morphological or physiological parameter that correlates with genome

size? It appears that such correlations can be found.

The best known correlation is that between nuclear DNA amount and

nuclear and cell size. Among closely related species with DNA amounts

that differ only slightly there is a great variability in cell sizes as

related to nuclear DNA amounts. The immediate factor determining nuclear

and cell size is the protein content of the nucleus or cell[14,15]. On a

wider scale, however, there is a clear relation between specific cell

size of homologous cells and the specific nuclear DNA content[1,16,17].

Figure 5 demonstrates this relationship for amphibian erythrocytes. It

should be noted that both cell size and DNA amount are drawn on

logarithmic scales. The slope of the resulting line is smaller than

unity. This result is typical for vertebrates[16,17] and several other

groups. The only exception seems to be a sample of higher plants[18] in

which a direct proportionality between nuclear volume and nuclear DNA

has been found. In all other cases, nuclear DNA increases faster than

cell size. Since the determination of nuclear and cell size by the

nuclear DNA amount seems to be mediated by the template function of DNA

in protein synthesis, larger genomes appear to be proportionally less

efficient as templates. This is true for algae as well as for

vertebrates[19]. It is tempting to speculate that this relatively lower

synthetic output of larger genomes is due to a relatively smaller pro-

portion of "producer genes" or "structural genes" and a relatively higher

proportion of controlling genes in larger genomes. Comparing total DNA

amounts, amounts of genes coding for ribosomal RNA, and cell size in

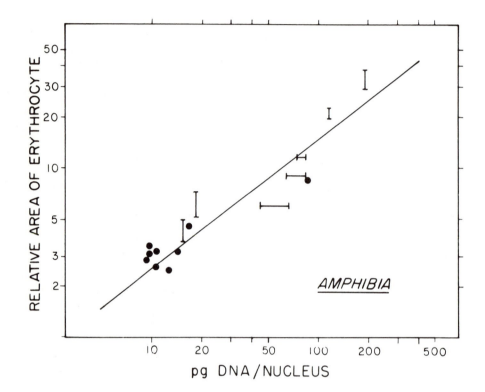

Fig. 5. Erythrocyte size and nuclear DNA amounts in amphibians.

several fish species, Pedersen found that ribosomal genes constitute a
relatively lower proportion of the larger genomes, and that cell size
parallels the amount of ribosomal genes much more closely than the total
DNA amount[20]. All this suggests some generalities for genome evolution
on a grand scale. Britten and Davidson have proposed a model of the
genome in which control genes play a crucial role[21]. Such a model would
allow for the observed effects.

Cell size is one of the visible correlates of the nuclear DNA
amount that has been implicated in genome evolution[22]. There are others,
for which the evidence is not as clear. (1) It has been claimed that
cell metabolism decreases relatively in larger cells with higher DNA
amounts[17,22]. There exist surprisingly few comparable data to indicate
the exact quantitative correlation between nuclear DNA amount and
metabolic activity of the cell. Preliminary data from our laboratory
indicate that the effect is considerably less than proportional (table 2)
and subject to much interspecific variation.

Table 2

Physiological correlations with genome size in toad and salamander

	diploid DNA amount	oxygen consumption of liver tissue	developmental time to stage 20 at 20°C
toad (Bufo terrestris)	11 pg	314 ± 18 ul O_2/hr/g	36 hours
newt (Notophthalmus viridescens)	86 pg	205 ± 13 ul O_2/hr/g	86 hours

Possibly a close correlation
can be found within certain groups. (2) A relation between nuclear DNA
amounts and the duration of the cell cycle is well established for plants
and a similar correlation may exist in vertebrates[23]. (3) We have
recently shown that tadpole development in anurans is slower in species

with higher DNA amounts[3]. The correlation is even closer when embryonic developmental rates are examined (fig. 6). If this correlation holds true generally for anurans, the nuclear DNA content of the developing oocyte rather than that of the somatic cells of the embryo would be the determining factor. The one frog species that does not fit the correlation of figure 6, _Ascaphus truei_, has a rather low DNA amount of about 7.5 pg, but its embryos develop much more slowly than any of the other species[24]. The oocyte of this species is exceptional in containing eight functional germinal vesicles throughout oogenesis[25]. In any case, this correlation between developmental rate and genome size is limited to the anurans. Salamanders, for instance, have both significantly higher DNA amounts and slower development than frogs and toads (table 2), but the quantitative details of the relation are different.

These correlations, incomplete as they are, suggest a tendency for large nuclear DNA amounts to be accompanied by large cell size, low relative metabolic rates, and slow development, that is, by attributes which may effectively limit the organism to a constant, uncompetitive environment. Adaptation to unstable, demanding, extreme, or competitive environments should favor the lowest DNA amount possible for the genome of a particular group. Several authors have noted that low DNA amounts appear to be a feature of advanced species rather than primitive ones, or even more convincingly, of specialized rather than generalized species[1,2,26,27]. Among the Anura, desert species of the genus _Scaphiopus_ have the lowest DNA amounts[3], among the teleosts, species with peculiar body shapes reflecting very specific adaptations exhibit the lowest DNA amounts[2]. In the same group, deep sea species may have considerably higher DNA amounts than close relatives living in the less stable and possibly more competitive environment of the surface waters[28]. We have discussed elsewhere the different evolutionary modes that might

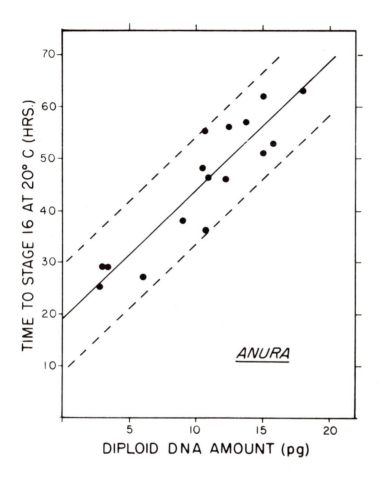

Fig. 6. Developmental time and nuclear DNA amounts in anurans.

be associated with DNA increase during the establishment of an evolutionary line and reduction of the nuclear DNA amount during the evolution of many specialized adaptive types from the ancestral stock[26]. We like to regard the evolution of the genome as an interplay between two tendencies: One is the increase in nuclear DNA by gene duplication which creates the material basis for new genes, new gene combinations and a greater genetic diversity. This increase has been documented in detail by Ohno[4,6,30]. The other tendency is a reduction of an ancestral genome once it is established. This reduction would involve selection of specific adaptive genomes from the ancestral one and a loss of redundant genes not necessary for these specific genomes. On the organismic level, it would correspond to the formation of new species all having features in common, but each restricted by its particular adaptive features.

The vertebrates are a particularly good group for further investigation of this hypothesis. Because vertebrates are able to form bone and teeth, hard structures that are resistant to decay and are readily fossilized, they have left a more adequate record of their evolutionary history than have most other groups of animals. Together with a vast amount of data on comparative anatomy and embryology, accumulated over more than a century of research, the fossil history of the vertebrates presents a rather reliable picture of their evolution based on independent evidence, against which we can check data and hypotheses on genome evolution. However, in spite of all the accumulated evidence, there are certain basic features of the evolutionary history of the vertebrates on which the evidence is incomplete and open to very diverse interpretation. Molecular data promise to provide the additional information needed in these cases to complete the picture. Certain molecular data, especially amino acid sequences in related proteins, have become powerful tools of evolutionary research, because the conservative nature of their evolution permits a rather unequivocal reconstruction of their ancestral states. This is not so for genome size, and we have to be very careful how we extrapolate back from the condition of present day forms to the ancestral state. A surviving species of an ancestral group which has kept many of the ancestral characters may still have become very specialized, in fact such specialization may have assured its survival.

In the following we shall attempt to present a tentative picture of vertebrate genome evolution.

HISTORY OF VERTEBRATE GENOME SIZE

It is believed that the pre-vertebrates evolved, probably by paedomorphosis, from a tunicate-like animal. "If, as seems surely to be the case, Paleozoic tadpoles of certain tunicates, or pretunicates, became sexually mature and no longer metamorphosed into sessile adults, a new mode of life opened up . . . Amphioxus . . . represents in slightly specialized fashion the stage in which sexual maturity of the tadpole has taken place, but not much progress toward higher evolutionary levels has occurred."[29] Two species of tunicates for which the DNA nuclear value has been determined have 0.2 and 0.4 pg/nucleus. Amphioxus has 1.2 pg/nucleus. These figures suggest that a three-fold increase in DNA amount may have occurred in the pre-vertebrate ancestor of the vertebrates[6].

It was presumably from an amphioxus-like ancestor that the first true vertebrates, members of the class Agnatha, evolved. The oldest known vertebrate fossils appear in Ordovician deposits nearly 500 million years old. These fossils are rare and fragmentary, but they have been shown to include representatives of two different groups of agnathans. The jawless fishes are represented in the modern fauna by the lampreys and hagfishes (Cyclostomata). Diploid DNA values reported for three species range from 2.7 to 5.5 pg. Again there appears to have been an increase in genome size at a critical stage of vertebrate evolution. It may well be that the agnathan line was initiated by tetraploidization of an amphioxus-like ancestor[30]. However, the differences between modern members of the class cannot be accounted for by polyploidy. The low value of 2.7 is for a lamprey (suborder Petromyzontoidei) and the high value (5.5) for a hagfish (suborder Myxinoidei). Six species of lampreys have diploid chromosome complements ranging from 60 to 156 and four species of hagfishes have complements of 46-52[31]. Whether evolution within the class was accompanied by serial duplications or by deletions of chromosomal segments cannot now be determined. Perhaps both processes were involved.

Vertebrate remains are rare throughout most of the Silurian but by late Silurian time, in addition to numerous agnathans, vertebrates with jaws had appeared.

The fish fauna of the Devonian was very diversified and by mid-Devonian times, roughly 375 million years ago, representatives of all the major stocks living today were present and clearly differentiated. Besides the agnathans and classes now extinct, we find sharks, ray-finned fishes, lungfishes, coelacanths, and rhipidistians, a group now extinct as fishes but ancestral to all the terrestrial vertebrates.

There is general agreement that the Chondrichthyes represent a lineage that has been distinct from the other fishes at least since the early Devonian. DNA values recorded for the Chondrichthyes range from 3 to 15 pg. The low value is for a chimaera (Hydrolagus) a member of the highly specialized Holocephali. Less highly specialized sharks have values of 5.5-6.5, close to the high value of the cyclostomes. The unspecialized Squalus has 15 pg. Evolution of the Chondrichthyes may thus have involved first an increase and then a loss of DNA.

The ray-finned fishes (Actinopterygii) appear to have been a distinct group since the Silurian. Evolution in these fishes also seems to have involved loss of nuclear DNA[1,2,6]. The more primitive chondrosteans and holosteans have DNA values from 2.4 to, possibly, 8.5 pg, while the modal value for the more advanced teleosts is about 1.7 pg. Moreover, within the teleosts, the more highly specialized species tend to have lower DNA values than the more generalized ones[2].

The remaining groups of fishes are of great interest because they include the ancestor of the terrestrial vertebrates. Paleontologists differ as to the proper classification of these fishes. Romer divides the class Osteichthyes into three subclasses, two with living representatives[32]. These are the ray-finned fishes (Actinopterygii) and the fleshy-finned fishes (Sarcopterygii). The latter includes the lungfishes (Dipnoi) and the cross-opterygians (=coelacanths + rhipidistians) (fig. 7).

Jarvik presents a different interpretation[33]. Following Stensiö[34] he places _Polypterus_ and _Calamoichthys_, two present day forms that Romer considers the most primitive surviving ray-finned fishes, in a separate class, Brachiopterygii. He states that the coelacanths, when they first appear in the fossil record in the Devonian, are quite distinct from the rhipidistians, but are probably more closely related to them than to the other Devonian fish groups. He believes that the lungfishes are not allied to either the rhipidistians or the coelacanths and suggests they may possibly be closer to the Chondrichthyes.

Schaeffer divides the Osteichthyes into three main stocks - Actinopterygii, Dipnoi, and Crossopterygii (=rhipidistians + coelacanths) and states that there is little unquestionable evidence that any two of them are more closely related to each other than either is to the third[35].

Nelson places the Brachiopterygii with the Sarcopterygii and says ". . . admittedly, the relationships of dipnoans, coelacanths, and

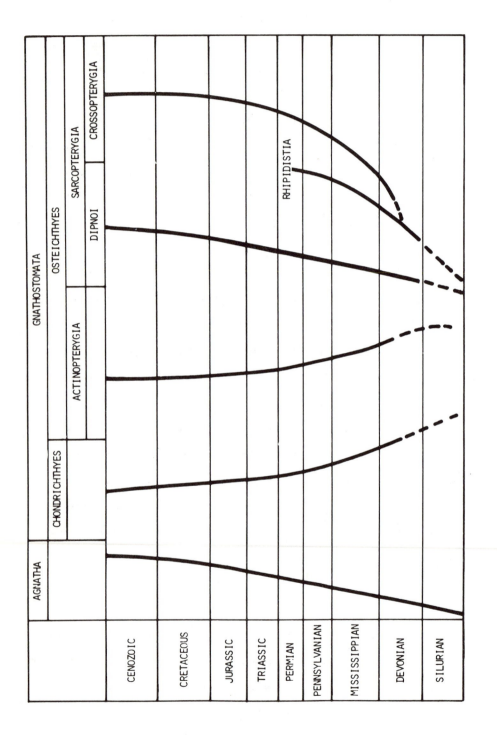

Fig. 7.   Classification of fishes.   After Romer, 1966.

brachopterygians are not very well established and all of these and the choanates (=rhipidistians + tetrapods) possibly do not form a monophletic group . . ."[36]

Denison has described the earliest known lungfish, from the Lower Devonian[37]. He points out that in many ways it is intermediate between the later lungfishes and the rhipidistians. This suggests that the lungfishes are more closely allied to the ancestors of the amphibians than most recent workers have assumed.

Ohno has discussed the enormous size (100-284 pg) of the genome of the lungfishes[6]. He believes that the increase in genome size resulted solely from tandem duplications and that it was necessarily accompanied by a great increase in cell size. The very large cells of the dipnoans require that all cell proteins be produced in great numbers. Ohno suggests that all the duplicated copies of the genes are needed to fulfill this requirement and that as a result it has become impossible for the lungfishes to eliminate genetic redundancy.

Thomson (pers. com.) says that cell size in primitive lungfishes was not markedly enlarged; the great increase in cell size (and DNA amount) took place some time after the divergence of the rhipidistian and dipnoan lines. It appears then that DNA values in lungfishes can tell us nothing about the values in the ancestors of the amphibians.

The single surviving coelacanth has a DNA value of 6.5, higher than the majority of actinopterygians, lower than the majority of amphibians. Latimeria represents a specialized sidebranch of the Sarcopterygii and, as Nelson indicates, its relationship to the rhipidistians is not clear[36].

A special interest attaches to the Brachiopterygii. They have not been traced back in the fossil record beyond the Tertiary, yet they are unquestionably very primitive. If the Sarcopterygii, including the Brachiopterygii, are a monophyletic group, then Polypterus may represent

the most primitive survivor of the ancestral stock from which the terrestrial vertebrates evolved. We have recently measured DNA in _Polypterus
palmas_ and found a diploid value of 12 pg. Lungfishes aside, this is the
highest value recorded for any fish. Does it indicate that a further
increase in genome size from the primitive gnathostome level (presumably
around 5-6 pg) took place in the line leading to the higher vertebrates?
And did this redundant genetic material provide the evolutionary flexibility that allowed the amphibians to invade the land?

The earliest known amphibian appeared in the Late Devonian.
Ichthyostega is very nearly an ideal intermediate between the rhipidistian fishes and the later amphibians. During the Carboniferous the
amphibians were numerous and diverse but most of them had died out by the
end of the Paleozoic and there is a gap in the fossil record between the
earliest known fossils of the modern orders and their Paleozoic ancestors.
Romer follows Parsons and Williams[38] in uniting the modern orders in the
subclass Lissamphibia, whose relationship to the Paleozoic amphibians
is unclear[32].

Brough and Brough divide the fossil orders of amphibians into class
Eobatrachia and class Eoreptilia and suggest that the two lineages may
have separated while they were still fishes[39]. The Broughs believe the
frogs evolved from the Eobatrachia and the salamanders, caecilians, and
reptiles from the Eoreptilia.

Jarvik believes the rhipidistian fishes were divided into two main
groups, Osteolepiformes and Porolepiformes[33]. The frogs, reptiles, and
probably caecilians evolved as separate lineages from the Osteolepiformes
and the salamanders from the Porolepiformes.

Schmalhausen rejects the idea that the amphibians are polyphyletic[40].
He believes there was a basic primitive stock from which the frogs, the
salamanders and caecilians, and the reptiles evolved as separate lineages

(Fig. 8).

Obviously there is little agreement between these workers on the degree of relationship of the orders of modern amphibians to each other on the one hand and to the reptiles on the other.

Salamanders parallel the lungfishes in having very high nuclear DNA values (36-200 pg) that seem to be due to serial duplications rather than to polyploidy. The disadvantages that attend large genome size may well explain why salamanders have been less successful as terrestrial animals than frogs. There are many fewer species of salamanders, they are less widely distributed, and they show less ecologic diversity. It is possible though that there has been some loss of redundant DNA during the evolution of the salamanders. The ones with the lowest DNA levels are members of the most advanced family, Plethodontidae. This is the largest family in the order and the only one that has been able to invade the tropics. Plethodontids entered South America after the establishment of the late Tertiary land bridge and there speciated rapidly. They may be in the process of freeing themselves from the evolutionary trap of too much DNA.

The modal value of DNA for frogs is 8.4 pg, somewhat above the level of the reptiles and mammals, but much lower than that of the salamanders. We agree with Ohno that the differences in pattern of chromosomal evolution shown by these two groups indicate a long separate evolutionary history and suggest the frogs may be closer to the reptiles than to the salamanders.

Shortly after they moved on to land in the late Devonian, the amphibians gave rise to the reptiles. This transition must have taken place in the Mississippian since the oldest reptile fossils known are from the Lower Pennsylvanian and by that time two distinct groups were present, pelycosaurs and captorhinomorphs. The pelycosaurs gave rise to the line leading to the mammals. Romer believes the captorhinomorphs

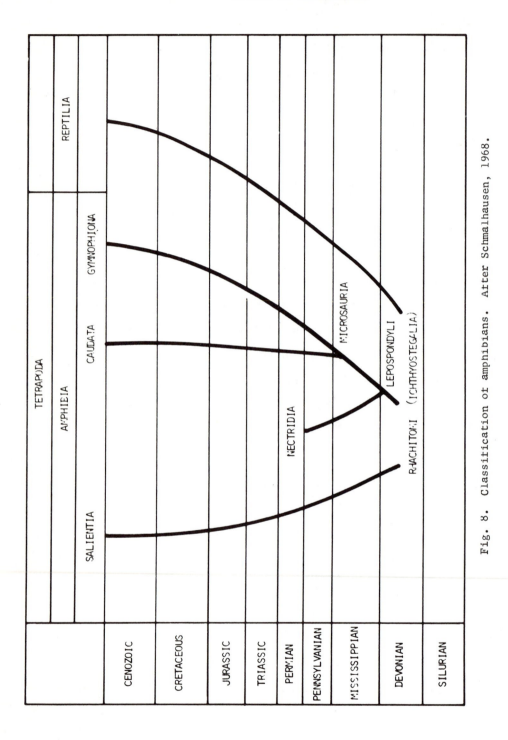

Fig. 8. Classification of amphibians. After Schmalhausen, 1968.

represent the basic reptilian stock from which the pelycosaurs had branched and which later gave rise to a number of other lines, including one leading to the turtles (Anapsida), one to the lizards, snakes, and tuatara (Lepidosauria), and the great archosaur stock from which evolved not only the dinosaurs but also the crocodilians and birds[32]. Carroll basically agrees with Romer but points out that the turtles are very isolated and that their relationship to the other groups is obscure[41]. Reig believes that the archosaur line diverged from the pelycosaurs rather than from the basic captorhinomorphs and thus that the crocodilians and birds are more closely allied to the mammals than they are to the lizards, snakes, and tuatara[42].

Turtles, crocodilians, and some lizards and snakes have DNA values of around 5 pg. Other members of the Squamata have values down to 3 pg and birds range from 1.5 to 3 pg. Ohno believes that the birds are more closely allied to the Squamata than they are to the crocodilians, partly because of the close approximation of DNA values of some members of the two groups, partly because microchromosomes are found in both, and partly because the Z-W sex determination mechanism present in birds is also found in some snakes. On the other hand, the paleontological evidence that birds and crocodilians are derived from archosaurs and that the Squamata represent a separate lineage is very strong. Morescalchi believes that some of the chromosomes of the crocodilians might be considered microchromosomes[13]. Crocodilians lack visible sex chromosomes, as do many lizards and snakes, while a number of lizards show male heterogamety[43,44]. Heterogamety must have developed several times late in the radiation of the reptiles, after the divergence of the snakes from the proto-lizards and after the divergence of the line leading to the birds from the line leading to the crocodilians. The presence of female heterogamety in birds and poisonous snakes and of male heterogamety in some lizards and mammals are examples of convergence. The low DNA values in the birds and some snakes again indicate a reduction in redundant genetic material with increasing specialization.

Mammalian genome size is relatively constant. It ranges from 5 to 10 pg, with most species having about 6-7 pg, slightly above the reptile level. It may be that the synapsid line retained somewhat more of the redundant DNA of the amphibians than did the other reptiles. Differences within the class apparently reflect tandem duplications and deletions, but the mammals show less tendency toward reduction of genome size than the reptiles or birds. This may be an indication that the archosaurs are more properly aligned with the other reptiles rather than with the synapsids.

Hotton has pointed out that the vertebrates may serve as a useful model in studying the evolution of higher categories simply because the origin of the vertebrate classes is closer to us in time than is that of most other major animal classes[45]. Thus all modern classes of Mollusca were present and well differentiated in the Cambrian. Their radiation must have taken place well over 600 million years ago and is not documented in the fossil record. The initial radiation of the fishes, on the other hand, apparently occurred during the Silurian or early Devonian, not much more than 400 million years ago, and the birds first appeared less than 150 million years ago.

Hotton points out that the origin of the classes of lower vertebrates took place rapidly. The time of origin of a new class is close (in the geologic sense) to the time of origin of the closest probable ancestor. The primitive fishes of the Silurian had given rise to the modern classes of fishes by mid-Devonian times. The rhipidistians appeared in the Lower Devonian and by the close of the period the amphibians had evolved. The divergence of the reptiles from the amphibians apparently took place

during the Mississippian. The origin of the mammals and birds presents
a different picture. Synapsids were present in the Lower Pennsylvanian
but not until the end of the Triassic did the mammals appear and they
did not begin their major radiation until the Cretaceous. Similarly the
archosaurs were present in the Upper Permian, Archaeopteryx not until the
Upper Jurassic. The evolution of a new class involves the invasion by a
group of a new major adaptive zone. The adaptive zones entered by the
fishes, amphibians and reptiles were empty of competitors, while the
zones eventually taken over by the birds and mammals were occupied by
the numerous, diverse, and well-adapted archosaurs[45]. Hotton believes
that the absence of competition allowed the rapid evolution of the
lower vertebrate classes, while the presence of potential competitors
delayed the appearance of the mammals and birds. In view of the evidence
presented above, we believe that another factor was also operative. It
is probable that polyploidization occurred in the chordate line leading
to the fishes and again in the sarcopterygian line leading to the
amphibians and reptiles. On the other hand, evolution of the birds from
the reptiles involved loss of DNA, while the mammals apparently either
retained the primitive reptile amount or increased it somewhat by tandem
duplication of chromosome segments. Ohno has stressed that polyploidization is by far the most effective means of making new genetic material
available for evolution[6]. This apparently is substantiated in the
evolution of the vertebrates. When redundant genetic material is
provided by polyploidization, major evolutionary shifts can take place
very rapidly. Major adaptive shifts without a closely preceding polyploidization are achieved much more slowly. Loss of redundant DNA
accompanies increasing specialization. Vertebrates with low DNA values
(teleosts, snakes, birds) speciate readily to fill many subzones but
they have lost evolutionary flexibility and are no longer capable of

major adaptive shifts. This seems to be the answer to why extreme specialization frequently leads to extinction and why new major categories arise from unspecialized members of preceding categories. The genome with the greatest evolutionary potential is one that has a moderate amount of redundant DNA that was introduced by polyploidy - not so much that it faces the restrictions laid on the lungfishes and salamanders, nor so little that it lacks flexibility.

REFERENCES

1. Mirsky, A. E. and Ris, H., *J. Gen. Physiol.* 34, 451 (1950).

2. Hinegardner, R., *Amer. Naturalist* 102, 517 (1968).

3. Goin, O. B., Goin, C. J. and Bachmann, K., *Copeia* 1968, 533 (1968).

4. Ohno, S., *Trans. Amer. Fisheries Soc.* 99, 120 (1970).

5. Beçak, W., *Genetics* Suppl. 61, 183 (1969).

6. Ohno, S., *Evolution by Gene Duplication* (Springer, New York, 1970).

7. Bogart, J. P., *Evolution* 22, 42 (1968).

8. Bachmann, K., *Chromosoma* 29, 365 (1970).

9. Ullerich, F., *Chromosoma* 18, 316 (1966).

10. Ullerich, F., *Chromosoma* 30, 1 (1970).

11. Miller, L. and Brown, D. D., *Chromosoma* 28, 430 (1969).

12. Walker, P. M. B., *Nature* 219, 228 (1968).

13. Moreschalchi, A., *Boll. Zool.* 37, 1 (1970).

14. Alvarez, M. and Cowden, R. R., *Z. Zellforschg.* 75, 240 (1966).

15. Bachmann, K. and Cowden, R. R., *Chromosoma* 17, 181 (1965).

16. Vialli, M., *Exp. Cell Res.* Suppl. 4, 284 (1957).

17. Commoner, B., *Nature* 202, 960 (1964).

18. Baetcke, K. P., Sparrow, A. H., Nauman, C. H. and Schwemmer, S. H., *Proc. Natl. Acad. Sci.* 58, 533 (1967).

19. Holm-Hansen, O., *Science* 163, 87 (1969).

20. Pedersen, R. A., Thesis, Yale (1970).

21. Britten, R. J. and Davidson, E. H., Science 165, 349 (1969).

22. Szarski, H., Nature 226, 651 (1970).

23. Van't Hof, J., Exp. Cell Res. 39, 48 (1965).

24. H. A. Brown, cited in Bachmann, K., Amer. Naturalist 103, 115 (1969).

25. Macgregor, H. C. and Kezer, J., Chromosoma 29, 189 (1970).

26. Goin, O. B. and Goin, C. J., Amer. Midland Naturalist 80, 289 (1968).

27. Bier, K. and Müller, W., Biol. Zentralbl. 88, 425 (1969).

28. Ebeling, A. W., Atkin, N. B. and Setzer, P. Y., Amer. Naturalist, in press.

29. Romer, A. S., Science 158, 1629 (1967).

30. Ohno, S., Wolf, U. and Atkin, N. B., Hereditas 59, 169 (1968).

31. Robinson, E. S. and Potter, I. C., Copeia 1969, 824 (1969).

32. Romer, A. S., Vertebrate Palaeontology (Chicago, 1966).

33. Jarvik, E., in Current Problems of Lower Vertebrate Phylogeny (Nobel Symposium 4), Ørvig, T., editor (1968).

34. Stensiö, E. A., Triassic Fishes from Spitzbergen (Vienna, 1921).

35. Schaeffer, B., in Current Problems of Lower Vertebrate Phylogeny (Nobel Symposium 4), Ørvig, T., editor (1968).

36. Nelson, G. J., Bull. Amer. Mus. Nat. Hist. 141, 477 (1969).

37. Denison, R. H., in Current Problems of Lower Vertebrate Phylogeny (Nobel Symposium 4), Ørvig, T., editor (1968).

38. Parsons, T. S. and Williams, E. E., Quart. Rev. Biol. 38, 26 (1963).

39. Brough, M. C. and Brough, J., Phil. Trans. Roy. Soc. London B 252, 107 (1967).

40. Schmalhausen, I. I., The Origin of Terrestrial Vertebrates (Academic Press, New York, 1968).

41. Carroll, R. L., Biol. Rev. 44, 393 (1969).

42. Reig, O. A., Bull. Mus. Comp. Zool. 139, 229 (1970).

43. Cole, C. J., Lowe, C. H. and Wright, J. W., Science 155, 1028 (1967).

44. Gorman, G. C. and Atkins, L., Copeia 1968, 159 (1968).

45. Hotton, N., North Am. Paleont. Convention, Chicago, Proc. H, 1146 (1970).

DISCUSSION

MELTON: I would like to reiterate a suggestion I made earlier in
this symposium that there may be a basic genome size for all vertebrates,
whereas you have offered a slightly different suggestion that each vertebrate class has its own characteristic minimal DNA value not the same
for all vertebrates. Yet, the spadefoot toad (which you mentioned) has
a genome size comparable to that of birds.

BACHMANN: Right. I prefer a more dynamic model of genome evolution, though. The "minimal genome" for any group may well be an
abstraction that either can't be defined precisely at all or is only a
tiny fraction of the DNA of any real living species. The smallest
genomes observed in any group in specialized forms may be more similar
to a medium-sized genome in the same group than to each other since we
think of genome evolution during specialization as irreversible loss of
parts of the genome not needed for the special adaptations of the
animal.

RICHMOND: You mentioned that there was evidence for evolutionary
"bursts" in polyploidy. Do you wish to speculate as to why or how
these increases occurred?

BACHMANN: Rather than speaking of evolutionary bursts in polyploidy, I would speak of polyploidy as an elegant way of achieving
bursts in the DNA amount. I don't want to speculate here on the
details of such an event. Perhaps you will allow me to refer you to
Ohno's new book.

MOSES: Is there any broad correlation (among groups) between DNA
content and chromosome number? In other words, if you treat chromosome
numbers as you have DNA values, can you show characteristics for each
group that may bear a relationship to DNA content? The point is that
if, as you have indicated, groups with smaller DNA amounts tend to be

more highly specialized, their larger numbers of chromosomes may also be an influential factor.

BACHMANN: As a general rule for all vertebrates, there seems to be no correlation between nuclear DNA amount and chromosome number. Both can vary rather freely and independently within limits. In many specific cases, though, clear correlations can be found. Reduction in chromosome number has been correlated with evolutionary advancement in some vertebrate groups, other advanced groups have large amounts of microchromosomes. Hinegardner has recently looked at DNA amounts, chromosome number and specialization in teleost fishes. Maybe he would like to comment on his results.

HINEGARDNER: In teleosts there is an overall increase in DNA with chromosome numbers. The correlation is significant at the 0.01% level. However, there is a wide spread of values.

GOIN: Rearrangements of chromosome numbers occur rather freely even in advanced forms. Lizards of the genus Cnemidophorus, for instance, show number changes by combining two acrocentrics into one metacentric or vice versa.

VERNICK: Does the relationship between specialization and DNA content described in vertebrates apply also to invertebrates?

BACHMANN: As far as we can tell from the data, yes. Hardly anything is known about invertebrate DNA, but insects seem to show a trend from high DNA values in generalized orders (Orthoptera) to low DNA values in advanced orders (Diptera, Coleoptera). Hinegardner has more invertebrate DNA data than anybody else and he sees this trend in all groups except in snails, if I understand right.

GALINSKY: Size difference and rate of development are dangerous criteria to use to correlate with differences in amount of DNA between groups. Dogs vary tremendously in size and growth rate but not in amount of DNA. It is important to keep quality and quantity of DNA in proper perspective.

BACHMANN: You are absolutely right. Pointing out correlations between DNA amount and any physiological parameter must not obscure the fact that all these parameters are very much dependent on the qualitative influence of the DNA. That is what "genetics of quantitative characters" is all about. The point that I am trying to make is that, amazingly, we can detect a quantitative influence of DNA in addition to and in spite of the qualitative determinants. Incidentally, body size in vertebrates, as I showed with the toads, has nothing to do with DNA amount and we don't know much yet about growth rate, just rate of embryonic development.

A SURVEY OF DNA CONTENT PER CELL AND PER CHROMOSOME

OF PROKARYOTIC AND EUKARYOTIC ORGANISMS:

SOME EVOLUTIONARY CONSIDERATIONS

A. H. Sparrow, H. J. Price, and A. G. Underbrink[*]

Biology Department, Brookhaven National Laboratory, Upton, N. Y. 11973

Abstract. A literature survey was made of DNA values for various
taxonomic groups of organisms. In addition, data are included representing the limits of variability found in DNA values estimated from
nuclear volume measurements of nearly 1000 plant species. Trends of
increasing DNA content with advancing organismic complexity exist
within prokaryotes and within eukaryotes, but cases of decreasing DNA
with advancing complexity or evolutionary specialization have apparently occurred in higher vertebrates and vascular plants. Large
differences in DNA content exist among species of comparable complexity
that can not be explained wholly by different levels of polyteny,
polyploidy or aneuploidy. In many instances, these variations probably
result from differences in the extent of differential gene redundancy.
Apparently, the amounts of DNA per cell and per chromosome have been
influenced in both increasing and decreasing directions by strong
selective pressures (see Summary).

I. INTRODUCTION

The pioneering work by Mirsky and Ris[1] reporting apparent

evolutionary trends in DNA content per cell in animals, aroused great

interest concerning the possible evolutionary changes in DNA content

in various animal and plant taxa.[2-8] Many more DNA values from

both prokaryotic and eukaryotic organisms have since been reported.

In this paper extensive data on DNA content per cell and the

average DNA content per chromosome are compiled, and the variations

Research carried out at Brookhaven National Laboratory under the
auspices of the U. S. Atomic Energy Commission.

[*]Present address: Department of Radiology, Radiological Research
Laboratories, Columbia University, New York, N. Y. 10032.

in these parameters within and among well recognized taxonomic groups are discussed in an evolutionary context.

II. MATERIALS AND METHODS

The nucleic acid contents per nucleus (or cell) of the organisms presented were obtained from the literature or from data collected in this laboratory. All nucleic acid values are presented as nucleotides x 10^n. In converting weight values, 1 gram of nucleic acid was estimated to contain 2.01×10^{21} nucleotides. For consistency in comparing haploid and diploid organisms, all DNA (or in some viruses, RNA) values are haploid (1n or gametic), except in the case of some protozoans which are bi- or multinucleated and/or endopolyploid. Since most of the data are from somatic tissue, the "haploid" value is one-half the somatic value.

Since nucleic acid values are in most cases relatively difficult to obtain by chemical analyses, relationships between nuclear volume and DNA content developed in this laboratory for higher plants[9-12] were utilized to obtain estimates of plant DNA values. The techniques of slide preparation and of obtaining nuclear volume measurements are given elsewhere.[9]   These relationships show that nuclear volume (NV) and interphase chromosome volume (ICV) are directly proportional to DNA content per cell and per chromosome, respectively. Therefore, when the nuclear volume of meristematic cells is known, an estimate of DNA content can be made. A similar relationship appears to hold also for organisms other than plants.[13]   ICV is obtained by dividing the average volume of the interphase nucleus by the somatic chromosome number and represents the volume occupied by an average chromosome at interphase, neglecting other nuclear components such as the nucleolus. ICV is useful in comparing average chromosome size among organisms. Using data from Baetcke et al.[12]   and Underbrink et al.,[13]

Conger[14]   has calculated the average volume for metaphase chromosomes
in eukaryotes to be about 0.35 to 0.38 times the ICV.

The data on ICVs and DNA values were originally obtained because
of a suspected relationship between these parameters and radiosensitivity.
It has now been clearly demonstrated that both ICV and DNA per chromosome are inversely related to radiosensitivity for a range of organisms
widely distributed taxonomically, as well as for higher plants.[9-13,15-17]

The approximate DNA content estimated from the nuclear volume is
usually reliable for meristematic cells of higher plants, but for the
various tissues of ferns and lower plants it is only tentative although
reasonable. The various factors influencing nuclear volume have been
discussed elsewhere.[15]

III. RESULTS AND DISCUSSION

Nucleic acid parameters used in this study are listed in Tables
1 to 13. The range of each parameter within major taxa is graphically
presented in Figure 1. The nuclear parameters of the various categories
of organisms are discussed below.

A. A Survey of DNA Content Per Cell and Per Chromosome of Major Taxa

1. Prokaryotes. The viruses are biochemically and structurally
the simplest life form and correspondingly have the lowest nucleic
acid contents (Table 1, Fig. 1). The lowest nucleic acid content is
1.3×10^3 nucleotides for the RNA containing tobacco necrosis.
satellite virus and the highest viral nucleic acid content is from the
double-stranded DNA pox viruses (5.3×10^5 nucleotides)--a range of
about 400-fold. The simplest viruses are considered to have about 3
to 5 genes.[18]

The bacteria all have more DNA than any known virus, but less
than most eukaryotic cells (Fig. 1). The DNA of bacterial cells is

in the form of a single DNA double helix and, therefore, the DNA per
cell (neglecting episomes) is equivalent to the DNA per chromosome.
The upper limit of bacterial DNA content overlaps the DNA contents per
cell only of the fungi and algae at their lower values. The algal
values overlap because of the low DNA content of the prokaryotic blue-
green alga Anacystis. However, not all prokaryotic algae have low DNA
contents; two species of Oscillatoria have at least 100 times as much
DNA per cell as do bacteria (Table 4).

Although most eukaryotes contain more DNA per cell than bacteria,
the smaller chromosomes of many higher taxa (fungi, algae, Arthropoda,
Urochordata, Agnatha, Teleostei, Aves) have on the average no more DNA
than do the largest bacterial cells (Fig. 1). Apparently the evolution
of DNA values per cell higher than ca. 6×10^7 nucleotides (the largest
bacterial value) was generally concomitant with the evolution of
multiple chromosome systems and their accessory achromatic structures.

2. Mitochrondria and Chloroplasts. In view of recent speculations
concerning the origin of mitochondria and chloroplasts from prokaryotic
organisms,[19, 20]   the DNA values for these organelles have been included (Table 2, Fig. 1b). The values for mitochrondria are similar to
those for larger 2-S DNA viruses. However, chloroplasts may have
appreciably more DNA and approach the maximum values found for bacteria
and the minimum for fungi and exceed the minimum for algae. Although
chloroplasts from only seven different species have been included, the
approximate 100-fold variation in DNA content exceeds that found in
most but not all of the large taxonomic groups so far studied.

3. Fungi. Most of the fungi studied tend to be characterized
by relatively low DNA contents, small nuclei, and very small chromosomes (Table 3, Fig. 1b, also see refs. 21,22). In addition to the

low DNA values, these small nuclei and chromosomes may be due partially to the reported lack of nuclear histones in some fungi.[23,24]

Saccharomyces cerevisiae, a yeast, has the lowest haploid DNA content (4.8×10^7 nucleotides) of any eukaryotic cell so far investigated, and apparently has only a very small repetitive DNA fraction.[25]   This suggests that the minimum amount of DNA needed to code for the structure and metabolism of eukaryotic cells is about 4 to 5×10^7 nucleotides. If the average gene is estimated to be 1500 base pairs,[18]   this is enough DNA to code for at least 13,000 genes.

4. Algae. Data for the algae show a range in DNA content extending over more than four orders of magnitude. One algal group with high DNA content is the Dinophyceae (Table 4). This group includes Gonyaulax polyedra which has a DNA content per cell exceeded only by that of certain Protozoa (Table 8) and Psilopsida (Table 5). The dinoflagellates also have some unique nuclear characteristics. The chromosomes are large permanently condensed, elongate structures lacking centromeres, RNA, histone and residual protein.[26-31]   The presence of a nuclear membrane but the lack of spindle fibers coupled with the above characteristics give this group a nuclear organization intermediate between prokaryotes and eukaryotes.

The DNA contents of the five species of green algae are between 8.0×10^7 and 7.0×10^9 nucleotides per cell (Fig. 1b, Table 4), therefore ranging over a factor of about 88. DNA values for other algal groups such as Rhodophyceae and Phaeophyceae are very much needed for a better understanding of any evolutionary significance of the large variability in DNA content in the algae.

5. Protozoa. The DNA content per cell found in Protozoa also extends over slightly more than four orders of magnitude (Table 8,

Fig. 1a). Polyploidy and multiple nuclei, however, are apparently responsible for a large amount of the variability.

The ciliates are characterized by a macronucleus and a varying number of micronuclei.[29] Although the micronucleus may be polyploid in some paramecia,[32] the macronucleus is always much larger and has been reported to be up to 860-ploid.[33,34] Woodard et al.[35] reported that _Tetrahymena pyriformis_ has 21.29 picograms (4.3 x 10^{10} nucleotides) of DNA per macronucleus and 0.86 picograms (1.7 x 10^{9} nucleotides) of DNA per micronucleus. The polyploidy of the ciliate macronucleus is achieved by endomitosis.[29]

Amoebae with high DNA contents per cell are _Chaos chaos_, _Amoeba proteus_, and _A. dubius_ (Table 8). _C. chaos_ is polynucleate[36,37] and _A. proteus_ is a polyploid having up to 500 chromosomes.[29] In addition, _A. proteus_ feeds on microorganisms and may have numerous symbionts living in its cytoplasm,[38] the DNA of which could contribute to the high DNA values per cell reported.

6. _Multicellular Invertebrates._ The multicellular invertebrates are poorly represented in number but the majority of those sampled have DNA contents similar to those of vertebrates (Table 9, Fig. 1a). The range in DNA content from the primitive sponge values to the highest arthropod value extends over two orders of magnitude (Fig. 1a). Mirsky and Ris[1] interpreted the lower DNA values found in sponges and coelenterates compared to values in echinoderms, crustaceans, and molluscs to represent a slight trend of increasing DNA content with advancing complexity. This trend was supported by the fact that the highly evolved squid has more DNA than the other more primitive molluscs.[1] The additional data from a nematode, five echinoderms, and three molluscs generally support this view, but the low values of some arthropods do not.

The data on DNA per chromosome so far available for multicellular invertebrates are similar to those of the vertebrates excluding urodele amphibians (Fig. 1a).

7. Chordata. DNA contents and phylogenetic trends are better known in the Chordata than for any other animal group. However, the number of DNA values for some groups, e.g., marsupials, monotremes, and some of the primitive taxa are highly inadequate.

The increasing DNA contents observed when comparing tunicates, the cephalochordate, and agnathans (Table 10, Fig. 1a) suggest that a large increase in DNA content (at least 6-fold) may have accompanied the evolution of the early vertebrates.[4,5] However, since very few species are represented in these groups, their true range of DNA values may not be known.

The line of vertebrate evolution that ended in teleost fish apparently acquired no large increase in nuclear DNA (Fig. 1a), and Hinegardner[3] suggested that a decrease in DNA content accompanied evolution and specialization within teleost fish. He observed that members of advanced fish families generally have less DNA than members of primitive ones.

A massive increase in DNA content per cell and per chromosome apparently took place in the ancestral vertebrate line from which dipnoans and the terrestrial vertebrates evolved.[1,4,5] The dipnoans have eleven times or more DNA per cell and at least thirty times more DNA per chromosome than do the other fishes surveyed. Very high DNA contents are also present in the primitive amphibian groups Apoda and Urodela (Table 1, Fig. 1a).[1,4,5,39] A reverse trend toward decreasing DNA content with evolutionary advancement is seen in the higher or more complex land vertebrates (Fig. 1a). The more specialized amphibians, the Anura, have less DNA per cell[4,39] and per chromosome than do the more primitive amphibians. At least a six-fold decrease in DNA content per cell is evident (Fig. 1a) in the amphibian-reptile-bird phylogenetic sequence.[1,4]

Unlike the apparent decrease in DNA values during the evolution of birds from reptiles, the evolution of mammals resulted in DNA levels similar to or slightly higher than those found in living reptiles (Fig. 1a).[4] Both birds and mammals have a restricted range of DNA content which seems to indicate that evolution within each group did not drastically affect the amount of DNA per nucleus.[40]

As seen in Tables 11 to 13, reptiles, birds and mammals generally have higher chromosome numbers than amphibians. The lower DNA contents in birds combined with their high chromosome numbers give rise to their low average DNA per chromosome (Table 12, Fig. 1a). Within the chordates equivalent or lower chromosomal DNA values are found only in fishes and some primitive forms. The low DNA per chromosome of some birds, fishes and lower chordates overlaps the higher values of bacteria (Fig. 1a).

Certain reptiles and many birds have several sets of macro- and many pairs of minute microchromosomes, $< 1 \mu$ long.[40-42] Such extreme bimodality of chromosome size is infrequent in other taxa but does occur in some holocephalian, chondrostean, and holostean fishes[43] and in some plants.[15]

For further discussions concerning evolution of DNA content in the vertebrates see Hinegardner,[3] Goin and Goin,[4] Ohno,[5] and Bachmann et al.[44]

8. _Terrestrial Plants_. It has been postulated that the land plants evolved from a green alga with isomorphic alternation of generations, that the primitive land plants probably had similar gametophytic and sporophytic generations, and that evolution proceeded in two

directions from this form.[45] One branch led to the bryophytes with a reduced sporophytic generation, and the other led to vascular plants with a dominant sporophytic generation.

a. Bryophyta.   Only four species of bryophytes have been studied and their estimated DNA contents per cell are relatively low but overlap the minimum values for vascular plants (Table 5, Fig. 1b). The DNA per chromosome is in the lower range for land plants (Fig. 1b). The minimum DNA per cell falls within, but the upper values exceed, the upper limit found in the green algae (Fig. 1b). Whether increasing DNA content is correlated with advancing complexity in comparing bryophytes with green algae will be known only when additional DNA values are available for both groups, but at present such a trend seems improbable.

b. Pteridophyta.   Psilopsida is clearly the most primitive vascular plant group. Two orders are recognized, Psilophytales and Psilotales.[46] The Psilophytales is an extinct order from which the Psilotales including Psilotum and Tmesipteris have descended. It is now generally agreed that the Psilophytales are ancestral or near-ancestral to all other vascular plant groups.[45] The data in Table 5 indicate that the living Psilopsida species examined have very large nuclear volumes and fairly large ICVs and, therefore, should have high DNA contents per nucleus and per chromosome. These species are, however, high polyploids (8x and 16x) with a base number x = 13.[47] Even if the estimated DNA contents per nucleus (Table 5) are considered on a genome basis (3.2 to 3.9 x 10^{10} nucleotides per genome), they are still relatively high compared with that of other vascular plants and more than three times that of any green alga or bryophyte so far studied (Fig. 1b). Although nuclear parameters are not known for extinct Psilopsida, it seems likely that a large increase in DNA content per nucleus and per chromosome may have accompanied the evolution of the first vascular plants, although it is possible that the increase occurred later.

The Lycopsida and Sphenopsida, two other ancient plant groups, arose during the Devonian, most likely from the Psilophytales. They reached a peak in the Paleozoic and rapidly declined in the Mesozoic.[48] Few genera are represented in the living flora. Except for Isoetes the species studied from both groups show estimated DNA contents per cell and per chromosome lower than the Psilopsida (Table 5, Fig. 1b). Our data, although very limited for these groups, suggest a loss of DNA with the evolution of these taxa from their psilophytalean ancestors. Stebbins[6] made a similar suggestion for the Lycopsida after comparing the nuclear size of the specialized Selaginella with those of the more primitive Lycopodium, Psilotum, and Tmesipteris.

The Pteropsida (ferns) which probably evolved from the Psilophytales through an intermediate group, the Primofilicales, include several thousand living species that are morphologically highly diversified.[46] Ferns are also cytologically highly variable in chromosome size and number with abundant polyploidy.[49-53] Relatively few species have been reported to have chromosome numbers less than n = 22, and some species such as Ophioglossum reticulatum L., n = 630, have very high numbers.[47] Ophioglossum petiolatum which has one of the highest chromosome numbers in plants (n = ca. 510), also has, as would be expected, a very large nuclear volume (NV = 4124 μ^3) and an estimated haploid DNA content of 2.8×10^{11} nucleotides (Table 5). This high value is exceeded appreciably only by a few other organisms (Fig. 1). The lowest estimated DNA per cell is that of hexaploid Salvinia (Table 5). Since diploid species of Salvinia have been reported to exist,[54,55] the lower limit of DNA content per cell is probably

lower by one-half to one-third than that shown in Figure 1b. Clearly
a wide range of DNA values exists in the extant fern species, but the
evolutionary trends are obscure.

c. Spermatophyta.

(i) Gymnospermae. The gymnosperms as a group have
relatively large nuclear volumes, high DNA contents per cell and per
chromosome and large ICVs (Table 6). The gymnosperms are cytologically
more stable with respect to chromosome number than either the ferns or
angiosperms and relatively few cases of polyploidy have been reported.[56]
The order Coniferales, except for the family Podocarpaceae,[57] is
extremely cytogenetically stable with chromosome numbers mostly n = 11,
12 and 13.[58]

Paleobotanical and morphological evidence suggests that the
gymnosperms are monophyletic in origin and descended from progymno-
sperms which probably evolved from psilophyte ancestry.[59,60] The
modern conifer families probably evolved from the Cordaitales through
now extinct transition forms.[61] The Pinaceae species have much higher
DNA values when compared to members of the other conifer families.[2,17,62]
However, since data on DNA content are lacking for fossil conifers, it
is not known whether the values for Pinaceae represent an increase or
the values in other families represent decreases in DNA content, or
whether both occurred in the evolution of conifer families.

(ii) Angiospermae. The angiosperms are the most recent,
most successful and most advanced plant group. Their immediate
ancestors are extinct and unknown, but they may have evolved from the
Lyginopteridales (pteridosperms), an extinct group of gymnosperms.[63]
They are cytologically much more variable than the gymnosperms and
polyploidy is abundant.[64] Our measurements of nuclear volumes, ICVs
(ca. 800 species; unpublished data), and chromosome numbers further

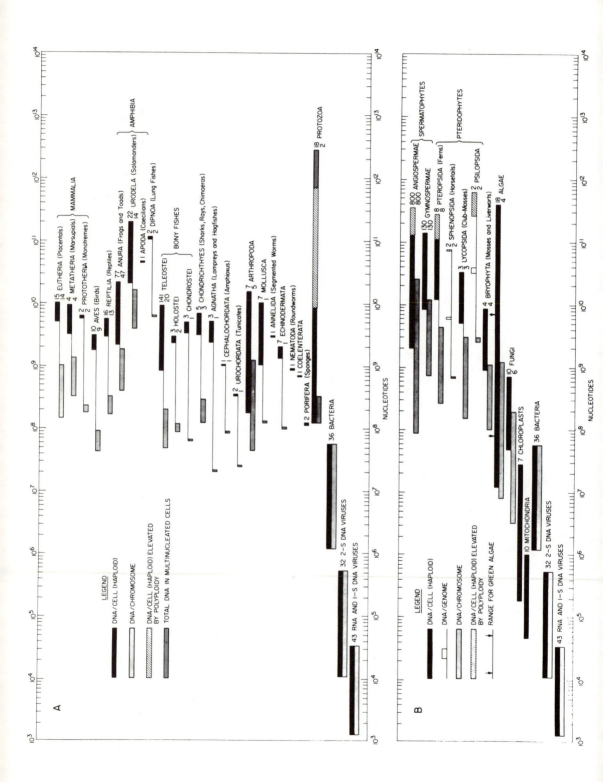

document the highly variable nature of these parameters in this group

(Table 7).

The lower limit of DNA content is less for angiosperms than that

of any other vascular plant group, and the higher estimated limit

represented by Sprekelia formosissima (Table 7) is exceeded only by

Tmesipteris. The angiosperms also have the widest range of DNA per

chromosome of any vascular plant group (Fig. 1b). Our extensive

unpublished data on nuclear volumes suggest that the majority of angio-

sperms investigated should have less DNA than most gymnosperms and

probably less than most ferns. This suggests that a reduction in

nuclear DNA has accompanied at least some angiosperm evolution.

Detailed analyses of nuclear parameters in angiosperm taxa are in

progress.

B. Relationship of DNA Content to Organismic Complexity

The significance of the large variation in DNA content per cell

among taxa is poorly understood. The data given in the preceding

section indicate a general trend in increasing DNA content with advanc-

ing complexity. They also show that the ranges of DNA contents are

wide within as well as among many major taxonomic categories (Fig. 1).

In view of this variation it seems illogical to assume that all the

differences in DNA content reflect differences in the number of unique

genes.[6,65] It is difficult to imagine, for instance, why some Amphibia

might need ten times as many individual genes as do the more highly

Figure 1. The ranges (horizontal logarithmic scale) of DNA (RNA for
some viruses) content per cell and per chromosome in major categories of
prokaryotic and eukaryotic organisms. The number of species represented
is to the right side of each entry. The vertical axes represent relative
evolutionary advancement. A. Ranges in nucleic acid values for viruses,
bacteria, and animals. B. Ranges in nucleic acid values for viruses,
bacteria, mitochondria, chloroplasts, and plants.

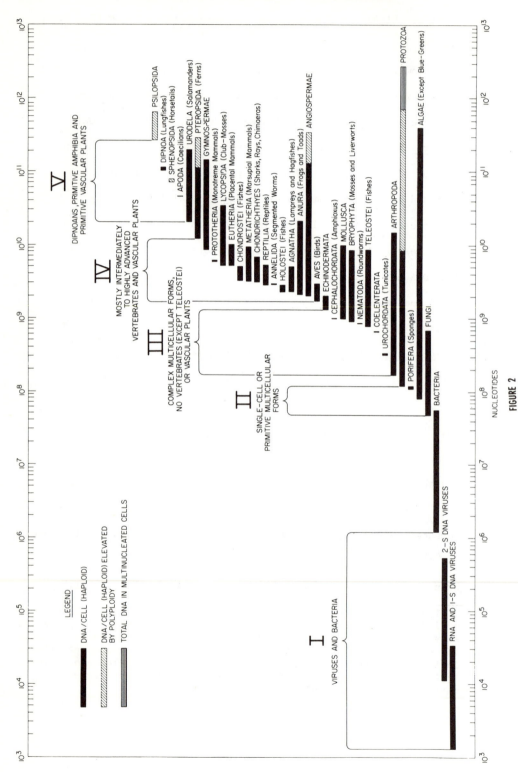

FIGURE 2

A. H. SPARROW, H. J. PRICE and A. G. UNDERBRINK

evolved and more structurally complex mammals or birds. It also seems

unlikely that Trillium luteum (Table 7) needs more than 60 and 1600

times as many unique genes respectively as does Arabidopsis (Table 7)

and the green alga Chlorella (Table 4). Therefore, to analyze

relationships between DNA content and structural complexity it seems

more prudent to compare the various major taxa on the basis of their

minimum DNA contents.[65]

Rearrangement of the taxa according to increasing minimum DNA per

cell (Fig. 2), independent of taxonomic or phylogenetic positions,

gives a more orderly picture than Figure 1. The taxa so arranged can

be grouped into five somewhat arbitrary divisions: I, viruses and

bacteria; II, single-celled and primitive multicellular forms; III,

taxa comprised of complex multicellular forms but no vascular plants

or vertebrates other than teleost fishes; IV, largely intermediate to

highly complex vertebrates and vascular plants; V, lungfishes, primitive

amphibians, and primitive vascular plants. The first four divisions

taken as groups suggest that the minimum amount of DNA per cell is

generally greater in the highly complex groups than in the less complex

groups, but many more data on DNA values are needed for poorly repre-

sented groups. The taxa of division V have minimum DNA contents per

cell much higher than might be expected by their relative structural

complexity, hence some other reason for these high values should be

sought.

C. Evolutionary Considerations

As discussed above, great differences in DNA content exist which

apparently cannot be explained by variation in the number of unique

genes. Polyploidy plays a role in the elevation of DNA content in some

---

Figure 2. The ranges of DNA (RNA for some viruses) content per
cell in major taxonomic categories arranged according to increasing
minimum DNA per cell.

groups, e.g. Protozoa, Pteridophyta and Angiospermae, but since large differences in DNA per chromosome and per cell exist quite independently of it, polyploidy alone clearly cannot be the basis of all the variability. However, polyploidy may be a pathway to the attainment of large nuclear volumes or duplicated gene copies which may have selective advantages under appropriate environmental conditions.

Stebbins[6,64] pointed out examples of plant families in which certain temperate genera have larger chromosomes and nuclei than tropical or subtroptical genera. Burley[66] found that the total length of the haploid chromosome complement of Picea sitchensis increased significantly and nuclear volume increased by a factor of 1.9 over 16 degrees of increasing latitude. Miksche's comparison of P. sitchensis seedlings from the southern to those of the northern parts of this range showed DNA increases per cell up to a factor of 1.9 in the northern representatives.[67]

Closely related diploid species having the same chromosome number often exhibit significant differences in chromosome size[64] and DNA content.[7,8] Rees and Hazarika[68] reported that the specialized inbreeding species of Lathyrus have less DNA per cell than the more primitive species. Variations in DNA content per nucleus have been found among species of Acrididae grasshoppers[69,70] and Amphibia[71] having the same chromosome numbers.

The above examples all suggest the existence of strong selective forces influencing the amount of DNA per nucleus.

Although the nature of the DNA selected for or selected against is not known, variation in the amount of redundant DNA sequences may account for much of the quantitative differences in DNA content observed in higher organisms.[6,65] Indeed, sequences of DNA from many higher organisms appear to be highly redundant.[25,72-74] Furthermore, the type of reiterated DNA has been shown to vary in related species

of Graminae,[75] Leguminosae,[76] and _Drosophila_;[77] the reiterated sequences

are more similar in closely than in distantly related species.

Robertson et al.[77] present evidence suggesting that the reiterated

fraction (ca. 10%) of the genome in _D. melanogaster_ and _D. simulans_ is

about half that in _D. funebris_. Bendich and McCarthy[75] proposed that

evolution of the more specialized species of the Graminae involved

loss of some repetitive sequences of DNA.

The functions of repetitive DNA are only partially understood.

Examples given below suggest that it serves a role of increasing the

number of gene copies coding for particular RNAs or proteins. A high

redundancy of gene copies coding for 28S and 18S ribosomal DNA exists

in _Drosophila melanogaster_[78,79] and _Xenopus laevis_.[80] In addition,

X. laevis has about 20,000 redundant cistrons for 5S rRNA[80] and _D._

melanogaster has a 13-fold redundancy in genes coding for each transfer

RNA.[81] An example of additional gene amplification is the differential

synthesis and release into the nucleoplasm of DNA complementary to rRNA

in some animal oocytes.[80,82]

Since polyteny exists in some tissues of insects and in the

suspensor cells of _Phaseolus_[83] and endopolyploidy occurs in tissues of

various animals[84-90] and plants,[91,92] it seems reasonable that other

types of mechanisms which increase the number of gene copies per

nucleus may be widespread. Differential tandem gene duplication may

be one of these mechanisms and genes other than those coding for

ribosomal and transfer RNAs may be highly redundant along the chromo-

somes. Indeed, Britten and Kohne[25] interpret the changing pattern of

hybridizable pulse-labeled RNA detected during differentiation of

amphibians,[93-96] sea urchins,[97,98] and mouse liver,[99] and during

liver regeneration[100] to be the expression of different families of

repeated sequences, since hybrids between RNA and nonrepeated DNA
sequences apprently did not occur in these experiments. In addition,
the DNA complements of 9S messenger RNA thought to code for histones
in sea urchins have been shown to be reiterated about 400-fold.[101]

Some reasons for variations in the extent of repetitive DNA and
consequent changes in the quantity of DNA within taxa can be postulated
in the following way: if a population of organisms is evolving in an
environment where its adaptability is hindered by a low production
of specific enzymes or proteins needed during a critical stage of its
members' life cycle, and messenger RNA is being transcribed and translated at its maximum rate, one way to increase protein production is
to increase the number of gene copies. Since enzyme activity has a
strong dependence on temperature (approximately doubling with each 10° C
rise up to an optimum) it could be one limiting factor with respect
to efficient growth at low temperatures in plants and cold-blooded
animals. Then, any organisms having duplications of rate-limiting genes
would have an immediate selective advantage, and the frequency of these
duplications in the population would be expected to increase in the
following generations. Unequal crossing-over and subsequent selection
could further increase the frequency and extent of redundant gene
copies. Duplications would be expected to accumulate until the gene
products in question were no longer limiting the adaptability of the
population.

Since the quantitative requirements for activity or amount of a
particular gene product probably vary with the environment, the extent
of redundancy of particular genes would be expected to decrease if it
were no longer of adaptive significance. Deletions of unneeded DNA
resulting from unequal crossing-over or other causes would be expected
to have an adaptive advantage for at least two reasons: 1) The cells

would be biochemically more efficient if nucleotides which are cofactors

in many biochemical reactions were not used to replicate surplus DNA,

for instance, the possible sequestering effects of DNA replication on

nucleotides has been discussed by Commoner;[102,103]   2) Since DNA

content and nuclear volume are proportional to mitotic cycle time in

diploid plants[10,104] and probably in animals,[39] the loss of function-

less DNA would permit a shorter mitotic cycle. This would be select-

ively advantageous to an organism undergoing selection for rapid

development, e.g., some annual plants growing under highly favorable

conditions of temperature and moisture,[6] or amphibians adapting to

temporary waters.[39]

Although admittedly oversimplified, the above discussion presents

ways in which some of the variations in DNA content per chromosome and

per nucleus that occurs so extensively in plants and animals can be

hypothetically explained in evolutionary terms. A better understanding

of the significance of the wide variation of DNA content in eukaryotic

cells will come when more is known about the organization of DNA in

the chromosome, regulatory mechanisms and processes, structural and

biochemical properties of eu- and heterochromatin, and the nature of

repetitive DNA sequences. However, much more data are needed particu-

larly for taxonomic groups now poorly represented either because of

paucity of species or a biased sampling of species.

IV. SUMMARY

An extensive literature survey was made of DNA values per chromo-

some and per cell for various taxonomic groups of prokaryotes and

eukaryotes. Listed, in addition, are data from plants representing

the limits of variability in DNA values estimated from nuclear volume

measurements of nearly 1000 plant species. The data considered in an

evolutionary context may be briefly summarized as follows:

1) Nucleic acid contents from the lowest viral RNA value to the
highest eukaryotic haploid nuclear value extend over eight orders of
magnitude (1.3×10^3 to 4.0×10^{11} nucleotides). However, nucleic acid
values in some cells may be higher due to multiple nuclei, endopolyploidy
or polyteny. With few exceptions, all prokaryotes have less DNA (or
RNA for some viruses) than eukaryotes. Average nucleic acid values per
chromosome extend over seven orders of magnitude (1.3×10^3 to $2.6 \times$
10^{10}). Small individual eukaryotic chromosomes have DNA contents
similar to those of larger bacteria. Apparently the evolution of DNA
values per nucleus higher than ca. 6×10^7 nucleotides was generally
concomitant with the appearance of multiple chromosome systems and
their accessory achromatic structures.

2) Within the multicellular invertebrates the range of DNA content
per nucleus extends over two orders of magnitude, and also indicates a
general trend of increasing DNA content with increasing complexity.
Likewise, in the chordate line at least a six-fold increase in nuclear
DNA apparently occurred during the evolution of the primitive vertebrates.
A further increase of at least 11-fold occurred in the vertebrate
line leading to lungfishes and primitive amphibians which have the
highest DNA values per haploid nucleus and per chromosome in the
animal kingdom. In higher vertebrates DNA decreases at least six-
fold with evolutionary advancement, from primitive amphibians (Urodela)
to birds, but increases slightly from reptiles to mammals.

3) A very strong trend of increasing DNA content with evolutionary
advancement exists when comparing major categories from viruses through
ferns. A large increase in DNA content apparently accompanied the
evolution of the first vascular plant group, the Psilopsida. Based
on the presumed genome values of living Psilopsida, further evolutionary

advancement of higher vascular plants was accompanied by both increases and decreases in DNA content.

4) Some of the increases in DNA content resulted from new genes added with advancing complexity. Large differences exist among species of comparable complexity, for instance, within algae, angiosperms, fishes and amphibians. This variation in DNA content cannot be wholly explained by different levels of polyteny, polyploidy or aneuploidy. In many cases, these variations probably do not reflect differences in the number of unique genes, but more likely are the result of differential gene redundancy.

5) It is speculated that redundancy may commonly exist in genes coding for proteins, and that increases in DNA content may represent, in part, the results of selection for duplications of genes whose products previously limited evolutionary adaptability. Evolutionary decreases in DNA values may represent losses of nonfunctional or over-reiterated DNA sequences which have lost their adaptive value. The data thus suggest that the amounts of DNA per cell and per chromosome have been influenced in both directions by strong selective pressures.

ACKNOWLEDGEMENTS

We thank Drs. H. H. Smith (BNL), A. D. Conger (Temple Univ.) and G. L. Stebbins (Univ. California, Davis) for suggestions concerning the manuscript. The authors also thank Dr. G. L. Stebbins for many valuable discussions. The assistance of Mrs. Anne Nauman, Miss Virginia Pond, Miss Susan S. Schwemmer, Mr. E. Eric Klug and Mr. C. Nauman is gratefully acknowledged.

Table 1. DNA (or RNA) content (nucleotides) per cell of viruses and bacteria

Organism	Nucleotides/cell	Ref.
RNA and 1-S DNA Viruses[†]		
Satellite virus of tobacco necrosis virus	1.3×10^3	13
Reovirus type 3	3.4×10^4	13
2-S DNA Viruses[‡]		
Simian vacuolating virus (SV40)	1.1×10^4	13
Pox virus	5.3×10^5	105,110
Bacteria		
Aerobacter aerogenes	5.7×10^7	13
A. aerogenes	4.0×10^6	106
Bacillus apiarius	1.3×10^7	106
B. cereus	2.6×10^7	106
B. medusa	4.2×10^7	106
B. megaterium	2.7×10^7	106
B. sotto	2.2×10^7	106
B. subtilis	1.3×10^7	13
B. thuringiensis	2.6×10^7	106
Bacterium aertrycke	2.4×10^7	13
B. lactis aerogenes	4.3×10^7	106
Chlamydia trachomatis	1.2×10^6	107
Diplococcus pneumoniae	4.0×10^6	13
Escherichia coli	9.0×10^6	107
E. coli B	1.3×10^7	13
E. coli B/r	3.4×10^7	13
Feline pneumonitis organism	4.7×10^6	13
Haemophilus influenzae	4.0×10^7	106
H. influenzae strain Rd4(3H)	3.2×10^6	13
Meningo pneumonitis organism	4.0×10^6	13
Micrococcus radiodurans	3.9×10^7	108
Mouse pneumonitis organism	5.1×10^6	13
Mycoplasm arthriditis	1.5×10^6	109
M. gallisepticum	4.0×10^6	109
M. hominis H39	1.7×10^6	109
M. laidlawii A	2.9×10^6	109
M. laidlawii B	2.7×10^6	109
Neisseria catarrhalis	4.6×10^6	107
N. flava	4.4×10^6	107
N. gonorrhoeae	3.0×10^6	107
N. meningitides	3.4×10^6	107
N. sicca	4.4×10^6	107
Proteus P18	3.7×10^7	106
P. P18 (L form)	1.2×10^6	106
Pseudomonas fluorescens	3.6×10^7	13
Rickettsia quintana	3.0×10^6	107
R. rickettsi	3.0×10^6	107
Salmonella typhimurium	2.6×10^7	13
Staphylococcus aureus	1.5×10^7	13

[†]The two values listed are the extreme values of 43 viruses taken from refs. 13 and 110.

[‡]The two values listed are the extreme values of 32 2-S DNA viruses taken from refs. 13 and 110.

Table 2. DNA content (nucleotides) of mitochondria and
chloroplasts

Source	DNA/mitochondrion	Ref.
Mitochondria		
Beef heart	5.2×10^4	111
Chicken liver	1.4×10^5	111
L cells (mouse fibroblasts)	1.8×10^5	111
Mung bean	1.0×10^6	111
Paramecium aurelia	7.4×10^5	112
Rat liver	$1.8-3.2 \times 10^5$	111
Sea urchin	4.6×10^4	111
Tetrahymena pyriformis	7.4×10^5	112
Turnip	1.0×10^6	111
Yeast	$1.2-7.4 \times 10^5$	111
	DNA/Chloroplast	
Chloroplasts		
Acetabularia mediterrranea	1.8×10^5	113
Antirrhinum majus	1.1×10^7	114
Brassica	1.0×10^6	115
Chlamydomonas reinhardi	2.0×10^7	116
Euglena gracilis	2.0×10^7	117
Ipomoea	1.0×10^6	115
Phaseolus vulgaris	1.0×10^6	115
	2.8×10^7	118

Table 3. DNA content (nucleotides) per cell and per chromosome of fungi

Organism	Chromosome no. (n)	Ref.	DNA/cell (haploid)	Ref.	DNA/chromosome
Ascobolus immersus	18	119	8.6×10^7	123	4.8×10^6
Aspergillus nidulans	8	120	8.8×10^7	123	1.1×10^7
A. sojae			1.5×10^8	123	
Blastocladiella emersonii			2.0×10^8	124	
Dictyostelium discoideum	7	121	7.0×10^8	121	1.0×10^8
Neurospora crassa	7	120	8.6×10^7	123	1.2×10^7
Ophiostoma multiannulatum			9.6×10^7	106	
Saccharomyces cerevisiae	15	13	4.8×10^7	123	3.2×10^6
Schizosaccharomyces pombe			4.8×10^7	123	
Ustilago maydis	2	122	3.8×10^8	123	1.9×10^8

Table 4. DNA content (nucleotides) per cell and per chromosome of algae

Organism	Chromosome no. (n)	Ref.	DNA/cell (haploid)	Ref.	DNA/chromosome
Bacillariophyceae					
Ditylum brightwelli			3.0×10^{10}	126	
Navicula pelliculosa			2.0×10^{8} †	126	
Skeletonema costatum			1.2×10^{9} †	126	
Thalassiosira fluviatilis			1.0×10^{10} †	126	
Chlorophyceae					
Chlamydomonas reinhardi	15	125	1.2×10^{8}	13	8.0×10^{6}
Chlorella ellipsoide			8.0×10^{7} †	127	
Dunaliella tertiolecta			1.2×10^{9} †	126	
Spirogyra sp.			5.8×10^{9}	128	
S. setiformis	ca.6	125	7.0×10^{9}	129	1.2×10^{9}
Chrysophyceae					
Monochrysis lutheri			2.0×10^{8} †	126	
Syracosphaera elongata			6.0×10^{9} †	126	
Cyanophyceae					
Anacystis nidulans			1.2×10^{7}	130	
Oscillatoria limosa			1.6×10^{10}	128	
O. princeps			1.7×10^{10}	129	
Dinophyceae					
Amphidinium carteri	25	28	6.0×10^{9} †	126	2.4×10^{8}
Chachonina niei			2.0×10^{10} †	126	
Gonyaulax polyedra			4.0×10^{11}	126	
Euglenophyceae (see Protozoa)					
E. gracilis	45	13	5.8×10^{9}	13	1.3×10^{8}

† Estimated from plotted data in ref. 126.

Table 5. NV, ICV and DNA content (nucleotides) per cell and per chromosome of bryophytes and pteridophytes

Organism	Chromosome no. (n)	Ref.	NV (μ^3)	Ref.	ICV (μ^3)	Ref.	DNA/cell (haploid)	Ref.	DNA/chromosome
Bryophyta (mosses and liverworts)									
Marchantia polymorpha (gametophyte)	9	13	13	13	1.4	13	9.0×10^8	128	1.0×10^8
							1.6×10^9	132	1.8×10^8
Mnium sp. (gametophyte, shoot apex)	ca.7	119	59	62	8.4	62	7.8×10^9	132	1.1×10^9
Riccia sp. (gametophyte, cells near apical notch)	8	62	64	62	8.0	62	8.6×10^9	132	1.1×10^9
Sphagnum sp. (gametophyte, central meristem of shoot apex)	19	131	20	62	1.1	62	2.6×10^9	132	1.4×10^8
Psilopsida									
Psilotum nudum (shoot apex, apical cell derivatives)	104	62	3844	62	18.5	62	2.6×10^{11}	132	2.5×10^9
Tmesipteris sp. (shoot apex, apical cell derivatives)	208	47	9366	62	22.5	62	6.3×10^{11}	132	3.0×10^9
Lycopsida (club mosses)									
Isoetes engelmannii (shoot apex)	11	131	498	62	22.6	62	3.3×10^{10}	132	3.0×10^9
Lycopodium lucidulum (shoot apex)	132	47	296	62	1.1	62	2.0×10^{10}	132	1.5×10^8
Selaginella kraussiana var. brownii (shoot apex)	10	119	77	62	3.8	62	5.1×10^9	132	5.1×10^8
Sphenopsida (horsetails)									
Equisetum arvense (nonapical cells of shoot apex)	108	47	1124	62	5.2	62	7.5×10^{10}	132	6.9×10^8
E. hyemale (nonapical cells of shoot apex)	108	47	1060	62	4.9	62	7.1×10^{10}	132	6.6×10^8
Pteropsida (ferns)									
Asplenium nidus[+]	144	47	280	62	1.9	62	3.7×10^{10}	132	2.6×10^8
Cystopteris fragilis[+]	84	47	447	62	5.3	62	6.0×10^{10}	132	7.1×10^8
Onoclea sensibilis[+]	37	62	842	62	22.8	62	1.1×10^{11}	132	3.0×10^9
Ophioglossum petiolatum (nonapical root cells)	ca.510	47	4124	62	4.0	62	2.8×10^{11}	132	5.5×10^8
Osmunda cinnamomea[+]	22	47	724	62	32.9	62	9.6×10^{10}	132	4.4×10^9
Pteridium aquilinum[+]	52	49	316	62	6.1	62	$\sim 4.4 \times 10^{10}$	132	8.5×10^8
Pteris longifolia[+]	58	62	451	62	7.8	62	6.0×10^{10}	132	1.0×10^9
Rumohra adiantiformis[+]	41	47	335	62	8.2	62	4.5×10^{10}	132	1.1×10^9
Salvinia sp. (apical cells)	27	62	176	62	3.2	62	1.2×10^{10}	132	4.4×10^8

[+] Gametophyte, 3-cell stage.

Table 6. NV, ICV and DNA content (nucleotides) per cell and per chromosome of gymnosperms

Organism	Chromosome no. (n)	Ref.	NV (μ^3)	Ref.	ICV (μ^3)	Ref.	DNA/cell (haploid)	Ref.	DNA/chromosome
Ephedra fragilis†	7	17	126	17	9.0	17	8.4×10^9	132	1.2×10^9
Juniperus virginiana	11	58	494	2	22.4	2	5.4×10^{10}	2	4.9×10^9
Larix laricina	12	2	840	2	35.0	2	6.0×10^{10}	2	5.0×10^9
Pherosphaera hookeriana†	13	17	140	17	5.4	17	9.4×10^9	132	7.2×10^8
Picea glauca	12	58	973	2	40.5	2	6.0×10^{10}	2	5.0×10^9
P. mariana	12	58	654	2	27.2	2	6.8×10^{10}	2	5.7×10^9
Pinus banksiana	12	58	1356	2	56.5	2	6.9×10^{10}	2	5.8×10^9
P. contorta	12	58	1861	2	77.5	2	7.5×10^{10}	2	6.2×10^9
P. resinosa	12	58	1907	2	79.4	2	1.4×10^{11}	2	1.2×10^{10}
P. rigida†	12	17	1516	17	63.2	17	1.0×10^{11}	132	8.3×10^9
P. strobus	12	58	1566	2	65.2	2	8.4×10^{10}	2	7.0×10^9
Thuja occidentalis	11	58	169	2	7.7	2	3.9×10^{10}	2	3.5×10^9
Tsuga canadensis	12	58	1413	2	58.9	2	7.6×10^{10}	2	6.3×10^9

†These three species represent the extremes in NV and ICV of 120 gymnosperm species reported in ref. 17.

Table 7. NV, ICV and DNA content (nucleotides) per cell and per chromosome of angiosperms[†]

Organism	Chromosome no. (n)	Ref.	NV (μ^3)	Ref.	ICV (μ^3)	Ref.	DNA/cell (haploid)	Ref.	DNA/chromosome
Allium cepa 'Excel'	8	12	628	12	39.2	12	5.4×10^{10}	12	6.8×10^9
Apocynum cannabinum	11	133	49	62	2.2	62	3.3×10^9	132	3.0×10^8
Arabidopsis thaliana	5	134	30	62	3.0	62	2.0×10^9	132	4.0×10^8
Brassica hirta	12	134	43	62	1.8	62	2.9×10^9	132	2.4×10^8
Chrysanthemum sp.	18	12					1.0×10^{10}	12	5.6×10^8
Chrysanthemum sp. (Diener's hybrid)	69	12	1396	12	10.1	12	7.9×10^{10}	12	1.1×10^9
C. lacustre	99	12	2247	12	11.3	12	1.4×10^{11}	12	1.4×10^9
C. nipponicum	9	12	318	12	17.7	12	4.4×10^{10}	12	4.9×10^9
C. yezoense	28	12	534	12	9.5	12	2.9×10^{10}	12	1.0×10^9
Clematis jackmannii	8	12	293	12	18.3	12	2.2×10^{10}	12	2.8×10^9
Gladiolus sp. 'Mansoer'	ca.30	12	393	12	6.6	12	6.0×10^9	12	2.0×10^8
Haemanthus katherinae	9	62	1872	62	10.4	62	1.2×10^{11}	132	1.3×10^{10}
Helianthus annuus 'Mammoth Russian'	17	12	353	12	10.4	12	1.6×10^{10}	12	9.4×10^8
Kalanchoe daigremontiana	17	12	67	12	2.0	12	1.1×10^{10}	12	6.5×10^8
Lilium longiflorum	24	12	3003	12	62.6	12	1.8×10^{11}	12	7.5×10^9
L. longiflorum 'Croft'	12	12	1278	12	53.2	12	1.1×10^{11}	12	9.2×10^9
L. pumilum	12	62	1760	62	73.3	62	1.2×10^{11}	132	1.0×10^{10}
Lycoris squamigera	3x=27	12	2069	12	76.6	12	8.6×10^{10}	12	9.6×10^9
Narcissus tazetta 'Paper White'	11	12					3.1×10^{10}	12	2.8×10^9
Nigella damascena	6	12	480	12	40.0	12	2.8×10^{10}	12	4.7×10^9
Raphanus sativus 'Cherry Belle'	9	12	91	12	5.1	12	5.0×10^9	12	5.6×10^8
Rumex longifolius	30	12	233	12	3.9	12	1.3×10^{10}	12	4.3×10^8
R. obtusifolius	20	12	154	12	3.8	12	8.0×10^9	12	4.0×10^8
R. sanguineus	10	12	106	12	5.3	12	5.0×10^9	12	5.0×10^8
R. stenophyllus	30	12	165	12	2.8	12	1.2×10^{10}	12	4.0×10^8
Scilla sibirica 'Alba'	6	12	995	12	82.9	12	7.3×10^{10}	12	1.2×10^{10}
Sedum album	34	135	50	62	0.7	62	3.3×10^9	132	9.7×10^7
S. alfredi 'Nagasakianum'	64	134	83	62	0.6	62	5.6×10^9	132	8.7×10^7
Sprekelia formosissima	ca.60	62	5292	62	44.1	62	3.5×10^{11}	132	5.8×10^9
Tradescantia sp. (clone 02)	6	12	634	12	52.8	12	5.9×10^{10}	12	9.8×10^9
T. blossfeldiana	36	12	538	12	7.5	12	4.0×10^{10}	12	1.1×10^9
T. paludosa	12	12	1330	12	55.4	12	1.2×10^{11}	12	1.0×10^{10}
T. paludosa (clone B2-2)	6	12	657	12	54.8	12	5.4×10^{10}	12	9.0×10^9
T. virginiana 'Purple Dome'	12	12	1381	12	57.5	12	1.2×10^{11}	12	1.0×10^{10}
Trillium luteum	5	133	1936	62	19.4	62	1.3×10^{11}	132	2.6×10^{10}
Tropaeolum majus	14	12	177	12	6.3	12	1.1×10^{10}	12	7.8×10^8
Tulipa sp. 'Golden Harvest'	12	12	1435	12	59.8	12	7.2×10^{10}	12	6.0×10^9
Vicia angustifolia	6	12	171	12	14.2	12	1.0×10^{10}	12	1.7×10^9
V. faba 'Sutton's Prolific Longpod'	6	12	585	12	48.8	12	4.4×10^{10}	12	7.3×10^9

[†]NVs and ICVs are representative data including the lower and upper limits of over 800 species studied in this laboratory.

Table 8. DNA content (nucleotides) per cell and per chromosome of protozoa

Organism	Chromosome no. (n)	Ref.	DNA/cell (haploid)	Ref.	DNA/chromosome
Ciliata					
Paramecium aurelia stock 51			3.2×10^{11}[†]	34	
P. caudatum			8.0×10^{11}[†]	106	
Tetrahymena pyriformis					
strain GL			2.7×10^{10}[†]	136	
strain W			6.1×10^{9}[†]	137	
Urostyla caudata			1.6×10^{12}[†]	138	
Flagellata					
Astasia longa			3.0×10^{9}	136	
Euglena bacillaris (5 strains)			$5.0\text{-}8.4 \times 10^{9}$	136	
Euglena gracilis	45	13	5.8×10^{9}	13	1.3×10^{8}
strain Z			$4.2\text{-}9.2 \times 10^{9}$	136	
Trichomonas gallinae			$6.0\text{-}8.0 \times 10^{8}$	136	
strain JB and YG					
T. vaginalis C_1			1.0×10^{9}	136	
Trypanosoma cruzi			3.4×10^{8}	139	
T. equiperdum			1.5×10^{8}	139	
T. evansi			4.0×10^{8}	136	
T. gambiense			1.5×10^{8}	139	
Sarcodina (amoebae)					
Amoeba dubias			7.0×10^{11}	36	
A. proteus	ca.250	29	8.0×10^{10}	140	3.2×10^{8}
Chaos chaos			2.8×10^{12}[†]	36	
Hartmannella sp.			4.0×10^{9}	141	
Sporozoa					
Plasmodium berghei			1.2×10^{8}	136	

[†]These species are multinucleate and/or endopolyploid and, therefore, are not haploid (gametic) values.

Table 9. DNA content (nucleotides) per cell and per chromosome of multicellular invertebrates

Organism	Chromosome no. (n)	Ref.	DNA/cell (haploid)	Ref.	DNA/chromosome
Porifera (sponges)					
Dysidea crawshagi			1.1×10^8	1	
Tube sponge			1.2×10^8	1	
Coelenterata					
Cassiopeia (jellyfish)			6.6×10^8	1	
Nematoda					
Panagrellus silusiae (round worm)			8.6×10^8	145	
Echinodermata					
Arbacia aequituberculata (sea urchin)			1.3×10^9	146	
Echinometria (sea urchin)			2.0×10^9	1	
E. mathaei			1.7×10^9	106	
Lytechinus pictus (sea urchin)			1.8×10^9	147	
Paracentrotus lividus (sea urchin)			1.4×10^9	146	
Stichopus diabole (sea cucumber)			2.0×10^9	1	
Strongylocentrotus purpuratus (sea urchin)	18	142	1.8×10^9	3	1.0×10^8
Annelida					
Nereid worm			2.9×10^9	1	
Mollusca					
Aplysia californica (sea hare)	ca.16	142	2.0×10^9	148	1.2×10^8
Chiton tuberculatus			1.3×10^9	1	
Fissurella barbadensis (limpet)			1.0×10^9	1	
Ilyanassa obsoleta (snail)			6.6×10^9	149	
Octopus vulgaris			1.0×10^{10}	150	
Squid			9.0×10^9	1	
Tectarius muricatus (snail)			1.3×10^9	1	
Arthropoda					
Chironomus tentans	4	143	5.0×10^8	143	1.2×10^8
Drosophila melanogaster (fruit fly)	4	144	1.7×10^8	151	4.2×10^7
Emerita analoga (crab)			5.8×10^9	152	
Goose barnacle	13	144	2.9×10^9	1	2.2×10^8
Gryllus domesticus (cricket)	11	142	1.1×10^{10}	106	1.0×10^9
Melanoplus differentialis (grasshopper)	12	144	1.5×10^{10}	153	1.2×10^9
Plagusia depressa (cliff crab)			3.0×10^9	1	

Table 10. DNA content (nucleotides) per cell and per chromosome of primitive chordates and fishes

Organism	Chromosome no. (n)	Ref.	DNA/cell (haploid)	Ref.	DNA/chromosome
Urochordata (tunicates)					
Ascidia atra			3.2×10^8	1	
Ciona intestinalis	14	154	3.4×10^8	156	2.4×10^7
Cephalochordata					
Amphioxus lanceolatus	12	144	$1.0 \times 10^{9\dagger}$	156	8.3×10^7
Agnatha					
Eptatretus stoutii (hagfish)	24	154	4.7×10^9	156	2.0×10^7
Lampetra planeri (brook lamprey)			$2.3 \times 10^{9\dagger}$	156	
Petromyzon (lamprey)			5.0×10^9	1	
Chondrichthyes (cartilaginous fishes)					
Carcharias longimanus (shark)			6.7×10^9	1	
C. obscurus (shark)			5.5×10^9	1	
Hydrolagus colliei (rat fish)	ca.24	43	3.0×10^9	5	1.2×10^8
Scyllium canicula (shark)	ca.40	142	6.7×10^9	4	1.7×10^8
Torpedo ocellata (shark)	ca.20	142	5.5×10^9	4	2.8×10^8
Chondrostei					
Acipenser stellatus			4.9×10^9	106	
A. sturio (sturgeon)			3.2×10^9	1	
Scaphirhynchus platorhynchus	ca.56	43	3.5×10^9	5	6.2×10^7
(shovelnose sturgeon)					
Holostei					
Amia (bowfin)			2.3×10^9	1	
A. calva (bowfin)	ca.23	43	2.5×10^9	5	1.1×10^8
Lepisosteus productus (spotted gar)	ca.34	43	2.8×10^9	5	8.2×10^7
Teleostei[‡]					
Corydoras aeneus			8.8×10^9	3	
Esox lucias (pike)	9	144	1.7×10^9	157	1.9×10^8
Pleuronichthys verticalis	24	155	$1.1 \times 10^{9\dagger}$	155	4.6×10^7
Tetradon fluiatilis			8.0×10^8	3	
Dipnoa (lung fishes)					
Lepidosiren paradoxa (South African lungfish)	19	155	1.1×10^{11}	147	5.8×10^9
Protopterus (African lungfish)	17	144	1.0×10^{11}	1	5.9×10^9

[†]Estimated from relative spectrophotometrically determined DNA values by comparing to a substituted value of 6.0×10^9 nucleotides for the human controls.

[‡]DNA values are the low and high taken from data in refs. 1, 3, 147, 155, 157, 158 totaling 141 species. Chromosome numbers were obtained from refs. 144, 155, 158.

Table 11. DNA content (nucleotides) per cell and per chromosome of amphibians

Organism	Chromosome no. (n)	Ref.	DNA/cell (haploid)	Ref.	DNA/chromosome
Apoda (Caecilians)					
Siphonops annulatus			$4.4 \times 10^{10+}$	71	
Urodela (salamanders)					
Ambystoma jeffersonianum	14	159	1.0×10^{11}	159	7.1×10^{9}
A. laterale	14	159	1.0×10^{11}	159	7.1×10^{9}
A. maculatum	14	160	8.8×10^{10}	163	6.3×10^{9}
A. mexicanum	14	142	7.6×10^{10}	163	5.4×10^{9}
A. tigrinum (tiger salamander)	14	144	8.3×10^{10}	163	5.9×10^{9}
Amphiuma means (Congo eel)	12	144	1.9×10^{11}	163	1.6×10^{10}
Desmognathus fuscus	12	144	3.6×10^{10}	163	3.0×10^{9}
D. monticola			2.0×10^{10}	39	
D. monticola			3.6×10^{10}	163	
D. ochrophaeus			3.6×10^{10}	163	
D. quadramaculatus			4.4×10^{10}	163	
Eurycea bislineata			7.1×10^{10}	163	
Gyrinophilus danielsi			2.0×10^{10}	39	
Necturus maculosus (mudpuppy)	12	160	$1.7 \times 10^{11+}$	155	1.4×10^{10}
Plethodon cinereus	12	144	5.3×10^{10}	163	4.4×10^{9}
P. glutinosus			8.6×10^{10}	163	
P. jordani			7.2×10^{10}	163	
Pleurodeles waltlii	12	71	$4.7 \times 10^{10+}$	71	3.9×10^{9}
Salamander salamander	12	161	$6.4 \times 10^{10+}$	161	5.3×10^{9}
Siredon mexicanum			9.7×10^{10}	165	
Triturus alpestris	12	161	$5.8 \times 10^{10+}$	161	4.8×10^{9}
T. cristatus	12	71	4.5×10^{10}	164	3.8×10^{9}
T. vulgaris	12	161	$5.5 \times 10^{10+}$	161	4.6×10^{9}
Anura (frogs and toads)					
Acris gryllus dorsalis			$8.2 \times 10^{9+}$	39	
Atelopus sp.			$2.7 \times 10^{9+}$	39	
Bombina bombina	12	161	$2.1 \times 10^{10+}$	161	1.8×10^{9}
B. variegata	12	161	$1.9 \times 10^{10+}$	161	1.6×10^{9}
Bufo bufo	11	142	$1.2 \times 10^{10+}$	166	1.1×10^{9}
B. calamita	11	162	8.9×10^{9}	162	8.1×10^{8}
B. crucifer	11	71	$1.0 \times 10^{10+}$	71	9.1×10^{8}
B. ictericus	11	71	$9.2 \times 10^{9+}$	71	8.4×10^{8}
B. marinus	11	162	8.7×10^{9}	162	7.9×10^{8}
B. paracnemis	11	71	$1.0 \times 10^{10+}$	71	9.1×10^{8}
B. pardalis	11	162	1.4×10^{10}	162	1.3×10^{9}
B. quercicus	11	142	$7.6 \times 10^{9+}$	39	6.9×10^{8}
B. t. terrestris	11	160	$8.7 \times 10^{9+}$	39	7.9×10^{8}
B. viridis	11	167	$8.8 \times 10^{9+}$	166	8.0×10^{8}
Ceratophrys calcarata	13	71	$6.0 \times 10^{9+}$	71	4.6×10^{8}
C. dorsata	51	71	$2.0 \times 10^{10+}$	71	3.9×10^{8}
Cycloramphus asper	13	71	$4.9 \times 10^{9+}$	71	3.8×10^{8}
C. dubias	13	71	$5.6 \times 10^{9+}$	71	4.3×10^{8}
Dermatonotus mülleri	11	71	$4.9 \times 10^{9+}$	71	4.4×10^{8}
Eleutheradactylus antillensis			$8.5 \times 10^{9+}$	39	
E. eneidae			$8.0 \times 10^{9+}$	39	
E. gossei			$5.3 \times 10^{9+}$	39	
E. martinicensis			$5.6 \times 10^{9+}$	39	
E. pantoni			$6.7 \times 10^{9+}$	39	
E. r. planirostris			$9.9 \times 10^{9+}$	39	
Eupemphyx nattereri	11	71	$5.4 \times 10^{9+}$	71	4.9×10^{8}
Gastrophryne c. carolinensis			$1.1 \times 10^{10+}$	39	
Gastrotheca sp.			1.0×10^{10}	39	

Table 11. Cont'd.

Organism	Chromosome no. (n)	Ref.	DNA/cell (haploid)	Ref.	DNA/chromosome
Hyla albomarginata	12	71	$8.4 \times 10^{9\dagger}$	71	7.0×10^{8}
H. arborea	12	161	8.8×10^{9}	161	7.3×10^{8}
H. brunnea			4.3×10^{9}	39	
H. cinera	12	142	$1.0 \times 10^{10\dagger}$	39	8.3×10^{8}
H. crepitans			$1.0 \times 10^{10\dagger}$	39	
H. crucifer bartramiana			8.5×10^{9}	39	
H. faber	12	71	8.7×10^{9}	71	7.2×10^{8}
H. femoralis			8.8×10^{9}	39	
H. fuscomarginata	12	71	$7.3 \times 10^{9\dagger}$	71	6.1×10^{8}
H. geographica			7.1×10^{9}	39	
H. heilprini			$8.2 \times 10^{9\dagger}$	39	
H. multilineata	12	71	$1.2 \times 10^{10\dagger}$	71	1.0×10^{9}
H. nana	15	71	6.0×10^{9}	71	4.0×10^{8}
H. ocularis			1.0×10^{10}	39	
H. parkeri	12	71	$7.4 \times 10^{9\dagger}$	71	6.2×10^{8}
H. polytaenia	12	71	7.5×10^{9}	71	6.2×10^{8}
H. pulchella prasina	12	71	$1.6 \times 10^{10\dagger}$	71	1.3×10^{9}
H. squirella			$1.0 \times 10^{10\dagger}$	39	
H. vasta			$5.5 \times 10^{9\dagger}$	39	
Leptodactylis fuscus	11	71	7.8×10^{9}	71	7.1×10^{8}
L. ocellatus	11	71	$4.8 \times 10^{9\dagger}$	71	4.4×10^{8}
Limnodystes tasmaniensis	12	166	4.8×10^{9}	166	4.0×10^{8}
Melanophryniscus moreirae	11	71	9.4×10^{9}	71	8.5×10^{8}
Odontophrynus americanus	22	71	$1.1 \times 10^{10\dagger}$	71	5.0×10^{8}
O. carvalhoi	11	71	$5.0 \times 10^{9+}$	71	4.5×10^{8}
O. cultripes	11	71	6.6×10^{9}	71	6.0×10^{8}
O. occidentalis	11	71	5.0×10^{9}	71	4.5×10^{8}
Oocormus microps	13	71	6.2×10^{9}	71	4.8×10^{8}
Phyllomedusa bicolor			$1.6 \times 10^{10\dagger}$	39	
Physalaemus fuscomaculatus	11	71	5.8×10^{9}	71	5.3×10^{8}
Pseudis paradoxa			3.3×10^{9}	39	
Pseudoacris brimleyi			9.5×10^{9}	39	
P. nigrita			8.1×10^{9}	39	
P. ornata			$7.8 \times 10^{9\dagger}$	39	
P. triseriata			$9.1 \times 10^{9\dagger}$	39	
Pseudopaludicola ameghini	10	71	7.7×10^{9}	71	7.7×10^{8}
Rana arvalis	12	166	1.1×10^{10}	166	9.2×10^{8}
R. capito			1.0×10^{10}	39	
R. catesbeiana	13	144	1.8×10^{10}	39	1.4×10^{9}
R. esculenta	13	166	1.3×10^{10}	166	1.0×10^{9}
R. grylio			1.0×10^{10}	39	
R. heckscheri			1.2×10^{10}	39	
R. pipiens	13	142	1.4×10^{10}	39	1.1×10^{9}
R. sphenocephala	13	142	1.2×10^{10}	39	9.2×10^{8}
R. temporaria	13	166	8.2×10^{9}	164	6.3×10^{8}
Scaphiopus hammondii			5.4×10^{9}	39	
S. h. holbrookii			2.1×10^{9}	4	
S. h. holbrookii			3.0×10^{9}	39	
Stombus appendiculatus	11	71	6.9×10^{9}	71	6.3×10^{8}
S. boiei	11	71	7.7×10^{9}	71	7.0×10^{8}

[†]Estimates of DNA content were calculated from relative spectrophotometrically determined values by comparing to substituted values. Bufo marinus (8.7×10^{9} nucleotides) and Triturus cristatus (4.5×10^{10} nucleotides) were used for converting values in refs. 39 and 71, respectively, and Rana temporaria (8.2×10^{9} nucleotides) was used for refs. 161 and 166. Relative values in ref. 155 were compared to the human control value (6.0×10^{9} nucleotides).

Table 12. DNA content (nucleotides) per cell and per chromosome of reptiles and birds

Organism	Chromosome no. (n)	Ref.	DNA/cell (haploid)	Ref.	DNA/chromosome
Reptilia					
Alligator mississipiensis	16	142	5.0×10^9	157	3.1×10^8
Amyda ferox (turtle)	33	169	$5.2 \times 10^9{}^\dagger$	169	1.6×10^8
Anolis carolinensis (chameleon lizard)	18	169	$3.9 \times 10^9{}^\dagger$	169	2.2×10^8
Boa constrictor amarali	18	169	$3.5 \times 10^9{}^\dagger$	169	1.9×10^8
Bothrops jararaca (South American jararaca)	18	169	$3.8 \times 10^9{}^\dagger$	169	2.1×10^8
Caiman sclerops (South American alligator)	21	169	$4.8 \times 10^9{}^\dagger$	169	2.3×10^8
Chelonia sp. (turtle)	28	142	5.3×10^9	157	1.9×10^8
Chelydra serpentina (snapping turtle)			5.0×10^9	157	
Clemmys insculpta (wood turtle)			4.9×10^9	157	
Coluber constrictor (black racer snake)			$2.8 \times 10^9{}^\dagger$	157	
Drymarchon corais couperi (gopher snake)	18	169	$3.6 \times 10^9{}^\dagger$	169	2.0×10^8
Elaphe absoleta (pilot snake)	18	168	4.3×10^9	157	2.4×10^8
Gerrhonotus multicarinatus (alligator lizard)	23	169	$3.8 \times 10^9{}^\dagger$	169	1.7×10^8
Gopherus agassizi (desert tortoise)	26	169	5.3×10^9	169	2.0×10^8
Natrix sp. (water snake)	20	144	5.0×10^9	157	2.5×10^8
Xenodon merremii (South American xenodon)	15	169	$4.1 \times 10^9{}^\dagger$	169	2.7×10^8
Aves					
Columbia livia (pigeon)	40	142	2.0×10^9	157	5.0×10^7
Duck	40	142	2.7×10^9	157	6.8×10^7
Gallus domestica (chicken)	39	142	2.3×10^9	157	5.9×10^7
Goose	ca.41	144	2.9×10^9	157	7.1×10^7
Meleagris gallopavo (turkey)	41	142	$1.9 \times 10^9{}^\dagger$	157	4.6×10^7
Melopsittacus undulatus (parakeet)	ca.26	169	2.3×10^9	169	8.8×10^7
Numida meleagris (guinea hen)	33	142	2.3×10^9	157	7.0×10^7
Passer sp. (sparrow)			1.9×10^9	157	
Phasianus sp. (pheasant)	41	144	1.7×10^9	157	4.1×10^7
Sirinus canarius (canary)	ca.40	169	$3.0 \times 10^{9+}$	169	7.5×10^7

$\dagger$Estimated from relative spectrophotometric DNA values by comparing to human (6.0×10^9 nucleotides) or chicken (2.3×10^9 nucleotides) values.

Table 13. DNA content (nucleotides) per cell and per chromosome of mammals

Organism	Chromosome no. (n)	Ref.	DNA/cell (haploid)	Ref.	DNA/chromosome
Prototheria (monotremes)					
Ornithorhynchus anatinus (platypus)	27	170	6.0×10^9†	171	2.2×10^8
Tachyglossus aculeatus (spiny anteater)	32	170	5.7×10^9†	171	1.7×10^8
Metatheria (marsupials)					
Macropus rufus (kangaroo)	10	172	3.1×10^9	172	3.1×10^8
Perameles nasuta (bandicoot)	7	172	9.2×10^9	172	1.3×10^9
Potorous tridactylus (potoroo)	6	171	4.9×10^9†	171	8.2×10^8
Trichosurus vulpecula (possum)	10	172	6.0×10^9	172	6.0×10^8
Eutheria (placentals)					
Bison bison (buffalo)			9.8×10^9	106	
Bos taurus (cattle)	30	144	ca. 6.5×10^9	106	2.2×10^8
Canis familiaris (dog)	39	169	5.6×10^9†	169	1.4×10^8
Cavia cobaya (guinea pig)	32	144	5.9×10^9	106	1.8×10^8
Equus caballos (horse)	32	144	5.8×10^9	157	1.8×10^8
Felis catus (cat)	19	144	7.1×10^9	106	3.7×10^8
Homo sapiens (man)	23	144	ca. 6.0×10^9	106	2.4×10^8
Mesocricetus auratus (golden hamster)	22	169	7.6×10^9†	169	3.4×10^8
Microtus oregoni (creeping vole)	9	169	5.4×10^9†	169	6.0×10^8
Mus musculus (mouse)	20	169	6.5×10^9	169	3.2×10^8
Orycteropus afer (aardvark)	10	173	1.1×10^{10}	173	9.5×10^8
Oryctolagus cuniculus (rabbit)	22	144	ca. 6.3×10^9	106	1.4×10^8
Ovis aries (sheep)	27	144	6.3×10^9	157	2.3×10^8
Rattus rattus (rat)	21	142	6.1×10^9	1	2.9×10^8
Sus scrofa (pig)	20	144	ca. 5.0×10^9	106	2.5×10^8

†Estimated from relative spectrophotometric or quantitative values by comparing to rat (6.1×10^9 nucleotides), human (6.0×10^9 nucleotides), or chicken (2.3×10^9 nucleotides) DNA values.

V. REFERENCES

1. MIRSKY, A. E. AND RIS, H., J. Gen. Physiol. 34, 451 (1951).

2. MIKSCHE, J. P., Can. J. Genet. Cytol. 9, 717 (1967).

3. HINEGARDNER, R., Am. Naturalist 102, 517 (1968).

4. GOIN, O. B. AND GOIN, C. J., Am. Midland Naturalist 80, 289 (1968).

5. OHNO, S., Evolution by Gene Duplication, Springer-Verlag, New York, 1970.

6. STEBBINS, G. L., Science 152, 1463 (1966).

7. REES, H., CAMERON, F. M., HAZARIKA, M. H., AND JONES, G. H., Nature 211, 828 (1966).

8. ROTHFELS, E., SEXSMITH, E., HEIMBURGER, M., AND KRAUSE, M. O., Chromosoma 20, 54 (1966).

9. SPARROW, A. H. AND MIKSCHE, J. P., Science 134, 282 (1961).

10. VAN'T HOF, J. AND SPARROW, A. H., Proc. Natl. Acad. Sci. U.S. 49, 897 (1963).

11. SPARROW, A. H., SPARROW, R. C., THOMPSON, K. H., AND SCHAIRER, L. A., Radiation Botany 5 (Suppl.), 101 (1965).

12. BAETCKE, K. P., SPARROW, A. H., NAUMAN, C. H., AND SCHWEMMER, S. S., Proc. Natl. Acad. Sci. U.S. 58, 533 (1967).

13. UNDERBRINK, A. G., SPARROW, A. H., AND POND, V., Radiation Botany 8, 205 (1968).

14. CONGER, A. D., Intern. J. Radiation Biol. 17, 381 (1970).

15. SPARROW, A. H., in Cellular Radiation Biology, p. 199, Williams and Wilkins, Baltimore, 1965.

16. SPARROW, A. H., UNDERBRINK, A. G., AND SPARROW, R. C., Radiation Res. 32, 915 (1967).

17. SPARROW, A. H., ROGERS, A. F., AND SCHWEMMER, S. S., Radiation Botany 8, 149 (1968).

18. WATSON, J. D., Molecular Biology of the Gene, W. A. Benjamin, New York, 1965.

19. COHEN, S. S., Am. Scientist 58, 218 (1970).

20. RAVEN, P. H., Science 169, 641 (1970).

21. OLIVE, L. S., Botan. Rev. 19, 439 (1953).

22. RAPER, J. R. AND ESSER, K., in The Cell, Vol. 4, p. 139, J. Brachet and A. E. Mirsky, Editors, Academic Press, New York, 1964.

23.   LEIGHTON, T. J., DILL, B. C., STOCK, J. J., AND PHILLIPS, C.,
      Proc. Natl. Acad. Sci. U.S. 68, 677 (1971).

24.   TONINO, G. J. M. AND ROZIJN, T. H., Biochim. Biophys. Acta 124,
      427 (1966).

25.   BRITTEN, R. J. AND KOHNE, D. E., Science 161, 529 (1968).

26.   CHUNOSOFF, L. AND HIRSHFIELD, H. I., J. Protozool. 14, 157 (1967).

27.   DODGE, J. D., J. Gen. Microbiol. 36, 269 (1964).

28.   DODGE, J. D., in The Chromosomes of the Algae, p. 96, M. B. E.
      Godward, Editor, Edward Arnold, London, 1966.

29.   GRELL, K. G., in The Cell, Vol. 6, p. 1, J. Brachet and A. E.
      Mirsky, Editors, Academic Press, New York, 1964.

30.   RIS, H., in The Interpretation of Ultrastructure, p. 69, R. C. J.
      Harris, Editor, Academic Press, New York, 1962.

31.   ZINGMARK, R. G., Am. J. Botany 57, 586 (1970).

32.   CHEN, T. T., J. Heredity 31, 175 (1940).

33.   WOODARD, J., BEATRICE, G., AND SWIFT, H., Exptl. Cell Res. 23,
      258 (1961).

34.   BEHME, R. J. AND BERGER, J. D., J. Protozool. 17, 20a (1970).

35.   WOODARD, J., GOROVSKY, M., AND KAMESHIRO, E., J. Cell Biol. 39,
      182a (1968).

36.   FRIZ, C. T., Comp. Biochem. Physiol. 26, 81 (1968).

37.   JAHN, H. E., How to Know the Protozoa, Wm. C. Brown, Dubuque,
      Iowa, 1949.

38.   ROTH, L. E. AND DANIELS, E. W., J. Biophys. Biochem. Cytol. 9,
      317 (1961).

39.   GOIN, O. B., GOIN, C. J., AND BACHMANN, K., Copeia 3, 532 (1968).

40.   BECAK, W., BECAK, M. L., NAZARETH, H. R. S., AND OHNO, S.,
      Chromosoma 15, 606 (1964).

41.   HAMMER, B., Hereditas 65, 25 (1970).

42.   RENZONI, A. AND VEGNI-TALLURI, M., Chromosoma 20, 133 (1966).

43.   OHNO, S., MURAMOTO, J., STENIUS, C., CHRISTIAN, L., AND KITTRELL,
      W. A., Chromosoma 26, 35 (1969).

44.   BACHMANN, K., GOIN, C. J., AND GOIN, O. B., See paper in this
      Symposium.

45.   CRONQUIST, A., Botan. Rev. 26, 425 (1960).

46. SPORNE, K. R., The Morphology of Pteridophytes, Hutchinson, London, 1962.

47. CHIARUGI, A., Caryologia 13, 27 (1960).

48. DITTMER, H. J., Phylogeny and Form in the Plant Kingdom, Van Nostrand, Princeton, N. J., 1964.

49. MEHRA, P. N. AND VERMA, S. C., Caryologia 13, 619 (1960).

50. BRITTEN, D. M., Am. J. Botany 40, 575 (1953).

51. BRITTEN, D. M., Can. J. Botany 42, 1349 (1964).

52. MANTON, I., Problems of Cytology and Evolution in the Pteridophyta, Cambridge Univ. Press, Cambridge, 1950.

53. MANTON, I. AND VIDA, G., Proc. Roy. Soc. (London) Ser. B. 170, 361 (1968).

54. D'AMATO-AVANZI, M. G., Caryologia 9, 373 (1957).

55. LOYAL, D. S., Current Sci. (India) 27, 357 (1958).

56. KHOSHOO, T. N., Evolution 13, 24 (1959).

57. HAIR, J. B. AND BEUZENBERG, E. J., Nature 181, 1584 (1958).

58. KHOSHOO, T. N., Silvae Genet. 10, 1 (1961).

59. BECK, C. B., Taxon 15, 337 (1966).

60. BECK, C. B., Biol. Rev. Cambridge Phil. Soc. 45, 379 (1970).

61. FLORIN, R., Acta Horti Bergiani 15, 285 (1951).

62. SPARROW, A. H., unpublished data.

63. CRONQUIST, A., The Evolution and Classification of Flowering Plants, Houghton Mifflin, Boston, 1968.

64. STEBBINS, G. L., Variation and Evolution in Plants, Columbia Univ. Press, New York, 1950.

65. BRITTEN, R. J. AND DAVIDSON, E. H., Science 165, 349 (1969).

66. BURLEY, J., Silvae Genet. 14, 127 (1965).

67. MIKSCHE, J. P., Chromosoma 32, 343 (1971).

68. REES, H. AND HAZARIKA, H., in Chromosomes Today, Vol. 2, p. 158, C. D. Darlington and K. R. Lewis, Editors, Plenum Press, New York, 1967.

69. JOHN, B. AND HEWITT, G. M., Chromosoma 20, 155 (1966).

70. KIKNADZE, I. I. AND VYSOSTSKAYA, L. V., Tsitologia 12, 1100 (1970).

71. BECAK, W., BECAK, M. L., SCHREIBER, G., LEVELLE, D., AND AMORIN, F. O., _Experientia_ 26, 204 (1970).

72. BRITTEN, R. J. AND KOHNE, D. E., _Carnegie Inst. Wash. Yearbook_ 65, 78 (1966).

73. BRITTEN, R. J. AND KOHNE, D. E., _Carnegie Inst. Wash. Yearbook_ 66, 73 (1967).

74. THOMAS, C. A., HAMKALO, B. A., MISRA, D. N., AND LEE, C. S., _J. Mol. Biol._ 51, 621 (1970).

75. BENDICH, A. J. AND McCARTHY, B. J., _Genetics_ 65, 545 (1970).

76. BENDICH, A. J. AND BOLTON, E. T., _Plant Physiol._ 42, 959 (1967).

77. ROBERTSON, F. W., CHIPCHASE, M., AND MAN, N. T., _Genetics_ 63, 369 (1969).

78. RITOSSA, F. M. AND SPIEGELMAN, S., _Proc. Natl. Acad. Sci. U.S._ 53, 737 (1965).

79. RITOSSA, F. M., ATWOOD, K. C., AND SPIEGELMAN, S., _Genetics_ 54, 819 (1966).

80. BROWN, D. D. AND DAWID, I. B., _Science_ 160, 272 (1968).

81. RITOSSA, F. M., ATWOOD, K. C., AND SPIEGELMAN, S., _Genetics_ 54, 663 (1966).

82. GALL, J. G., _Genetics_ 61 (Suppl. 1, part 2), 121 (1969).

83. NAGL, W., _Nature_ 221, 70 (1969).

84. SWIFT, H. H., _Physiol. Zool._ 23, 169 (1950).

85. COGGESHALL, R. E., YAKSTA, B. A., AND SWARTZ, F. J., _Chromosoma_ 32, 205 (1970).

86. NAORA, H., _J. Biophys. Biochem. Cytol._ 3, 949 (1957).

87. LEUCHTENBERGER, C., _Science_ 120, 1022 (1954).

88. SWARTZ, F. J., _Chromosoma_ 8, 53 (1956).

89. FLOYD, A. D. AND SWARTZ, F. J., _Exptl. Cell Res._ 56, 275 (1969).

90. HERMAN, C. J. AND LAPHAM, L. W., _Science_ 160, 537 (1968).

91. D'AMATO, F., _Caryologia_ 17, 41 (1964).

92. HALLET, M. J., _Compt. Rend._ 271, 2110 (1970).

93. DENIS, H., _J. Mol. Biol._ 22, 269 (1966).

94. DENIS, H., _J. Mol. Biol._ 22, 285 (1966).

95. CRIPPA, M., DAVIDSON, E. H., AND MIRSKY, A. E., *Proc. Natl. Acad. Sci. U.S.* 57, 885 (1967).

96. BROWN, D. D. AND GURDON, J. B., *J. Mol. Biol.* 19, 399 (1966).

97. GLISIN, V. R., GLISIN, M. V., AND DOTY, P., *Proc. Natl. Acad. Sci. U.S.* 56, 285 (1966).

98. WHITELEY, A. H., McCARTHY, B. J., AND WHITELEY, H. R., *Proc. Natl. Acad. Sci. U.S.* 55, 519 (1966).

99. CHURCH, R. B. AND McCARTHY, B. J., *J. Mol. Biol.* 23, 477 (1967).

100. CHURCH, R. B. AND McCARTHY, B. J., *J. Mol. Biol.* 23, 459 (1967).

101. KEDES, L. H. AND BIRNSTIEL, M. L., *Nature* 230, 165 (1971).

102. COMMONER, B., *Am. Scientist* 52, 365 (1964).

103. COMMONER, B., *Nature* 202, 960 (1964).

104. VAN'T HOF, J., *Exptl. Cell Res.* 41, 274 (1966).

105. JOCLIK, W. K., *J. Mol. Biol.* 5, 265 (1962).

106. SHAPIRO, H. S., in *Handbook of Biochemistry*, p. H52, H. A. Sober and R. Harte, Editors, Chemical Rubber Co., Cleveland, Ohio, 1968.

107. KINGSBURY, D. T., *J. Bacteriol.* 98, 1400 (1969).

108. DRIEDGER, A. A., *Can. J. Microbiol.* 16, 1136 (1970).

109. MOROWITZ, H. J., BODE, H. R., AND KIRK, R. G., *Ann. N. Y. Acad. Sci.* 143, 110 (1967).

110. HORSFALL, F. L. AND TAMM, I. (Editors), *Viral and Rickettsial Infections of Man*, 4th ed., J. B. Lippincott, Philadelphia, 1965.

111. NASS, M. M. K., *Science* 165, 25 (1969).

112. SUYAMA, Y. AND PREER, J. R., JR., *Genetics* 52, 1051 (1965).

113. GIBOR, A. AND IZAWA, M., *Proc. Natl. Acad. Sci. U.S.* 50, 1164 (1963).

114. RUPPEL, H. G. AND WYK, V., *Z. Pflanzenphysiol.* 53, 32 (1965).

115. SUYAMA, Y. AND BONNER, W. D., JR., *Plant Physiol.* 41, 383 (1966).

116. SAGER, R. AND ISHIDA, R., *Proc. Natl. Acad. Sci. U.S.* 50, 725 (1963).

117. BRAWERMAN, G. AND EISENSTADT, J. M., *Biochim. Biophys. Acta* 91, 477 (1964).

118. GYLDENHOLM, A. O., *Hereditas* 59, 142 (1968).

119. ORNDUFF, R., (Editor), Index to Plant Chromosome Numbers for 1967, (Regnum Vegetabile 59), Intern. Bureau Plant Taxonomy and Nomenclature, Utrecht, Netherlands, 1969.

120. AINSWORTH, G. C. AND SUSSMAN, A. S., The Fungi, Academic Press, New York, 1965.

121. DEERING, R. A., SMITH, M. S., THOMPSON, B. K., AND ADOLF, A. C., Radiation Res. 43, 711 (1970).

122. CHRISTENSEN, J. J. AND RODENHISER, H. A., Botan. Rev. 6, 389 (1940)

123. ESSER, K. AND KUENEN, R., Genetics of Fungi, Springer-Verlag, New York, 1967.

124. DEERING, R. A., Radiation Res. 34, 87 (1968).

125. GODWARD, M. B. E., in The Chromosomes of the Algae, p. 1, M. B. E. Godward, Editor, Edward Arnold, London, 1966.

126. HOLM-HANSEN, O., Science 163, 87 (1969).

127. IWAMURA, T., Ann. N. Y. Acad. Sci. 175, 488 (1970).

128. ISHIDA, M. R., Cytologia 26, 359 (1961).

129. SHINKE, N., ISHIDA, M. R., AND UEDA, K., Cytologia 22 (Suppl.), 156 (1957).

130. ASATA, Y. AND FOLSOME, C. E., Genetics 65, 407 (1970).

131. ORNDUFF, R., (Editor), Index to Plant Chromosome Numbers for 1966, (Regnum Vegetabile 55), Intern. Bureau Plant Taxonomy and Nomenclature, Utrecht, Netherlands, 1968.

132. DNA content estimated from nuclear volume (see Materials and Methods).

133. DARLINGTON, C. D. AND WYLIE, A. P., Chromosome Atlas of Flowering Plants, George Allen and Unwin, London, 1955.

134. UHL, C. H., personal communication.

135. MOORE, R. J., (Editor), Index to Plant Chromosome Numbers for 1968, (Regnum Vegetabile 68), Intern. Bureau Plant Taxonomy and Nomenclature, Utrecht, Netherlands, 1970.

136. MANDEL, M., in Chemical Zoology, Vol. 1, p. 541, M. Florkin, B. T. Scheer and G. W. Kidder, Editors, Academic Press, New York, 1967.

137. IVERSON, R. M. AND GIESE, A. C., Exptl. Cell Res. 13, 213 (1957).

138. PIGON, A. AND EDSTROM, J. E., Exptl. Cell Res. 16, 648 (1959).

139. RIOU, G. AND PAUTRIZEL, R., J. Protozool. 16, 509 (1969).

140. COHEN, A. I., Ann. N. Y. Acad. Sci. 78, 609 (1959).

141. LITTLE, J. B. AND CHANG, R. S., Radiation Res. 35, 132 (1968).

142. MAKINO, S., A Review of the Chromosome Numbers in Animals, Hokuryuka, Tokyo, 1956.

143. DANEHOLT, B. AND EDSTROM, J. E., Cytogenetics 6, 350 (1967).

144. ALTMAN, P. L. AND DITTMER, D. S., (Editors), Biology Data Book, Federation Am. Soc. for Experimental Biology, Washington, D. C., 1964.

145. SIN, W. C. AND PASTERNAK, J., Chromosoma 32, 191 (1970).

146. WHITE, M. J. D., The Chromosomes, John Wiley and Sons, New York, 1961.

147. MARKERT, C. L., J. Cell Physiol. 72 (Suppl. 1), 213 (1968).

148. LASEK, R. J. AND DOWER, W. J., Science 172, 278 (1971).

149. COLLIER, J. R. AND McCANN-COLLIER, H., Exptl. Cell Res. 27, 553 (1962).

150. PACKARD, A. AND ALBERGONI, V., J. Exptl. Biol. 52, 539 (1970).

151. KURNICH, N. B. AND HERKOWITZ, I. H., J. Cell. Comp. Physiol. 39, 281 (1952).

152. VAUGHN, J. C. AND LORY, R. D., J. Histochem. Cytochem. 17, 591 (1969).

153. SWIFT, H. AND KLEINFELD, R., Physiol. Zool. 26, 301 (1953).

154. TAYLOR, K. M., Chromosoma 21, 181 (1967).

155. OHNO, S. AND ATKIN, N. B., Chromosoma 18, 455 (1966).

156. ATKIN, N. B. AND OHNO, S., Chromosoma 23, 10 (1967).

157. BRAWERMAN, G. AND SHAPIRO, H. S., in Comparative Biochemistry, Vol. 4, p. 107, M. Florkin and H. S. Manson, Editors, Academic Press, New York, 1962.

158. OHNO, S., MURAMOTO, J., CHRISTIAN, L., AND ATKIN, N. B., Chromosoma 23, 1 (1967).

159. MacGREGOR, H. C. AND UZZELL, T. M., Science 143, 1043 (1964).

160. GOIN, C. J. AND GOIN, O. B., Introduction to Herpetology, W. H. Freeman, San Francisco, 1962.

161. ULLERICH, F. H., Chromosoma 30, 1 (1970).

162. GRIFFIN, C. S., SCOTT, D., AND PAPWORTH, D. G., Chromosoma 30, 228 (1970).

163. BACHMANN, K., Histochemie 22, 289 (1970).

164. WALKER, P. M. B. AND YATES, H. B., _Proc. Roy. Soc. (London) Ser._
     _B._ _140_, 274 (1952).

165. EDSTROM, J. E. AND KAWIAK, J., _J. Biophys. Biochem. Cytol._ _9_,
     619 (1961).

166. ULLERICH, F. H., _Chromosoma_ _21_, 345 (1967).

167. WICKBOM, T., _Hereditas_ _31_, 241 (1945).

168. BECAK, W. AND BECAK, M. L., _Cytogenetics_ _8_, 247 (1969).

169. ATKIN, N. B., MATTINSON, G., BECAK, W., AND OHNO, S., _Chromosoma_
     _17_, 1 (1965).

170. BICK, Y. A. E. AND JACKSON, W. D., _Nature_ _214_, 600 (1967).

171. BICK, Y. A. E. AND JACKSON, W. D., _Nature_ _215_, 192 (1967).

172. RENDEL, J. M. AND KELLERMAN, G. M., _Nature_ _176_, 829 (1955).

173. BENIRSCHKE, K., WURSTER, D. H., AND LOW, R. J., _Chromosoma_ _31_,
     68 (1970).

DISCUSSION

BASKIN: How was the interphase chromosomal volume determined?

SPARROW: ICV is obtained by dividing the average volume of the
interphase nucleus by the somatic chromosome number. References to
techniques of slide preparation and of obtaining nuclear volume measurements are given in the materials and methods section of the
manuscript.

MELTON: In your survey of the literature on DNA values, what
kinds of methodological differences were encountered? In particular,
how frequently did you find cytophotometric values in agreement with
biochemical extraction results?

PRICE: Methodological differences encountered in the literature
were varied and at times particular DNA values estimated by different
procedures, or the same procedure from different laboratories, were,
as you might expect, not always in agreement. The DNA of eukaryotic
cells is most commonly extracted and estimated by the diphenylamine

reaction, or estimated cytologically by a variety of microspectrophotometric techniques. When DNA contents have been determined by both of
these methods, the results are generally in good agreement.

HINEGARDNER: DNA contents have been reported in different units,
why did you use number of nucleotides?

SPARROW: We chose nucleotides to be consistent with previous
reports from this laboratory dealing with radiosensitivity and target
theory. We have found that DNA values for prokaryotes are often given
as nucleotides while those from eukaryotes are often in picograms.
Conversion from one to the other is easy. Perhaps it would be desirable
at times to use double scales when DNA values are presented in graphic
form.

MILKMAN: Since evolution does occur in small steps, and since
there is impressive variation within, for example, the genus _Pinus_, I
wonder if variations in DNA content have been observed over relatively
short periods of time or among strains of a species?

SPARROW: The best example of intraspecific variation in DNA content that we are aware of, excluding polyploidy or aneuploidy, is that
in _Picea sitchensis_ reported by J. P. Miksche and cited in the discussion of this paper. Avanzi, et al. (_Mutation Res._ 3, 426 (1966))
have shown by a cytophotometric analysis that _Triticum durum_ var.
Aziziah has a DNA content per nucleus 6% higher than variety Cappelli.
More investigations on variability of DNA content within species are
indeed needed.

STEVENSON: Comment: I can't recall the source, but I believe there
is recent evidence for interpreting the extant Psilopsids, the
Psilotales, as degenerate ferns rather than direct descendants of the
extinct forms the Psilophytales. This would, I believe, accommodate
them into your scheme in a more consistent pattern.

EVOLUTION OF PREFERENTIAL TRANSMISSION MECHANISMS

IN CYTOPLASMIC GENETIC SYSTEMS

Ruth Sager
Hunter College
City University of New York

This session brings together some considerations of
cytoplasmic genes in Chlamydomonas and of episomes in
bacteria. In terms of evolution we might ask whether the
control of episomal DNA in the bacterial system provides
a useful model for untangling the complexities of interactions between organelle and nuclear genomes in
eukaryotic cells, In my own experience, episomal
genetics, especially lambda genetics, has provided a
model of elegant experimentation and analysis to
emulate in studies of cytoplasmic genes.

I wish now to present a few remarks on an
evolutionary theme concerning the preferential
transmission of cytoplasmic genes.

A fundamental property of all cytoplasmic genetic
systems so far investigated is preferential transmission
to the progeny of cytogenes, i.e., cytoplasmic genes
from one of the parents. Different mechanisms have been

proposed to control this process in different organisms.
These differences in mechanism are closely related to
the special features of the sexual life cycle in each
organism. When the transmission mechanisms are compared,
they are seen to have one outstanding feature in common:
they all lead to greatly reduced opportunities for
genetic recombination.

To show how preferential transmission and limited
recombination are achieved, I have summarized the
patterns of transmission of cytogenes in the principal
classes of organisms in relation to their life cycles.
The patterns of segregation of cytogenes in meiosis
and mitosis and the consequences with respect to uniparental inheritance and to recombination are presented
in the tables.

Since preferential transmission is such a pervasive
phenomenon, it seems evident that it is the result of
powerful selective pressures operating in evolution. I
have been interested in this problem both from the
theoretical side: what is the advantage to the organism
of preferential transmission; and from the experimental
side: what is the mechanism of this process at the
molecular level.

Concerning the molecular basis of preferential
transmission, Chlamydomonas is the only organism in which

this process is being investigated. Chloroplast DNA
can be readily examined in $CsCl_2$ gradients where by
virtue of its high AT content, chloroplast DNA it is
well separated from nuclear DNA.[1] We have been studying
the fate of chloroplast DNAs from male and female
parents in zygotes, using density and radioisotope
labels to distinguish the two. We have found that
chloroplast DNA from the male or mt^- parent is pre-
ferentiallly destroyed during zygote maturation while that
of the mt^+ parent is replicated and transmitted to the
progeny.[2,3] The details of the process are very complex,
suggesting a mechanism analogous to bacterial host
modification and restriction. Thus Chlamydomonas is
equipped with an elaborate enzymatic system for
achieving maternal transmission of chloroplast genes.
Chiang[4] has also studied this system and proposed a
somewhat different interpretation.

On the theoretical side, some time ago I suggested
that the cytoplasmic genetic system operates to restrict
recombination,[5] and the data in the tables support that
hypothesis. Lewontin[6] has provided new mathematical
evidence in polymorphic populations that highly restricted recombination maximizes fitness at the population
level. This idea in a more qualitative form has been
discussed by many geneticists and goes back at least as
far as R. A. Fisher[7].

Lewontin asks: "Why does not the genome congeal into a single nonrecombining supergene?" His answer: the need for adaptation to a varying environment; and the use of chiasmata for the orderly distribution of chromosomes. In the organelle systems, apparently the opportunity for recombination has been preserved, but delegated to a small fraction of the population. Whether physical exchange plays a role in distribution of organelle DNAs is not known, but so long as the molecules are identical the consequences of exchange go undetected in nature.

The finding of elaborate mechanisms to restrict recombination of cytogenes is particularly interesting because it further emphasizes the importance to the organism of these cytoplasmic genomes.

Preferential Transmission of Cytogenes in Meiosis

Chlamydomonas

(A) Maternal transmission to tetrads. Mechanism:
Exclusion of parental cytogenes initially present in
zygote.

(B) Biparental transmission in tetrads. Mechanism:
Inhibition of parental exclusion process. Transmission of
both parental cytogenomes to all zoospores. Recombination
occurs in vegetative growth of zoospore clones.

Yeast

(A) Uniparental (4:0) transmission to tetrads. Mechanism:
Sporulating diploids are cytoplasmic homozygotes before
sporulation.

(B) Biparental: irregular segregation in tetrads.
Mechanism: sporulating diploids are cytohets and segregation in meiosis is irregular. Segregation mechanism
unknown.

Fungi

(A) Uniparental (4:0) transmission to tetrads. One parental
genome excluded in ascus. Mechanism unknown. Recombination unknown.

Higher Plants

(A) Maternal. Parental cytogenome excluded. Mechanism
unknown.

(B) Biparental. Both maternal and parental cytogenes
transmitted through zygote, usually with maternal
preference. Mechanism unknown. Recombination unknown.

Preferential Transmission of Cytogenes in Mitosis

Chlamydomonas

Segregation occurs in vegetative clones originating from cytohet zoospores. Segregation patterns:

a. 1:1 by reciprocal exchange.

b. average 1:1 by non-reciprocal "conversion" events.

Segregation mechanism: recombination of DNA molecules at 4-strand stage. No preferential transmission or replicative dominance seen in mitosis.

Yeast

Segregation in vegetative zygote clones not 1:1. Segregation patterns:

a. preferential transmission (polarity) determined
   by mitochondrial sex factor .

b. replicative dominance of some mitochondrial DNAs
   (suppressiveness).

Segregation mechanism unknown.

Fungi

Preferential transmission (replicative dominance) in mycelial growth. Recombination unknown.

Higher Plants

Preferential segregation with various ratios from bi-parental zygotes. Segregation mechanisms unknown. Recombination unknown.

Consequences of Preferential Transmission of Cytogenes

Chlamydomonas   (Chloroplast genes)

Rare opportunity for recombination provided by spontaneous (10^{-2} to 10^{-4}) biparental transmission in zygotes,
exception to the rule of maternal inheritance. Recombination and segregation of new types occurs in vegetative growth of cytohet progeny from these exceptional
zygotes.

Yeast (Mitochondrial genes)

Recombination occurs in zygotes, limited by preferential
transmission and by rapid segregation in diploid clones.

Fungi

Uniparental transmission occurs in asci.

Preferential tranmission occurs in cytohet mycelia.

Recombination unknown.

Higher Plants

Maternal inheritance predominant.

Recombination unknown.

1. R. Sager and M. R. Ishida, <u>Proc. Natl. Acad. Sci.</u> 50, 725 (1963).

2. R. Sager and D. Lane, <u>Fed. Proc.</u> 38, 347 (1969).

3. D. Lane and R. Sager, ms. in prep.

4. K. S. Chiang, <u>Proc. Natl. Acad. Sci.</u> 60, 194 (1968).

5. R. Sager, <u>Autonomy and Biogenesis of Mitochondria and Chloroplasts</u> (North-Holland, Amsterdam, 1971), p.250.

6. R. Lewontin, <u>Proc. Natl. Acad. Sci.</u> 68,984 (1971).

7. R. A. Fisher, <u>The Genetical Theory of Natural Selection</u> (Clarendon Press, Oxford, 1930).

INTERACTIONS BETWEEN NUCLEAR AND ORGANELLE GENETIC SYSTEMS

R. P. Levine

The Biological Laboratories, Harvard University

Abstract. Experiments with Chlamydomonas reinhardi are described which reveal that the alga's chloroplast genetic system
as well as certain components of its photosynthetic apparatus
are of hybrid origin. Their specification at the transcriptional level is determined by information contained both in
nuclear and chloroplast DNA and at the translational level by
events that occur on both chloroplast and cytoplasmic ribosomes.

INTRODUCTION

The unicellular green alga Chlamydomonas reinhardi is an

attractive organism for the study of organelle genetic systems,

particularly the genetic system of the chloroplast. In a re-

cent review (1) we described several of the features of the

chloroplast system in C. reinhardi, including its possession

of ribosomes that have a similar sedimentation coefficient

(68S) to those of bacteria, and its sensitivity to the anti-

biotics rifampicin, spectinomycin, and chloramphenicol, inhi-

bitors of transcription and translation that are also effective

against bacteria. We also presented evidence that the chloro-

plast is by no means genetically autonomous, since many of its

components appear to be specified by nuclear genes and synthe-

sized in the cytoplasm.

In this paper I should like to summarize some recent

experiments, performed in my laboratory and the laboratories
of others, that have served to extend our understanding of
how the chloroplast genetic system of C. reinhardi is organized, operates, and interacts with the other genetic systems
of the cell.

THE FOUR TYPES OF DNA IN C. REINHARDI

Four species of DNA, each characterized by its buoyant
density, have been identified in C. reinhardi. Nuclear DNA,
representing 85% of the total, has a density of 1.723 (2).
The density of chloroplast DNA is 1.695, and it represents
14% of the total DNA (2). A third species of DNA, the so-
called γ-band (2), is now believed to be an amplification of
the genes specifying the RNA of cytoplasmic ribosomes (K.S.
Chiang, personal communication). It represents 1% of the
total and its density is 1.715 (2). M-band DNA is detected
in zygotes (2); its density, 1.715 (2), is indistinguishable
from that of γ-band DNA. However, its cellular location and
function have not yet been determined. A species of DNA as-
cribable to the mitochondrion has not yet been identified in
C. reinhardi.

The mode of replication of nuclear, γ-band, and chloroplast
DNA is semiconservative (2), and in cells from synchronous
cultures their replication times are distinct (2).

THE ORGANIZATION OF C. REINHARDI CHLOROPLAST DNA

When thin sections of C. reinhardi are examined with the
electron microscope, it is not uncommon to observe within the
chloroplast a number of widely-separated regions of DNA, suggesting that the organelle may be polyploid (U. Goodenough,
unpublished observations). Although certain considerations

have led to the suggestion that the chloroplast DNA of C. reinhardi is diploid (3), a recent study of the renaturation kinetics of chloroplast DNA (4) has indicated a level of redundancy of some 50-fold in vegetative cells. Whether this redundancy derives from repeated sequences within a single chromosome, multiple chromosome copies of a single sequence, or a combination of the two has not yet been determined.

Despite these uncertainties, some understanding of the arrangement of genes within the chloroplast DNA has been obtained by Surzycki and Rochaix (unpublished data) using the technique of transcriptional mapping (5,6). This technique allows one to alter the extent of expression of the various genes within a transcriptional unit by perturbing the transcribing activity of the DNA-dependent RNA polymerase. The technique permits conclusions as to whether specific genes are indeed transcribed together, and it allows one to determine their relative order within the transcriptional unit. In the experiments with C. reinhardi, the synthesis of the 16S and 23S species of chloroplast ribosomal RNA by genes that appear to lie within chloroplast DNA (6a), was perturbed in 3 ways: 1) by using actinomycin D to block transcription at random points within the genes; 2) by using rifampicin to prevent initiation of transcription, such that only those polymerases bound to the template at the start of the experiment would complete the transcription of ribosomal RNA genes; and 3) by depleting genes of polymerases before the synchronous restart of transcription commences in synchronous culture.

A number of possible arrangements of the genes for the 16S and 23S chloroplast ribosomal RNA's can be envisaged, as shown

in Fig. 1. For each case, one can predict the effects that
the 3 kinds of perturbations mentioned above will have on the
amount of ribosomal RNA synthesized in a given time. The
mathematical analysis (due principally to Rochaix) by which
the predicted values were obtained will not be described here
except to say that it makes three·assumptions: 1) The DNA-
dependent RNA polymerases are equally spaced on any one gene
in the act of being transcribed, an assumption based on electron micrographs of transcriptional complexes in both eukaryotes and prokaryotes (7). 2) The polymerases travel at a constant speed on a given operon during transcription, an assumption that is supported by the fact that the essentially equal
spacing of polymerases along the ribosomal RNA genes that has
been observed (7) is only possible if the enzymes are travel-
ing at equal speed along the transcriptional unit. 3) The
genes for the 23S ribosomal RNA are twice as long as those
for the 16S ribosomal RNA, an assumption in accord with the
fact that the ratio of the molecular weights of the two chloroplast ribosomal RNA's is close to two (8).

Each of the arrangements of ribosomal RNA cistrons gives
a set of predicted values for the ratio of inhibition of
synthesis of 23S:16S RNA in the presence of actinomycin D and
the ratio of 23:16S RNA synthesized after the addition of
rifampicin. These ratios are a function of the number and
sequence of gene pairs (e.g. 16S16S, 23S23S, 16S23S,etc.) per
transcriptional unit and, with the exception of a purely random arrangement of genes (arrangement 1 in Fig. 1), they have
the values shown in Figs. 2 and 3. Figures 2 and 3 also
give sets of experimentally determined values, which were

Possible arrangements of chloroplast rRNA genes

Numbers 16 and 23 are used to designate genes for the small and large rRNA respectively. Sp refers to Spacer DNA.

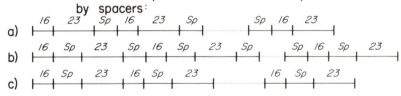

Arrangement 1: The 16S and 23S genes are located in independent transcriptional units:

a)

b)

c)

In arrangement Ic, the 16S and the 23S genes are separated from their promoters by a spacer.

Arrangement 2: Pairs of 16S and 23S genes are tandemly arranged within one or more transcriptional units with the 16S gene first:

Arrangement 3: The same as *2* but with 23S gene first:

Arrangement 4: The 16S and 23S genes are located in one transcriptional unit and are separated by spacers:

a)

b)

c)

Arrangement 5: The 16S and 23S genes are tandemly arranged within one or more transcriptional units with all 16S gene tandems first:

Arrangement 6: The same as *5* but with all 23S gene tandems first:

Figure 1.

obtained in the following way. Cells were exposed to radioactive precursors of RNA and either rifampicin or actinomycin
D, and the amount of radioactively-labelled 16S and 23S RNA
in the cells was then determined. The ratio of inhibition
of RNA synthesis in the presence of actinomycin D or the ratio
of RNA synthesis in the presence of rifampicin is then calculated, using the values obtained from several experiments.
It is found that the ratio of inhibition of 16S and 23S RNA
synthesis obtained in the presence of actinomycin D is 1.57
± 0.09 (the value encompassed by the cross-hatched bar in
Fig. 2), and that the ratios obtained for the remaining synthesis of the two RNA's in the presence of rifampicin is
between 1.47 and 1.53 (the cross-hatched bar in Fig. 3).

By comparing the predicted with the experimental results,
one can clearly eliminate those arrangements in which the gene
for the 23S ribosomal RNA preceeds the gene for the 16S ribosomal RNA (arrangements 3 and 6 in Fig. 1). Arrangement 5,
in which there are tandems of 16S and of 23S ribosomal RNA
genes within a transcriptional unit but with all 16S tandems
preceeding 23S tandems, can also be eliminated. We are left
then with either a random arrangement of genes (arrangement 1)
or an arrangement in which a 16S ribosomal RNA gene precedes
a 23S ribosomal RNA gene (arrangements 2 and 4). Arrangements
1a and 1b both give a predicted inhibition ratio with actinomycin D (or synthesis ratio with rifampicin) of 2. Since the
experimental values are clearly different, these arrangements
can be rejected. Arrangement 1c is compatible with any ratio
depending on the length of the spacers (lengths of DNA that
are not transcribed), but this arrangement is unlikely because

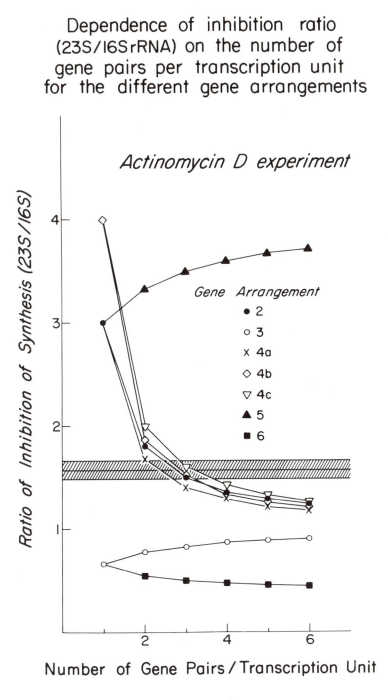

Figure 2.

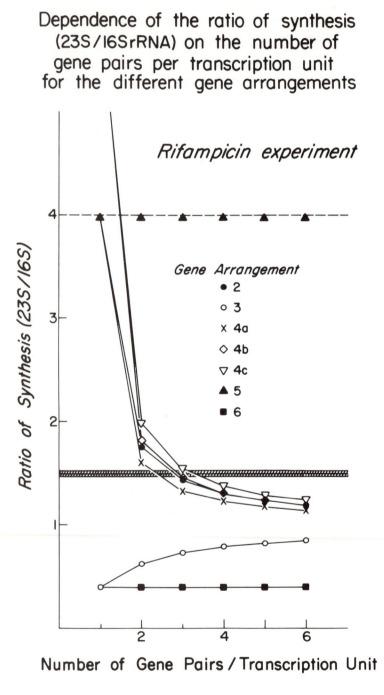

Figure 3.

it requires that each gene be separated from its promoter by
a considerable distance; moreover, the length of the 16S
spacer would need to be markedly longer than the 23S spacer.

We are thus left with arrangements 2 and 4, where the
predicted values are most nearly in accord with the experimental values. Both call for tandem sets of 16S and 23S gene
pairs, with the 16S gene preceding the 23S gene for each pair.
The values, moreover, indicate that the most likely arrangement is a sequence of 2 or 3 pairs in each tandem set. The
data obtained from studies of the hybridization between chloroplast DNA and RNA also indicate that there are 2 to 3 gene
copies per DNA copy in the chloroplast of C. reinhardi (K.S.
Chiang, personal communication). Whether spacer DNA is involved (as in arrangement 4) is currently under investigation
in my laboratory by Rochaix.

THE TRANSCRIPTIONAL AND TRANSLATIONAL SITES FOR THE SPECIFI-
CATION OF THE CHLOROPLAST GENETIC SYSTEM OF C. REINHARDI

The endosymbiont theory for the origin of cell organelles
(see ref. 9 for a recent revival of this theory) holds that
a pre-historic, free-living prokaryote infected a primitive
cell and established a stable relationship with its host. If
such events indeed occurred, then it would be not surprising
to find that the host had acquired certain means to regulate
the genetic activities of the invading organism. Whether or
not the theory is correct, it is becoming clear that the
nuclear-cytoplasmic system of the eukaryotic cells does indeed exert control over the construction of a complete chloroplast genetic system in C. reinhardi, as will become apparent below.

Since the conclusions I will present rest heavily on the

use of rifampicin, a word must be said about its use. Its
effect in C. reinhardi is only maximally observed in vivo,
and certainly only in vitro, if the cells to be treated are
taken from synchronously growing cultures prior to the onset
of their chloroplast RNA synthesis. At this stage in the
cycle, the chloroplast DNA-dependent RNA polymerases are free
to bind to the antibiotic and thus become inactivated. I
should also point out that the chloroplast enzymes from
C. reinhardi and Euglena (10) are apparently different from
the enzymes found in higher plants (11), where rifampicin has
no apparent affinity for any polymerase present and is without
effect on chloroplast ribosomal RNA synthesis.

Previous studies (6a, 12, and experiments cited in the
preceding section) have shown that the genes for chloroplast
ribosomal RNA do reside in chloroplast DNA in C. reinhardi.
This result has been confirmed by electron microscopy (13)
where it is shown that the chloroplasts of cells grown for 4
generations in the presence of rifampicin are virtually devoid
of ribosomes, the cytoplasmic ribosome levels remaining nearly
normal. The use of rifampicin should allow one to determine
whether specific transfer RNA species are also encoded in
chloroplast DNA. The report that the synthesis of certain
transfer RNA species is stimulated by light in Euglena (14)
certainly enhances our interest in seeking possible transfer
RNA genes in the chloroplast DNA of C. reinhardi.

The sites for the transcription of information for the
proteins of chloroplast ribosomes are not known, and there
are only fragmentary data regarding the site where this
information is translated. Cells of C. reinhardi can grow

heterotrophically in the presence of either spectinomycin or chloramphenicol, inhibitors of translation on C. reinhardi chloroplast ribosomes (15, 16), and although such cells have lost the capacity to synthesize certain components of the photosynthetic apparatus (13, 17), they contain chloroplast ribosomes at normal levels, even after 10 cell generations (13). This observation suggests that chloroplast ribosomes can be synthesized under conditions where translation on chloroplast ribosomes is prevented and hence that the site for the translation of the proteins of these ribosomes may be on cytoplasmic ribosomes [a similar suggestion has been made for the site of synthesis of the ribosomal proteins of certain fungal mitochondria (18 - 21)]. We do not yet know whether the chloroplast ribosomes formed under these conditions have a protein complement identical to normal ribosomes; it is certainly possible that a few ribosomal proteins are not synthesized when chloroplast protein synthesis is inhibited, and that the ribosomes formed under these conditions may be defective in their ability to translate messenger RNA. It nonetheless appears likely that at least the bulk of the chloroplast ribosomal proteins are synthesized outside the chloroplast whereas the chloroplast ribosomal RNA is synthesized within the chloroplast. At this key level, therefore, the two genetic systems of the cell seem to take part in a complex kind of interaction.

A similar picture presents itself with regard to the synthesis of the chloroplast-located DNA-dependent RNA polymerase and to the synthesis of the enzymatic apparatus required for chloroplast DNA replication. During long-term

growth experiments in the presence of chloramphenicol or
spectinomycin, chloroplast ribosomal RNA continues to be
synthesized; moreover, neither rifampicin, chloramphenicol,
nor spectinomycin has any apparent effect on the ability of
cells to synthesize chloroplast DNA. In this same series of
experiments, on the other hand, chloroplast membrane organization and photosynthetic capacity are greatly disrupted in
every case; thus the antibiotics are clearly exerting an
effect. Again, we do not yet know whether the nucleic acids
that are synthesized under these conditions are "normal" in
all respects, but the fact that synthesis appears to continue
after several generations of growth suggests that the polymerases (and associated enzymes) are synthesized on cytoplasmic
ribosomes; furthermore, the rifampicin experiments imply that
information for the synthesis of the DNA replicating apparatus
of the chloroplast is encoded in nuclear DNA.

Should other lines of experimentation confirm these
results with antibiotics, then the chloroplast genetic system
is indeed hybrid in nature. How such joint participation in
the construction process could have evolved is, to my mind,
a fascinating but as yet unsolved question.

THE GENETIC SPECIFICATION OF COMPONENTS OF THE PHOTOSYNTHETIC
APPARATUS OF C. REINHARDI

Given the endosymbiont theory for the origin of the
chloroplast of eukaryotic photosynthetic organisms, it is
particularly intriguing to analyze the photosynthetic apparatus
from the point of view of the genetic events of transcription
and translation that specify the different components of the
photosynthetic electron transport chain and photosynthetic
carbon metabolism. In one sense such an analysis is an

exercise in producing a catalogue of the genetic events that
pertain to the chloroplast, but in another, and perhaps more
important sense, the analysis reveals the degree to which
these events rely upon an interaction between the genetic
system of the chloroplast and that of the nucleus and cytoplasm.

The idea behind the analysis is quite simple: it is to
ask which components are made and which are not made when
limitations are placed upon transcription in the chloroplast
or upon translation on chloroplast and cytoplasmic ribosomes.
Two types of experimental approach have been used in this
laboratory. One takes advantage of ac-20, a mutant strain
of C. reinhardi having a deficit in the production of chloroplast ribosomes (22, 23). The other takes advantage of
synchronously growing cultures in which the photosynthetic
capacity doubles prior to cell division and in which the
effect of the inhibitors of transcription and translation on
this doubling can be monitored (17). I shall describe these
two approaches in turn.

The analysis with ac-20

Many of the general properties of ac-20 have been
described elsewhere (22 - 24), and a few will be described
here in order to facilitate an understanding of the analysis.
In C. reinhardi, heterotrophic growth (in the dark) or mixotrophic growth (in the light) occurs in the presence of
acetate as a source of reduced carbon. A photosynthetically
active chloroplast is synthesized under these conditions.
Phototrophic growth occurs in a minimal medium with CO_2 as
the sole carbon source. When as-20 cells undergo mixotrophic

growth, they come to possess levels of chloroplast ribosomes
that are 5 to 10 per cent of those of the wild type. When
ac-20 cells are cultured in minimal medium, on the other hand,
the number of chloroplast ribosomes is perhaps 25 per cent of
the wild type. If ac-20 cells are transferred from an
acetate-supplemented medium to a minimal medium, the number
of ribosomes will increase to perhaps 40 per cent of the
wild-type level. This increase occurs relatively slowly if
the transferred cells are maintained in minimal medium in the
dark, requiring at least 12 hours; in contrast, when the
transfer is carried out in the light, the increase takes place
within a few hours (22).

The in vivo photosynthetic capacity of ac-20 cells
cultured in acetate-supplemented medium is at best 4 per cent
of wild type, whereas for phototrophically growing cells it
is at least 25 per cent. When cells are transferred from
acetate-supplemented to minimal medium, the photosynthetic
capacity increases only after the chloroplast ribosome
increase has occured, and then only if the cells are placed
in the light. For example, the number of chloroplast ribosomes
can be allowed to increase in the dark, as described above,
but there is no concomitant increase in photosynthetic
capacity; when such cells are placed in the light, photosynthetic rates rise rapidly and reach a steady state within
a few hours. On the other hand, if cells are placed in the
light immediately after transfer, there is a lag in the onset
of increased photosynthetic capacity which corresponds in
duration to the time taken for the number of chloroplast
ribosomes to increase (22).

Clearly, the photosynthetic capacity of these cells seems to be correlated with the level of chloroplast ribosomes present, and thus it can be reasonably proposed that the photosynthetic capacity is dependent upon translational steps that take place on these ribosomes. Our analysis (24, 25, and Levine and Armstrong, unpublished data) has revealed that four components of the photosynthetic electron transport chain are affected, namely the ascorbate reducible cytochrome 559, and cytochromes 552, and 563, and Q, plus one enzyme of the photosynthetic carbon reduction cycle, ribulose 1,5-diphosphate carboxylase. Each of these undergoes an increase in amount when cells are transferred from acetate-supplemented medium to minimal medium. The proposition that these components require translational steps on chloroplast ribosomes for their synthesis is supported by the experiments that will be described in the following section.

We have studied (23) the effects that rifampicin, spectinomycin, chloramphenicol, and cycloheximide have on the recovery of photosynthetic capacity under experimental conditions where chloroplast ribosome recovery has already occurred. The study described here focuses on levels of photosynthetic electron transport as measured by the Hill reaction, and thus upon the recovery of cytochromes 559 and Q. Transfer experiments are performed as described above; the cells are incubated in the dark in minimal medium to allow chloroplast ribosome levels to recover (a light-to-dark transfer) and the culture is then divided into several parts. One serves as a control, and the others receive appropriate aliquots of either rifampicin, spectinomycin,

chloramphenicol, or cycloheximide. The recovery of the Hill
reaction rates in the ensuing light period is then measured.
In the control, as previous experiments had shown (24), full
recovery is achieved 9 to 12 hours in the light. In the presence of rifampicin, chloramphenicol, or spectinomycin, on the
other hand, no increase in Hill activity occurs during this
period. In contrast, cycloheximide has no effect on recovery.
These results are summarized in Figs. 4-7.

Figures 4-7 also indicate the effects of the antibiotics
when they are administered at the time cells are transferred
to minimal medium and placed directly in the light (a light-
to-light transfer), that is, at a time when ribosome recovery
has not yet begun. The same results are obtained under these
experimental conditions, namely that cycloheximide does not
block recovery whereas the other three antibiotics do.

These results, in accord with other studies of C. reinhardi, 22-26 support the contention that both transcription
and translation within the chloroplast are essential for the
formation of a functional chloroplast.

Chloroplast membranes are disorganized and the pyrenoid
is deformed in cells of ac-20 that are cultured under mixotrophic growth conditions. Upon transfer of these cells to
a minimal medium and after the number of chloroplast ribosomes
has attained the "phototrophic level", chloroplast membranes
come to have a more normal organization and a more normal
pyrenoid appears within the chloroplast.

Similar kinds of experiments in which the recovery of
chloroplast membrane organization in ac-20 is monitored after
the administration of antibiotics also indicate that trans-

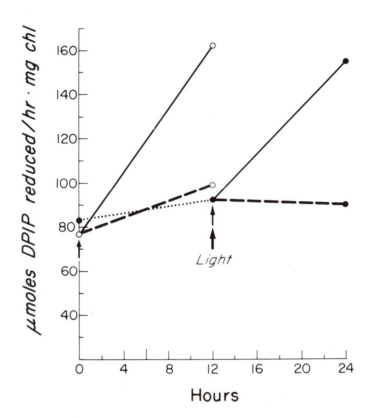

Figure 4. Effect of 250 µg/ml riframpicin on recovery of Hill activity in mixotrophic *ac-20* transferred to minimal medium at 0 hr. Open symbols represent values from light-to-light transfer experiment, closed symbols represent values from light-to-dark transfer experiment. Control values are indicated by circles and solid lines. Values from rifampicin-treated cells are symbolized by circles and dashed lines. For the light-to-dark transfer experiment, reaction rates at the beginning and at the end of the dark incubation period are given; the two values are connected by a dotted line. The time that light is provided in the light-to-dark transfer experiment is indicated. Rifampicin was added to the experimental cultures at times indicated by the thinner arrows.

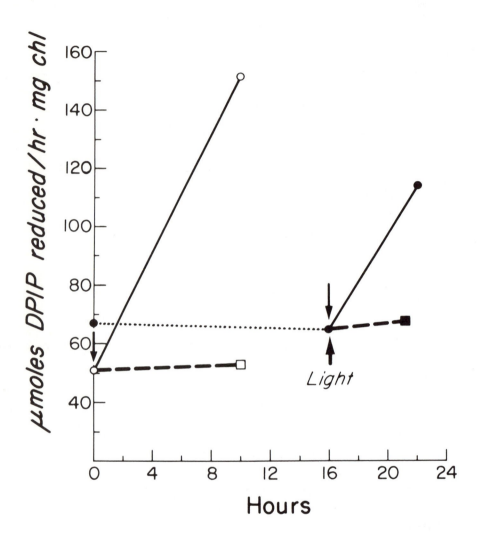

Figure 5. Effect of 100 μg/ml chloramphenicol on recovery of Hill activity in mixotrophic *ac-20* transferred to minimal medium at 0 hr. Symbols as described for *Fig. 4* except that values from chloramphenicol-treated cells are symbolized by squares, and thinner arrows indicated times of additions of chloramphenicol.

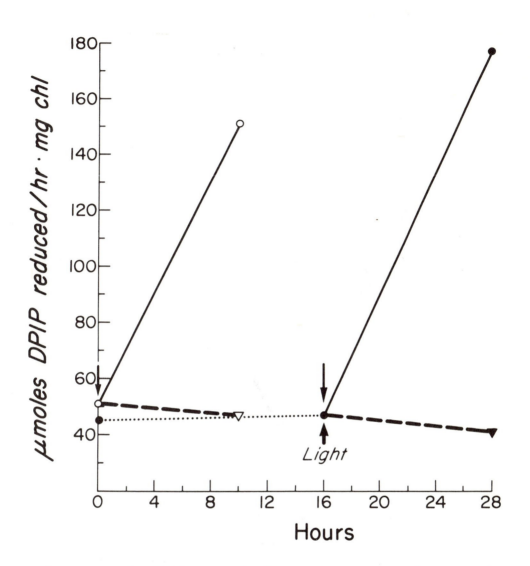

Figure 6. Effect of 3 μg/ml spectinomycin on recovery of Hill activity in mixotrophic *ac-20* transferred to minimal medium at 0 hr. Symbols as described for *Fig. 4* except that values from spectinomycin-treated cells are symbolized by triangles, thinner arrows indicate times of addition of spectinomycin.

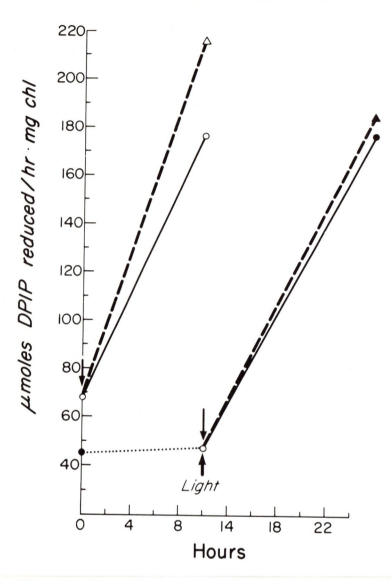

Figure 7. Effect of 1 μg/ml cyclohex-imide on recovery of Hill activity in mixotrophic *ac-20* transferred to minimal medium at 0 hr. Symbols as described for *Fig. 4* except that values from cycloheximide-treated cells are symbolized by triangles, thinner arrows indicate times of addition of cycloheximide.

lational steps on chloroplast ribosomes are essential for the
construction of a normally constructed system of chloroplast
membranes in C. reinhardi.

Finally, the experiments with ac-20 have suggested the
interesting possibility that light may have a stimulatory effect on the transcription of chloroplast DNA. As mentioned
earlier, the recovery of photosynthetic capacity requires both
the increase in the level of chloroplast ribosomes, an event
that can occur in the dark, and the exposure of such "ribosome-
recovered" cells to light. If this exposure takes place in
the presence of rifampicin, no photosynthetic recovery occurs.
Thus the light appears to have the effect of triggering a
rifampicin-sensitive transcriptional event; whether this
triggering is direct or indirect remains unknown.

The analysis with synchronous cultures

Cells of C. reinhardi can be cultured in synchrony under
phototrophic conditions by exposing them to a 12 hour light/
12 hour dark cycle (27). The amounts and activities of
chloroplast components increase in the light at a time when
the number of ribosomes per cell is not increasing and the
cells are not dividing. Cell division occurs during the
dark period. The time when the chloroplast components
increase can be utilized to ask the following question: which
components and activities depend for their increase on transcriptional events within the chloroplast and translational
events on chloroplast or cytoplasmic ribosomes? This question
can be answered by exposing the cells to the appropriate
inhibitor of protein synthesis during the period of time when
the component or activity normally increases (17). Figure 8

shows the change in photosynthetic and respiratory rates in a
control culture and in cultures exposed to the antibiotics.
It can be seen that the expected increase in photosynthetic
oxygen evolution is inhibited by rifampicin, spectinomycin,
and cycloheximide, whereas the increase in the rate of
respiration is affected only by cycloheximide. These results
have been interpreted (17) to mean that the doubling of
chloroplast components prior to cell division requires transcriptional and translational events within the chloroplast
and translational events in the cytoplasm. (We have as yet
found no way to selectively inhibit transcription in the
nucleus.)

When a similar analysis is carried out to determine which
of several readily identifiable chloroplast components are
affected by the three antibiotics, results are obtained that
are summarized in Table 1. Chloroplast DNA appears to contain
information for the synthesis of cytochromes 553 and 564
since their increase is prevented by rifampicin. Cytochrome
553 poses some interesting questions. A mutant strain,
ac-206, is known that lacks cytochrome 553 (28), and since
the mutant gene shows a Mendelian pattern of inheritance, the
possibility exists that both the nucleus and the chloroplast
contain information for the synthesis of this cytochrome (see
discussion in ref. 17).

The synthesis of chlorophyll, ferredoxin, ferredoxin NADP
reductase, phosphoribulokinase, and ribulose 1,5-diphosphate
carboxylase (17) as well as that of the carotenoids (Sirevag,
unpublished observations) does not appear to require transcriptional events within the chloroplast, since their increase

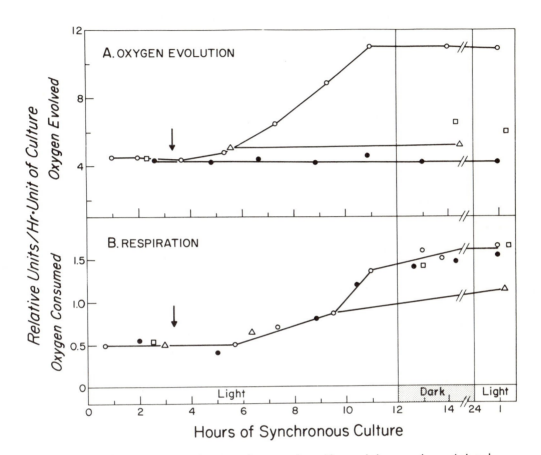

Figure 8. Effects of spectinomycin, rifampicin, and cyclohexi-mide on the increase of (A) photosynthetic oxygen evolution and (B) respiration in synchronous cultures of <u>C. reinhardi.</u> ∘, control; •, 3 μg/ml spectinomycin; ▫, 250 μg/ml rifampicin; △, 1 μg/ml cycloheximide. Spectinomycin and rifampicin were added 50 min before zero time; cycloheximide was added at the time indicated by the arrow.

is insensitive to rifampicin. This kind of conclusion,
however, is valid only if the lifetime of a given species of
chloroplast messenger RNA is assumed to be shorter than one
cycle of growth. If not, then rifampicin would not be expected
to have a detectable effect on the synthesis of that messenger
RNA's protein product. Whether or not long-lived chloroplast
messenger RNA does indeed exist in C. reinhardi could be
investigated by the use of actinomycin D in synchronous
cultures.

Translational events required in the formation of chloroplast components are shared between the chloroplast and cytoplasmic ribosomes. Table 1 shows that the synthesis of
chlorophyll, ferredoxin, ferredoxin NADP reductase,
phosphoribulokinase (17), the carotenoids, β-carotene,
neoxanthin, and trollein (Sirevag, unpublished observations)
depends upon active cytoplasmic ribosomes but is independent
of chloroplast ribosomes. Cytochrome 552, however, shows the
reverse dependence since its synthesis is prevented by
spectinomycin but not cycloheximide. The synthesis of the
carotenoids lutein and violaxanthin is unaffected by either
spectinomycin or cycloheximide (Sirevag, unpublished
observations). This result could be explained by assuming
that the enzymes responsible for the biosynthesis of these
carotenoids are present in sufficient amounts so that neither
antibiotic exerts any recognizable effect. This explanation
is certainly no satisfactory. Unfortunately too little is
known of the enzymes required for the biosynthesis of these
carotenoids.

Two components, cytochrome 563 and ribulose 1,5-diphos-

phate carboxylase, show a dual dependence in that translational
events on both cytoplasmic and chloroplast ribosomes appear to
be required for their synthesis. In the case of ribulose
1,5-diphosphate carboxylase, it has been shown in Criddle's
laboratory (29) that the synthesis of polypeptides of one of
the enzyme's subunits is inhibited in barley by chloramphenicol
whereas the synthesis of the polypeptides of the other subunit
is inhibited by cycloheximide. Thus dual control over the
polypeptides of a single complex enzyme has apparently evolved
both for C. reinhardi and for a higher plant.

TABLE 1

Chloroplast Components of C. reinhardi Whose Synthesis
in Synchronous Cultures is affected by Inhibitors
of Transcription and Translation

Chloroplast Component	Rifampicin	Spectinomycin	Cycloheximide
Chlorophyll	–	–	+
Carotenoids			
β-carotene	–	–	+
lutein	–	–	–
violaxanthin	–	–	–
neoxanthin	–	–	+
trollein	–	–	+
Cytochrome 553	+	+	–
Cytochrome 563	+	+	+
Ferredoxin	–	–	+
Ferredoxin NADP reductase	–	–	+
Phosphoribulokinase	–	–	+
Ribulose 1,5-diphosphate carboxylase	–	+	+

+ = antibiotic inhibits synthesis

– = antibiotic does not inhibit synthesis

CONCLUSIONS

The chloroplast genetic system and the photosynthetic components of the chloroplast of C. reinhardi are of hybrid origin, their specification at the transcriptional level being determined by information both in nuclear and chloroplast DNA and their translation occuring on both chloroplast and cytoplasmic ribosomes or, in certain cases, on both classes of ribosomes. An attempt to summarize the minimal sites of genetic specification is given in Table 2. It is essential to keep in mind in examining this table that the effects of the inhibitors reveal only that there are requirements for site-specific transcription or translation; they do not necessarily reveal the nature of these requirements. For example, the inhibition of the synthesis of cytochrome 553 by rifampicin could mean that the chloroplast DNA contains information for the primary structure of the protein moiety of the molecule or, alternatively, for the synthesis of an enzyme that in turn is required for the biosynthesis of the heme moiety of the cytochrome. The fact that both cycloheximide and spectinomycin inhibit the synthesis of cytochrome 563 implies only that translational steps on both cytoplasmic and chloroplast ribosomes are required. These steps could be in the synthesis of proteins upon which cytochrome synthesis in turn depends.

TABLE 2

A Tentative Assignment of Sites of Transcription and
Translation Required for the Synthesis of Some
Chloroplast Components of C. reinhardi

Chloroplast Component	Transcriptional Site(s)	Translational Site(s)
DNA polymerase and other enzymes involved in DNA replication	probably nucleus	cytoplasm
DNA-dependent RNA polymerase	probably nucleus	cytoplasm
Ribosomal RNA (5s, 16s, 23s)	chloroplast	-
(Most) ribosomal proteins	?	cytoplasm
Phosphoribulokinase	probably nucleus	cytoplasm
Ribulose 1,5-diphosphate carboxylase	probably nucleus	chloroplast and cytoplasm
Ferredoxin	probably nucleus	cytoplasm
Ferredoxin NADP reductase	probably nucleus	cytoplasm
The ascorbate reducible cytochrome 559	probably nucleus	chloroplast
Cytochrome 553	nucleus and chloroplast	chloroplast
Cytochrome 563	chloroplast	chloroplast and cytoplasm
Chlorophyll	probably nucleus	cytoplasm
β-carotein	probably nucleus	cytoplasm
Lutein	probably nucleus	?
Violaxanthin	probably nucleus	?
Neoxanthin	probably nucleus	cytoplasm
Trollein	probably nucleus	cytoplasm

The danger in summarizing data as in Table 2 is that the summary implies that questions have been answered. In fact, our research has left us very impressed by the complexity of the interactions between the nuclear-cytoplasmic and the chloroplast genetic systems, and the recent work of Surzycki and Gillham (30) indicates that a third genetic system, apparently localized in mitochondria, also interacts with the first two systems in C. reinhardi. Thus I close by emphasizing our ignorance of how these genetic systems actually control their closely interdependent synthetic activities. Nothing is known, for example, about how a protein synthesized in the cytoplasm is selectively taken up by an organelle, nor whether the products of any regulatory genes in one genetic system exert an effect on the other. Until such fundamental questions can be answered, it is difficult to speculate profitably on how the present interdependence might have evolved.

ACKNOWLEDGEMENT

    Research in the author's laboratory is supported by a grant from the National Science Foundation (GB 1866).

REFERENCES

1. R. P. Levine and U. W. Goodenough, Ann. Rev. Genetics 4, 47 (1970).

2. K. S. Chiang, Autonomy and Biogenesis of Mitochondria and Chloroplasts. 1971. Boardman, N. K. Linnane, A. W., and Smillie, R. M. eds. Amsterdam, North-Holland, 235.

3. R. Sager and Z. Ramanis, Symp. Soc. Exp. Biol. 24, 401 (1970).

4. D. Bastia, K. S. Chiang, H. Swift, and P. Siersma, Proc.
    Natl. Acad. Sci. U.S. (1971). In press.

5. M. Bleyman, M. Kondo, N. Hecht, and C. Woese. J. Bact.
    99, 535 (1969).

6. M. Bleyman and C. Woese, Proc. Natl. Acad. Sci., U. S.
    63, 532 (1969).

6a. S. J. Surzycki, Proc. Natl. Acad. Sci. U. S. 63, 1327
    (1969).

7. O. L. Miller and B. R. Beatty, Science 164, 955 (1969).

8. U. E. Loening, Symp. Soc. Gen 1 Microbiol. 20, 77 (1970).

9. H. Ris and W. Plaut, J. Cell Biol. 13, 383 (1962).

10. R. D. Brown, D. Bastia, and R. Haselkorn, First Lepetit
    Colloquium, Florence, Italy (1969).

11. W. Bottomley, D. Spencer, A. M. Wheeler, and P. R. Whit-
    feld, Arc. Biochem. Biophys. 143, 269 (1971)

12. S. J. Surzycki, U. W. Goodenough, R. P. Levine, and J. J.
    Armstrong, Symp. Soc. Exp. Biol. 24, 13 (1970).

13. U. W. Goodenough, J. Cell Biol. (1971). In press.

14. W. E. Barnett, C. J. Pennington, and S. A. Fairchild,
    Proc. Natl. Acad. Sci., U.S. 63, 1261 (1969).

15. W. G. Burton, J. Cell Biol. 47, 28a (1970).

16. J. K. Hoober and G. Blobel, J. Mol. Biol. 41, 121 (1969).

17. J. J. Armstrong, S. J. Surzycki, B. Moll, and R. P.
    Levine, Biochemistry 10, 692 (1971).

18. P. J. Davey, R. Yu, and A. W. Linnane, Biochem. Biophys.
    Res. Commun. 36, 30 (1969).

19. H. Kuntzel, Nature 222, 142 (1969).

20. W. Neupert, W. Sebald, J. Schwab, A. Pfaller, and T.
    Bucher, Eur. J. Biochem. 10, 585 (1969).

21. W. Neupert, W. Sebald, A. J. Schwab, P. Massinger, and
     T. Bucher, Eur. J. Biochem. 10, 589 (1969).

22. U. W. Goodenough and R. P. Levine, J. Cell Biol. 44,
     547 (1970).

23. U. W. Goodenough and R. P. Levine, J. Cell Biol. (1971).
     In press.

24. R. K. Togasaki and R. P. Levine, J. Cell Biol. 44, 531
     (1970).

25. R. P. Levine and A. Paszewski, J. Cell Biol. 44, 540
     (1970).

26. J. K. Hoober, P. Siekevitz, and G. E. Palade, J. Biol.
     Chem. 244, 2621 (1969).

27. J. Kates and R. F. Jones, J. Cell Comp. Physiol. 62,
     157 (1964).

28. D. G. Gorman and R. P. Levine, Plant Physiol.

29. R. S. Criddle, B. Dau, B. Kleinkopf, and R. C. Huffaker,
     Biochem. Biophys. Res. Commun. 41, 621 (1970).

30. N. W. Gillham and S. J. Surzycki, Proc. Natl. Acad.
     Sci., U. S. (1971). In press.

DISCUSSION

MELTON: (1) Are the chloroplasts permeable to actinomycin? (2) Is there any selective probe available for the
role of nuclear transcription?

LEVINE: (1) Apparently, for Surzycki and Rochaix have
found that it inhibits the synthesis of Chlamydomonas
chloroplast DNA. (2) I know of none so far for Chlamydomonas.

EDMUNDS: Have you found any evidence for cyclic changes in photosynthetic capacity in <u>non-dividing</u> cells (e.g., stationary cultures) under constant illumination?

LEVINE: We have not studied the photosynthetic capacity of cells from stationary cultures.

Episomes in Evolution

ALLAN CAMPBELL
Department of Biological Sciences, Stanford University
Stanford, California 94305

It may appear superfluous to discuss the role of episomes in evolution when cogent reasons have been presented for abandoning the term "episome" altogether.[1] Terminology being largely a matter of convenience, time should not be consumed in defending archaic nomenclature. Nevertheless, I feel that the objects originally included in this category have special evolutionary potentialities and constraints not completely shared by anything else.

EPISOMES AND PLASMIDS

In 1958, Jacob and Wollman[2] proposed the term "episome" for genetic elements such as bacteriophage lambda or the fertility factor F that could replicate in two alternative modes--either as part of the bacterial chromosome or as autonomous, extrachromosomal elements. The term has been useful in encouraging a unified approach to the manner in which the chromosome and extra element associate; but this purpose has now been served. The older term "plasmid" includes all autonomous extrachromosomal elements, whether or not chromosomal integration is demonstrable. It is more useful than "episome" in many contexts because it includes objects obviously related to known episomes but which integrate rarely if at all. Bacterial sex factors, for example, include

534

both the classical F factor, whose integration is sufficiently frequent

to have been discovered in the course of investigations on the mechanism

of mating, and others such as resistance transfer factors, which promote

rare transfer without integration.

Of course, the term "episome" does not denote a phylogenetic taxon.
Neither do "plasmid," "sex factor" or "virus." The utility of all these
categories must be sought either in terms of operational convenience or
of unified treatment of problems common to different members of the
category.

Hayes' remarks are directed at the category, not at the word
"episome." They cannot be answered by re-labeling episomes as "integrating plasmids," for instance. The problem is that, since different
elements integrate at vastly different rates and probably by various
mechanisms, no clear operational distinction is feasible between integrating and non-integrating elements. I postponed the same problem
elsewhere[3] by stipulating that the term episome should ultimately apply
only where ability to integrate is the product of natural selection
rather than happenstance. In principle, this approach still seems
logical to me: If there were unequivocal evidence that the ability of
F to integrate had been the object of direct selection, I doubt that
anyone would object to including it in a category of "integrating
elements." On the more practical level of devising a working nomenclature to classify present knowledge, I can't say whether the ability of
a plasmid to integrate at any rate whatever merits special attention.

Integration of an episome has two requirements: (1) The element
must be inserted into the chromosome. (2) The resulting structure must
survive and replicate with sufficient hereditary stability to be recognizable as a biotype. The diversity of recombinational mechanisms
assures that insertion at some rate should be possible for any element.

Present knowledge of replication control does not permit assessment of how restrictive the second requirement may be.

COMPOSITE REPLICONS

Figure 1 diagrams insertion of the lambda prophage into the chromosome of its host Escherichia coli. It is a formal genetic scheme which can be interpreted literally with respect to the behavior of DNA molecules. It can be extended to other cases where two circular elements, initially separate, fuse to form one. We know of no other way that two elements can become joined.

Note the symmetry of the basic event. We say the prophage is inserted into the bacterial chromosome. A statement that the bacterial chromosome is inserted into the prophage would describe Figure 1 equally well.

One feature not explicit in Figure 1 is that each element--bacterial chromosome and phage genome--can replicate as such. The chromosomes of both λ and E. coli seem to have the properties that Jacob et al.[4] suggested a replicon should have. Each element contains genes encoding diffusible products that initiate replication at a specific site on the DNA of that element. Replication then proceeds in one or both directions from that site.

After fusion, the product (lysogenic chromosome) bears all the information--both specific genes and specific initiation sites--of both replication systems. I will call any such element formed by the integration of two replicons and bearing the complete information for initiation at more than one distinct site (whether or not this information is expressed) a composite replicon. We would like to know whether both systems are in fact active in any one composite replicon; and if not, why one of them is non-functional.

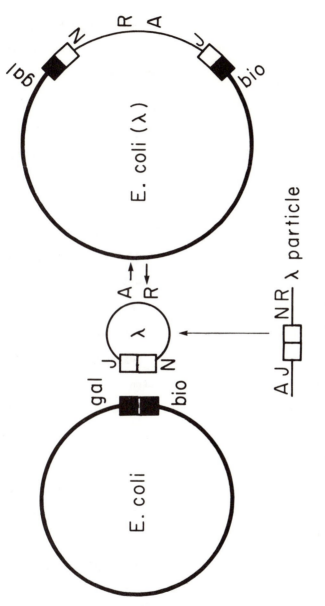

Figure 1. Insertion of λ prophage into the chromosome of its host. A, J, N, and R are genes of the phage. gal and bio are genes of the host.

Replication of Composite Replicons

Replication of lambda is inhibited by a repressor encoded by the phage gene cI. Both synthesis of specific replication proteins (O and P) and ability of the initiation site to serve as template for them are inhibited by repressor. In those cells that survive viral infection and become lysogenic, the cI gene is turned on and lambda replication is turned off. In the lysogenic chromosome, only the bacterial replication system is expressed. Prophage DNA is passively replicated as part of the bacterial chromosome.

Insertion of F into the bacteria chromosome proceeds in a manner similar to that shown in Figure 1. In the resulting Hfr chromosome, measurements of origin and direction of replication show that the bacterial replication system is expressed while the F-specific system ordinarily is not.[5] A plausible explanation for the shutoff of F replication has been given by Pritchard et al.[6] They suppose that F replication initiation is triggered by the replication act itself. Subsequent initiation does not take place until the inhibitor concentration falls, due to dilution with growth. An autonomous F element can then seek its own time for replication within the cell cycle; but in an Hfr chromosome, bacterial replication of F can maintain the inhibitor concentration above the threshold value at all times. Little direct evidence relevant to this ingenious hypothesis is available.

If the bacterial initiation system is rendered thermolabile by mutation, an inserted F element can allow growth at high temperature, apparently by providing an alternative replication system.[7] Some Hfr derivatives of this same bacterial strain are still thermosensitive, indicating that not every inserted F can function in this manner.

Another class of complex replicons arise from fusions of extra-chromosomal elements with each other rather than with the bacterial

chromosome. An example is lysogenization of an F' element. F' elements
are variants of F in which small segments of the bacterial chromosome
are inserted. If the bacterial segment includes the lambda attachment
site, lambda can insert there. In cells carrying such lysogenic F' elements, lambda repressor shuts off the phage replication system so that
the composite element replicates as an F.

A less contrived example may be the R agents, which confer drug
resistance and mating ability on cells harboring them. Such agents can
apparently form by integration of a sex factor similar to F and a plasmid that alone confers resistance but not ability to conjugate.[8] Without detailing the variations of individual systems upon this theme (and
without the evaluation of individual systems on which interpretation
must ultimately stand or fall) I have sketched in Figure 2 one concep-
tion of R agents. For some R isolates, molecular species corresponding
to the circles of Figure 2 have been identified.[9] Rapid dissociation of
the composite element into components occurs in certain host strains.
Replication control of the components and the complex is not well
understood.

There is no known composite replicon in which two alternative
replication systems are expressed simultaneously. Perhaps such an element would not survive. If that is so, the combinations of elements
that can integrate are limited. If integration has contributed to
survival, selection should have favored compatibility of replication
systems.

Insertion and Excision

Laboratory experiments cannot prove the significance of a process
in nature, but may provide relevant insights. If insertion and excision
are important to natural phage development, their occurrence in the life
cycle might be regulated rather than haphazard. In fact, the lambda

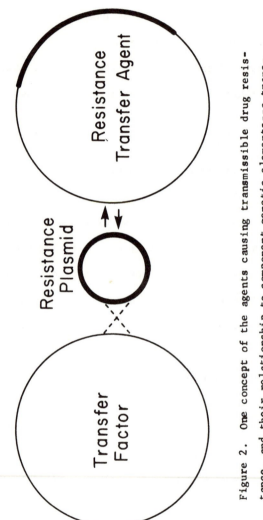

Figure 2. One concept of the agents causing transmissible drug resistance, and their relationship to component genetic elements--a transfer factor, which causes bacterial conjugation; and a resistance
plasmid, that replicates extrachromosomally and causes drug resistance.

phage determines a highly specific, elaborate and precisely controlled
system that serves no obvious purpose beyond getting the phage in and
out of the bacterial chromosome at appropriate times during its life
cycle.

The insertion act depicted in Figure 1 comprises a reciprocal
recombination between two unique sites on the chromosomes of phage and
bacterium respectively. These two sites have no detectable homology
with each other. Recombination between them requires the product of a
phage gene (int), specific to this combination of sites. Another phage
(21), related to lambda, inserts at a different site on the bacterial
chromosome. The 21 int gene product will not substitute for that of
lambda.[10]

Insertion forms a lysogenic chromosome. To regenerate virus from
this complex, the phage chromosome must be excised. Excision, like
insertion, requires the int gene product. The product of a second phage
gene, xis, is also required for excision but not insertion.[11]

A difference in the catalytic requirements for excision and insertion makes sense only in conjunction with the non-equivalence of phage
and bacterial sites of recombination. If the two sites were identical,
then insertion and excision would be chemically indistinguishable.

If the site on the bacterium is denoted B.B' and that on the phage
P.P', the requirements for insertion and excision are described as

$$(1) \quad P.P' + B.B' \xrightleftharpoons[int + xis]{int} P.B' + B.P'$$

The apparent difference in catalytic requirements for forward and
reverse directions of the same reaction must be rationalized with the
second law of thermodynamics. Two suggestions have been made. Dove[12]
proposed that the combination P.B' + B.P' has a lower energy content
than P.P' + B.B' and that the equilibrium of the reaction as written is

therefore to the right. The product of the _xis_ gene functions by pumping energy into the reaction to force it to the left.

A second possibility[3] is that neither insertion nor excision is the
simple exchange reaction implied by equation (1), but that in both directions the complete reaction is essentially irreversible. The reactions
might be, for example,

$$(2) \quad P.P' + B.B' + ATP \xrightarrow{\text{int}} P.B' + B.P' + ADP + iP$$

$$(3) \quad P.B' + B.P' + ATP \xrightarrow{\text{int + xis}} P.P' + B.B' + ADP + iP$$

This is tantamount to saying that energy is "pumped into" the
reaction in both directions. It satisfies the thermodynamic requirements for microscopic reversibility, because (2) and (3) do not constitute reverse reactions.

Equations (2) and (3) more closely describe the chemical models
commonly proposed for recombination in general than does equation (1).
Equation (1) portrays the catalyzed breaking and rejoining of two DNA
molecules in both strands as a simple exchange reaction, leaving no
nicks that require resynthesis or covalent bond formation. It is only
under this literal interpretation that the relative energy contents of
the various nucleotide sequences need determine the ratio of reactants
to products at the steady state.

Whatever the chemical mechanism turns out to be, the biological
implications are noteworthy. Since P.P' and B.B' are non-identical, no
detectable recombination between them is catalyzed by host recombinases.
The phage can control the time of insertion or excision by regulating
the activities of the _int_ and _xis_ genes. The different catalytic
requirements for insertion and excision allow regulation of the direction as well as the extent of the reaction.

Like most phage genes, _int_ and _xis_ are repressed in the lysogenic cell. This guarantees stability of the lysogenic state. Both genes are expressed following infection of non-lysogenic cells, or following induced derepression of lysogenic cells. This allows prophage insertion after infection, and excision after induction. There is no firm evidence that the phage utilizes the obvious possibility of controlling the direction by differential regulation of _int_ and _xis_ gene transcription. Rather these two adjacent genes are thought to be expressed coordinately.

However, Weisberg and Gottesman[14] have shown that _int_ function is much more stable than _xis_, where both functions are assayed by an ingenious _in vivo_ test. The greater stability of _int_ provides a sufficient explanation for the prevalence of insertion rather than excision following infection. In those cells that survive infection and can be recovered as lysogens, survival requires establishment of repression before phage development has proceeded too far. Once repression is established, it is assumed that _int_ and _xis_ genes are not further expressed. The greater _in vivo_ stability then assures that insertion is ultimately favored.

Phage P2 uses a different method to stabilize the inserted state. As in λ, the _int_ gene is adjacent to the _att_ site of this phage. Unlike λ, the P2 _int_ gene is not repressed in lysogenic bacteria but rather is unexpressed because the insertion event separates it from its normal promoter.[15]

Insertion of the F agent into the bacterial chromosome in a manner similar to that shown in Figure 1 produces an Hfr chromosome. In contrast to λ, insertion and excision occur at low frequency (estimated at 10^{-4} per generation for insertion). The genetic control of F insertion is not known.

Significance of Rare Integration and of Gene Transfer

In λ and P2, insertion and excision are determined by specific gene products whose synthesis and activity are precisely regulated. When these products are fully active, insertion and excision are very frequent. Dissociation of R factors into their components also seem to be frequent under appropriate circumstances. Insertion and excision of F are rare and perhaps unregulated.

Where these reactions are rapid and controlled, it is hard to doubt that they serve some function in the life cycle of the element concerned. The exact reason why it pays λ to insert can be debated, but the specialized machinery required for the process can hardly have developed and survived by accident.

On the other hand, the converse conclusion is unwarranted: Just because F integration appears rare and haphazard, we cannot assert that integration is insignificant in the natural history of F or its hosts. The observed rate may be sufficient to the need; and rare events need not be unimportant.

One purpose that F insertion may serve is to effect transfer of host genes in conjugation. The significance of the "sex factor" function of F has itself been debated on similar grounds.

F was discovered during studies of bacterial gene transfer. Further work showed that F primarily promotes its own intracellular transfer, and only rarely that of other genes. Conjugation serves mainly to disseminate F among hosts, and gene transfer seems to be a minor side reaction.

This places the relation of F to gene transfer in proper perspective: Just as viruses can sometimes transduce host genes, F allows their occasional transfer. Both agents raise the same general questions, common to most host-parasite relationships. In the long run, both host and parasite should evolve so that the complex has an advantage over parasite-free competitors. Recombinational potential is a reasonable guess for the contribution that F or λ make to the welfare of E. coli.

Bodmer[16] has summarized the best evidence for the evolutionary role of recombination in prokaryotes, together with some theoretical arguments for its relative unimportance in this group of organisms. He concludes that the optimal situation may be a predominantly asexual mode of replication, with recombination kept in reserve for circumstances when rapid evolution is needed.

Additional theoretical analysis addressed to the question of whether its contribution to recombinational potential can maintain a fertility factor in the bacterial population should help define the possibilities. Another question is whether the selective value conferred on fertility factors by their ability to mediate host recombination has played an important role in their survival.

The drug resistance factors may be instructive in this regard. As they are the only example of their kind, the facts are subject to diverse generalizations. I shall emphasize one. This is the only instance on record where natural bacterial populations have been subjected to radical large scale alterations in selective conditions and the genetic basis of their response examined in detail.

The alteration was the widespread introduction by man of antibiotics into humans, animals, and their environment. The response of the coliform bacteria in the intestines of these subjects was consistent throughout the world: Although antibiotic-resistant variants can arise through mutation in pure laboratory cultures, such variants do not predominate when the experiment is done in the wild. Instead, selection favors the acquisition of drug resistance plasmids.

My inclination is to generalize from this one example and say that natural bacterial populations may frequently respond to drastic changes in their environment by recombination rather than by fresh mutation. The alternative hypothesis that there is something special about drug resistance is equally tenable. But if we ask, "How do natural bacterial populations evolve?" we probably should not ignore the one case where the process has been observed. In speaking here of evolution by recombination, we use the term in its broadest sense, to denote selection of individuals whose genetic information derives from two parental sources, whether or not homologous replacement is involved.

Whereas increasing numbers of previously undetected plasmids and transfer factors are coming to light, in nature any one factor may have a restricted range. Most natural isolates of E. coli, for example, although susceptible to infection with the classical F factor, do not harbor it. This suggests that selection favors F$^-$ bacteria, perhaps merely because of the metabolic expense of maintaining F and its products.

Maintenance of F in E. coli might then be attributed to a balance between infection, that converts F$^-$ to F$^+$, and selection, favoring F$^-$ One way that such a balance might be stabilized is by continued rare contact of E. coli with cells of some other species that harbor F and provide a permanent reservoir. In the reservoir species, F would necessarily be selectively advantageous or at most neutral.

Drug resistance is determined by a complex replicon in which transfer element and resistance plasmid have joined. Whether these resistance genes were generally plasmid-borne in their species of origin or became so during transfer cannot be specified. At any rate, addition of antibiotics to the human environment allowed widespread dissemination only of those transfer agents that could associate with resistance

determinants. To be effective, the occasional strong selection for
integrative ability must counterbalance mutational pressure favoring its
loss.

Bacterial recombination, despite its haphazard appearance at the
individual level, seems marvelously well adapted to its purpose. Not
only is the capacity for sexual reproduction frequently held in abeyance,
but the determinants of recombination themselves can be perpetuated in a
small fraction of the bacterial population and thus impose a minimal
metabolic load.

FUSIONS BETWEEN DIFFERENT REPLICONS

Integration of two replicons into one composite structure creates
further evolutionary possibilities. In a lysogenic bacterium, for
example, deletion or rearrangement can produce structures from which
excision is no longer possible, so that genes originally belonging to
one replicon become permanently welded into the other. No novel
principles are involved. A few illustrative examples, taken for convenience from work in our laboratory on bacteriophage lambda, will be
given here. I will first describe them as laboratory creations and
return later to their possible significance.

Incorporations of Bacterial Genes into the Lambda Virus

Figure 1 shows the normal lysogenic cycle of lambda. Regardless
of chemical detail, excision results in exact genetic reversal of insertion. This fact attests to the precision of site recognition by the
specific phage-coded recombination enzymes. Artificial selection for
lambda particles derived by imprecise excision reveal that they too
occur, but at a much lower rate--about 1 in 10^5 for lambda particles
that carry the neighboring genes for galactose catabolism. These

abnormal particles do not arise from imprecise specific excision, but
rather from other pathways whose genetic control is unknown.

Galactose-transducing (λgal) particles or of two types:

The first type can undergo a complete lysogenic cycle. They are
defective (unable to form plaques) and generate infectious progeny only
when a normal (non-defective) lambda supplies protein components of the
virions. Replication of λgal prophage does not require a helper phage;
like normal λ prophage, the viral DNA is passively replicated by the
host. Each λgal isolate of this type can therefore be propagated
indefinitely and engender any desired number of identical progeny.

Genetic and physical studies agree that λgal's of this type arise
by abnormal excision, as diagrammed in Figure 3. The exact location of
the breakpoints is unique to each isolate; but every λgal contains a
connected segment of the lysogenic chromosome from which it derived.
The molecular mechanism is perhaps identical to the genesis of ordinary
deletion mutations.

Particles of the second class cannot undergo a complete cycle, even
with normal λ helper. They cannot lysogenize, and transduce only by
recombinational replacement of the gal gene of the recipient with its
homolog from the donor. Little and Gottesman[17] showed that some of
these particles have the structure indicated in Figure 4. They cannot
lysogenize because the left end of normal lambda is needed for circularization.

Incorporation of Viral Genes into the Host

Since the prophage of a lysogenic bacterium is under host control
rather than vice versa, lysogenization is properly considered incorporation of virus into the host. Because of specific excision, this incorporation is reversible. Deletions can create irreversible unions of
host and viral genes.

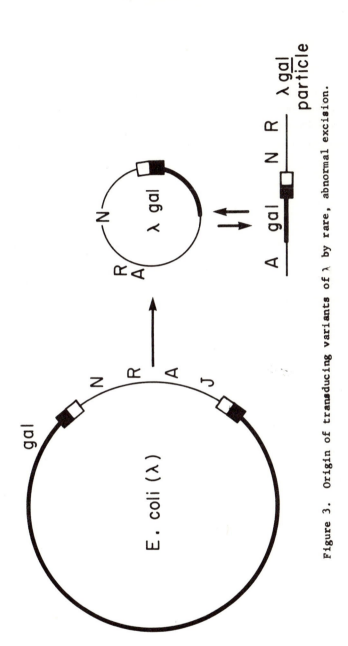

Figure 3. Origin of transducing variants of λ by rare, abnormal excision.

Figure 4. Molecular structure of DNA from one type of transducing particle found in λ lysates, as elucidated by Little
and Gottesman.[17] The right end of the molecule as drawn is
normal for λ, with a single stranded region 12 nucleotides
long. The left end of the molecule is bacterial DNA. Absence
of the normal λ left end precludes pairing of the ends and
intracellular circularization. Therefore the structure shown
cannot reproduce itself and is detectable only by recombination with other structures bearing homologous genes.

Deletions penetrating the lambda prophage can be obtained by selection for chlorate resistance.[18] The prophage insertion site lies between two bacterial genes (chlD and chlA) that specify components of nitrate reductase. Cells containing nitrate reductase are sensitive to chlorate under anaerobiosis, probably because the chlorite produced by reduction is toxic. Some chlorate resistant mutants are deletions of chlA or chlD together with adjacent genes, including all or part of the lambda prophage.

Compound Aberrations

Figure 5 shows the origin of transducing phages and deletion prophages. One diagram serves both purposes because the two constitute reciprocal products of the same type of heterologous exchange. (Reciprocity at the individual level is undemonsted and irrelevant to the point.) This mode of origin imposes restrictions on the nature of the product: A single event can generate a transducing phage in which gal and all bacterial genes to its right are inserted between a terminal phage segment of variable extent to the left. Likewise, a single event may give rise to bacterial deletions in which any terminal prophage segment is fused to bacterial genes.

More complicated monsters can be created either by a succession of single fusions or by recombination between parents with different aberrations. Transducing phages have been isolated from a lysogen in which the gal genes had been fused with the lambda Q gene.[19] Figure 6 shows some of the transducing phages obtained. Lysogenization by one such phage, followed by homologous recombination, produced the structure shown in Figure 7, where a tiny segment of phage genome is fused into the bacterium.[20]

One intriguing property of structurally abnormal phages is the tendency for further aberrations to accumulate. λgal derived from

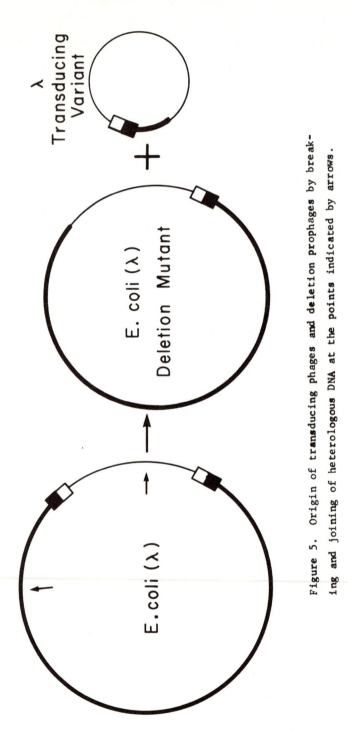

Figure 5. Origin of transducing phages and deletion prophages by breaking and joining of heterologous DNA at the points indicated by arrows.

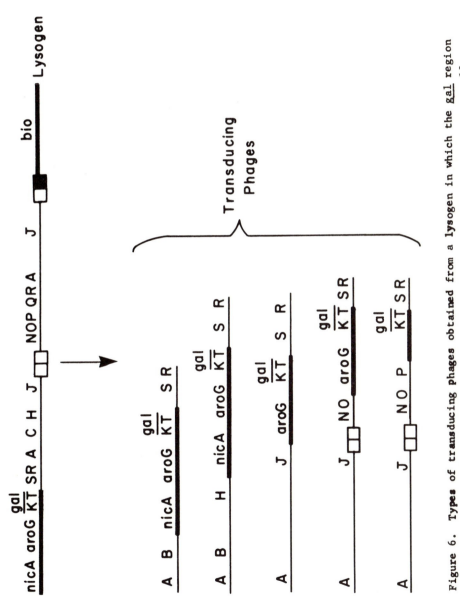

Figure 6. Types of transducing phages obtained from a lysogen in which the gal region of E. coli was fused with the Q gene of λ by deletion. After Sato and Campbell.[19]

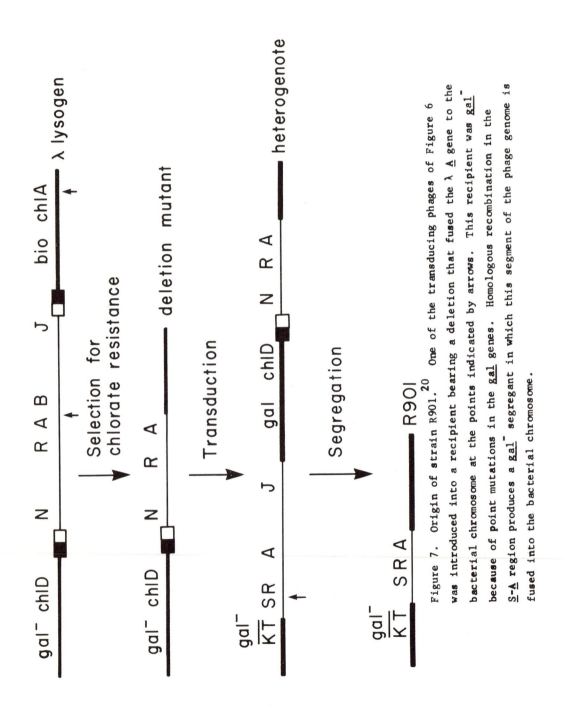

Figure 7. Origin of strain R901.[20] One of the transducing phages of Figure 6 was introduced into a recipient bearing a deletion that fused the λ A gene to the bacterial chromosome at the points indicated by arrows. This recipient was gal⁻ because of point mutations in the gal genes. Homologous recombination in the S-A region produces a gal⁻ segregant in which this segment of the phage genome is fused into the bacterial chromosome.

normal lysogens are quite stable. These phages have all the lambda
genes and specific sites concerned with intracellular replication and
recombination and lack only those involved in virion construction.
Because they are defective, they are generally replicated as prophages.

With other transducing phages, secondary aberrations are more
common. In a very thorough study of seven plaque-forming derivatives
of lambda that transduce the bioA gene, Manly[21] found that the genetic
content of five of these obeyed the rules prescribed by Figure 3. The
other two could only have resulted from additional or more complicated
changes occurring prior to isolation.

The phage at the bottom of Figure 6 does not form plaques because
it lacks part of the Q gene of lambda. A plaque-forming mutant (λqin)
selected from it proved to be a substitution of bacterial DNA of unknown
origin for the gal gene formerly present.[22,23] Starting from a λbio
phage that fails to form plaques because the N gene of lambda is absent,
Court and Sato[24] isolated a mutant (nin) able to form plaques. All nin
mutations selected from this λbio phage have entailed density changes,
and therefore changes in DNA content. The one case analyzed by electron
microscopy proved to be a simple deletion[25] that presumably alters the
regulation of the adjacent Q gene, whose expression is normally controlled by N. The interesting point here is that these nin mutants have
been obtained only from this λbio stock, not from point mutants of N;
although the latter are suppressed by nin when N^-nin recombinants are
made.[26]

Whether the effect of the bio substitution on generation of nin
mutations is inductive or purely selective has not been rigorously
tested. The relevance to the present discussion is that, once a single
aberration has occurred, selection can favor types with further

structural changes, so that the system evolves rapidly into a form radically different from the original.

Recombination and gene fusions in nature

The above examples of specific mutations of a particular temperate virus illustrate some of the ways in which a host-episome complex can change. Extensive occurrence of aberrations fusing host genes into the episome and vice versa could inextricably merge the evolution of the two elements. Transferable episomes provide means whereby host genomes can not only recombine by exchange of homologous information but also expand by addition of information from heterologous sources. Present knowledge does not permit any useful assessment of the actual extent of these processes in natural evolution.

Of the temperate phages that grow on the same hosts as lambda, several can recombine genetically with it. Generally, homology is restricted to a few small regions of the chromosome.[27,28] Whereas each of these phages makes a different specific repressor and contains operators responding to it, some of them (such as lambda and 434) insert at the same chromosomal site. More "missing links" must be found before the phylogenetic relationships of these viruses can be determined.

In considering viral evolution, the possibility of host gene incorporation must be borne in mind at all times. Viruses, as we now understand them, have two essential properties: (a) Their nucleic acid can replicate autonomously. (b) They form infectious particles that transport nucleic acid from one cell to another. The temperate phages have a third property: ability to integrate into the chromosome. All these properties are in principle separable, and various combinations have been observed. It seems unlikely that nature has ignored completely the possibility of constructing new viruses and episomes by fusing DNA segments that previously each determined only one of these properties.

Formation of infectious, extracellular particles of defined morphology is perhaps the most uniquely viral characteristic in the minds of
many investigators. It is a complex property, which can be further
subdivided into particle formation, liberation from cell of origin, and
uptake by recipient cell. "Virus-like" particles have frequently been
reported where no infectivity was demonstrable. Some authors have been
so struck with their similarity to known infectious agents as to dub
them "viruses" with no qualifications. Extensive packaging of heterogeneous fragments of host DNA into such particles (with no demonstrated
preference for the DNA segment determining their formation) was observed
in one instance.[29] Naturally occurring defective viruses could equally
well be either descendants of past viruses or potential ancestors of
future ones (or, of course, both).

More specifically, if recombination is useful to the host, evolution of gene transfer mechanisms as a strictly host function, later
exploited by adventitious plasmids, is plausible. Some years ago,
Lwoff[30] noted that, whereas viruses might in principle be related to
normal cell constituents, they in fact do not resemble known organelles.
A morphological distinction between viruses and non-viruses is less
defensible if we include the possibility that the normal counterparts of
viruses are to be found outside, rather than inside, the cell.

Figure 8 shows the lambda map with genes classified according to
function. The clustering of genes with related functions may simplify
regulation of virus development. It also means that a phage such as
lambda could have been constructed in a few steps by heterologous fusion
between a non-viral plasmid and host genes determining DNA packaging,
autolysis, and recombination, respectively.

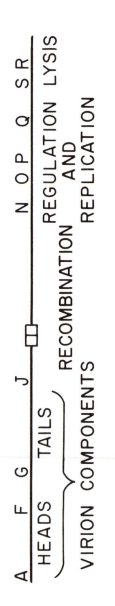

Figure 8. Distribution of genetic functions on the λ map.

EPISOMES AND EUKARYOTES

Discussion of episomes in eukaryotes seems necessarily confined to enumeration of a few genetic elements that might be episomes and speculation as to what the episomes might be. I will comment here only on the origin of eukaryotes.

The development of the eukaryotic cell from its prokaryotic ancestors perhaps occurred by fusion of several different prokaryotes, each ancestral to the determinants of specific subcellular structures.[31] According to this very appealing picture, a eukaryotic cell comprises a community of elements, each descended from a free-living ancestor, that coexist in intimate proximity. The novelty of this "great step forward" in evolution appears in wider perspective if we view the prokaryotic cell as already comprising a community of separable, potentially autonomous genetic elements.

Acknowledgment. Some of the experimental work reported here was supported by Grant AI-8573 of the National Institutes of Health, U. S.

REFERENCES

1. HAYES, W., in Bacterial Episomes and Plasmids, CIBA Foundation Symposium, pp. 4-11. G.E.W. Wolstenholme and Maeve O'Conner, Editors. J.A. Churchill, 1969.

2. JACOB, F. AND WOLLMAN, E.L. Compt. rend. 247, 154 (1958).

3. CAMPBELL, A. Episomes, Harper and Row, New York, 1969.

4. JACOB, F., BRENNER, S., AND CUZIN, F. Cold Spring Harbor Symp. Quant. Biol. 28, 329 (1963).

5. CARO, L.G. AND BERG, C.M. Cold Spring Harbor Symp. Quant. Biol. 33, 559 (1968).

6. PRITCHARD, R.H., BARTH, P.T., AND COLLINS, J. Symp. Soc. Gen. Microbiol. 19, 263 (1969).

7. NISHIMURA, Y., CARO, L., BERG, C.M., AND HIROTA, Y. J. Mol. Biol. 55, 441 (1971).

8. ANDERSON, E.S., in Bacterial Episomes and Plasmids, CIBA Foundation Symposium, pp. 102-119. G.E.W. Wolstenholme and Maeve O'Conner, Editors. J.A. Churchill, 1969.

9. NISIOKA, T., MITANI, M., AND CLOWES, R.C. J. Bacteriol. 103, 166 (1970).

10. ZISSLER, J. AND CAMPBELL, A. Virology 37, 318 (1969).

11. GUARNEROS, G. AND ECHOLS, H. J. Mol. Biol. 47, 565 (1970).

12. DOVE, W.F. J. Mol. Biol. 47, 585 (1970).

14. WEISBERG, R. AND GOTTESMAN, M., in The Bacteriophage λ. A.D. Hershey, Editor. Cold Spring Harbor Laboratories, New York, 1971 (in press).

15. BERTANI, L.E. Proc. Natl. Acad. Sci. U.S. 65, 331 (1970).

16. BODMER, W. Symposia Soc. Gen. Microbiol. 20, 279 (1970).

17. LITTLE, J. AND GOTTESMAN, M., in The Bacteriophage λ. A.D. Hershey, Editor. Cold Spring Harbor Laboratories, New York, 1971 (in press).

18. ADHYA, S., CLEARY, P., AND CAMPBELL, A. Proc. Natl. Acad. Sci. U.S. 61, 956 (1968).

19. SATO, K. AND CAMPBELL, A. Virology 41, 474 (1970).

20. CAMPBELL, A., ADHYA, S., AND KILLEN, K., in Bacterial Episomes and Plasmids, CIBA Foundation Symposium, pp. 12-31. G.E.W. Wolstenholme and Maeve O'Conner, Editors. J.A. Churchill, 1969.

21. MANLY, K. Virology 42, 138 (1970).

22. HRADNECA, Z., FIANDT, M., SATO, K., CAMPBELL, A., AND SZYBALSKI, W. Summarized by Fiandt et al., in The Bacteriophage λ. A.D. Hershey, Editor. Cold Spring Harbor Laboratories, New York, 1971 (in press).

23. HERSKOVITZ, I. Personal communication.

24. COURT, D. AND SATO, K. Virology 39, 348 (1969).

25. FIANDT, M., COURT, D., CAMPBELL, A., AND SZYBALSKI, W. Summarized

     by Fiandt et al., in The Bacteriophage λ. A.D. Hershey, Editor.

     Cold Spring Harbor Laboratories, New York, 1971 (in press).

26. COURT, D. Thesis, University of Rochester (1970).

27. LIEDKE-KULKE, M. AND KAISER, A.D. Virology 32, 465 (1967).

28. FIANDT, M. AND SZYBALSKI, W., in The Bacteriophage λ. A.D. Hershey,

     Editor. Cold Spring Harbor Laboratories, New York, 1971 (in press).

29. HAAS, M. AND YOSHIKAWA, H. J. Virol. 3, 248 (1969).

30. LWOFF, A., in The Viruses, Vol. 2. F.M. Burnst and W.M. Stanley,

     Editors. Academic Press, New York, p. 187, 1957.

31. SAGAN, L. J. Theoret. Biol. 14, 225 (1967).

DISCUSSION

JACOB: In all cases of circular DNA there exists two forms. The open circular and the twisted covalently-closed molecules. What is the form that you would select for your integration model?

CAMPBELL: I have no preference. In fact, lambda DNA seems to be covalently closed when it lysogenizes.

LEE: In microbial systems, there is a strong tendency for tandem duplications to be reduced. What is the driving force for this tendency? Further, animal systems keep these repeating sequences in large amounts. Could you explain this?

CAMPBELL: In both higher organisms and microbes, tandem duplications tend to be eliminated. The stability of those redundancies observed in higher organisms may either be because of reverse tandem arrangements, or of sufficient differentiation between similar regions so that homology is reduced.

SIMON: Are there not cases in the lambda virus where after enough DNA has been deleted, multiple copies of some lambda genes are generated?

CAMPBELL: Yes. N. Franklin at Stanford found that certain deletion derivatives of lambda accumulate density increases. R. Baldwin at
Stanford has shown that many of these increases are attributable to
tandem duplications of phage DNA.

STRICKBERGER: There are at least two selective pressures which
would tend to cause the permanent incorporation of at least a section
of the lambda genome into the bacterial chromosome; (1) the advantage
to the bacterium to maintain a locus that produces repressor, preventing
further infection of lambda particles; (2) the advantage to the viral
chromosome to be replicated via the bacterial system, at least as long
as the bacterial clone survives. Is there evidence of such incorporation, by discovery of lambda strains which have made this kind of
transition, or of E. coli chromosomes bearing "relict" lambda genes?

CAMPBELL: I know of no case where a natural isolate of E. coli
carried a defective lambda prophage. Defective lysogens are found in
the laboratory, generally following mutagenesis. Natural defective
relatives of other viruses have frequently been encountered. Some of
these may differ from active virus by a few point mutations, but
probably many of them have large deletions or substitutions similar to
those I have described.

Abata, J. J.
  Division of Radiation Therapy
  Meadowbrook Hospital
  East Meadow, New York 11554

Ahmed, Asad
  Department of Genetics
  University of Alberta
  Edmonton, Alberta, Canada

Arnheim, N.
  State University of New York
  Stony Brook, New York 11790

Atwood, K. C.
  Columbia University
  630 West 168th Street
  New York, N. Y. 10032

Bachmann, K.
  Department of Biology
  University of South Florida
  Tampa, Florida 33620

Baker, W. K.
  Department of Biology
  University of Chicago
  1103 East 57th Street
  Chicago, Illinois 60637

Baskin, L. S.
  Stevens Institute of Technology
  Castle Point Station
  Hoboken, New Jersey 07030

Bass, A. I.
  Department of Biology
  State University of New York
  Albany, New York 12203

Beebe, D. C.
  Biology Department
  University of Virginia
  Charlottesville, Virginia 22904

Bendich, A. J.
  Department of Botany
  University of Washington
  Seattle, Washington 98105

Berg, C. M.
  Department of Genetics and
    Cell Biology
  University of Connecticut
  Storrs, Connecticut 06268

Berlyn, M.
  Department of Biology
  Yale University
  New Haven, Connecticut 06520

Boguslawski, G.
  Department of Genetics
  Cornell University
  Ithaca, New York 14850

Bonventre, P.
  Department of Microbiology
  University of Cincinnati
  Cincinnati, Ohio 45219

Bores, R. J.
  Brookhaven

Bottino, P. J.
  Brookhaven

Boyer, S. H.
  Johns Hopkins Hospital
  933 Traylor Building
  Baltimore, Maryland 21205

PARTICIPANTS

Britten, R. J.
  Carnegie Institution of
    Washington
  5241 Broad Branch Road, N. W.
  Washington, D. C. 20015

Bronk, B. V.
  Brookhaven

Brooks, J. S.
  Oklahoma State University
  Stillwater, Oklahoma 74074

Brooks, M.
  Oklahoma City University
  Stillwater, Oklahoma 74074

Burholt, D. R.
  Brookhaven

Camargo, A. M.
  Brookhaven

Campbell, A. M.
  Department of Biological Sciences
  Stanford University
  Stanford, California 94035

Carlson, P. S.
  Wesleyan University
  Middletown, Connecticut 06457

Cassidy, D. M.
  Section of Genetics, Development
    and Physiology
  Cornell University
  Ithaca, New York 14850

Chaleff, R.
  Department of Biology
  Yale University
  New Haven, Connecticut 06520

Chao, E. S-E.
  Michigan State University
  East Lansing, Michigan 48823

Chourey, P. S.
  Brookhaven

Coggins, J. R.
  Brookhaven

Coggins, L. F. W.
  Brookhaven

Cohen, E. H.
  Kline Biology Tower
  Yale University
  New Haven, Connecticut 06520

Colbert, D. A.
  Department of Biomedical Sciences
  Brown University
  Providence, Rhode Island 02912

Collins, A. M.
  Department of Genetics
  Ohio State University
  Columbus, Ohio 43201

Combatti, N. C.
  Brookhaven

Conger, A. D.
  Radiation Biology
  Temple University School of
    Medicine
  Philadelphia, Pennsylvania 19140

Costello, R. C.
  Brookhaven

Crockett, W. C.
  Brookhaven

Crotty, W. J.
  Department of Biology
  New York University
  Washington Square
  New York, N. Y. 10003

Curley, D. M.
  Biology Department
  Long Island University
  385 Flatbush Avenue Extension
  Brooklyn, New York 11201

Daly, K. R.
  Department of Zoology
  University of Texas
  Austin, Texas 78712

Dancis, B.
  Department of Biological Chemistry
  Harvard Medical School
  Boston, Massachusetts 02115

Davidson, D.
  Biology Department
  McMaster University
  Hamilton, Ontario, Canada

Dean, W. W., Jr.
  Biological Research Laboratories
  Syracuse University
  Syracuse, New York 13210

PARTICIPANTS

Donaldson, D. W.
Academic Faculty of Genetics
Ohio State University
Columbus, Ohio 43210

Drozdowicz, B. Z.
Section of Genetics, Development
and Physiology
Cornell University
Ithaca, New York 14850

Dudock, B.
State University of New York
Stony Brook, New York 11790

Dunn, H.
Biochemistry Department
University of Illinois
Urbana, Illinois 61801

Dunn, N.
Biochemistry Department
University of Illinois
Urbana, Illinois 61801

DuPraw, E. J.
Stanford University
Stanford, California 94305

Duvick, D. N.
Department of Plant Breeding
Pioneer Hi-Bred International
Johnston, Iowa 50131

Eckhardt, R. A.
Department of Biology
Yale University
New Haven, Connecticut 06520

Edmunds, L.
State University of New York
Stony Brook, New York 11790

Elkind, M. M.
Brookhaven

Farquhar, M. N.
Department of Biochemistry
University of Washington
Seattle, Washington 98105

Fasy, T. M.
The Rockefeller University
New York, N. Y. 10021

Ferguson, M. L.
Department of Biological Sciences
Kent State University
Kent, Ohio 44242

Fink, G. R.
Section of Genetics, Development
and Physiology
Cornell University
Ithaca, New York 14850

Fitch, W. M.
Department of Physiological
Chemistry
University of Wisconsin
Madison, Wisconsin 53706

Flickinger, R.
Department of Biology
State University of New York
Buffalo, New York 14214

Floyd, B. M.
Brookhaven

Fox, T.
Section of Genetics, Development
and Physiology
Cornell University
Ithaca, New York 14850

Freeling, M.
Department of Botany
Indiana University
Bloomington, Indiana 47401

Friedlander, G.
Brookhaven

Fussell, C. P.
Pennsylvania State University
McKeesport, Pennsylvania 15132

Gage, L. P.
Department of Embryology
Carnegie Institution of Washington
115 West University Parkway
Baltimore, Maryland 21210

Galinsky, I.
Biology Department
Hofstra University
Hempstead, New York 11550

Gavazzi, G.
Institute of Genetics
University of Milan
Milan, Italy

Gaydos, M. D.
University of Virginia
Charlottesville, Virginia 22904

Geisbusch, W. J.
Brookhaven

PARTICIPANTS

Gerstel, D. U.
  Department of Crop Science
  North Carolina State University
  Raleigh, North Carolina 27607

Goin, C. J.
  Department of Zoology
  University of Florida
  Gainesville, Florida 32601

Goin, O. B.
  Department of Zoology
  University of Florida
  Gainesville, Florida 32601

Goldhaber, M.
  Brookhaven

Gonsalves, N. I.
  Rhode Island College
  Providence, Rhode Island 02908

Gonzalez, F. W.
  Brookhaven

Gravel, R.
  Biology Department
  Yale University
  New Haven, Connecticut 06520

Greene, L. J.
  Brookhaven

Grist, E.
  Brookhaven

Grist, K. L.
  Brookhaven

Gross, S.
  Institute for Muscle Disease, Inc.
  515 East 71st Street
  New York, N. Y. 10021

Guimaraes, R. C.
  Department of Genetics and
    Cell Biology
  University of Connecticut
  Storrs, Connecticut 06268

Guy, O.
  Brookhaven

Halaban, R.
  Brookhaven

Hall, C. A. S.
  Brookhaven

Hart, G. E.
  Texas A & M University
  College Station, Texas 77840

Haut, W. F.
  Biology Department
  Hofstra University
  Hempstead, New York 11550

Heinig, K.
  Department of Biological Sciences
  State University of New York
  Albany, New York 12203

Hillman, W. S.
  Brookhaven

Hinegardner, R.
  Division of Natural Sciences
  University of California
  Santa Cruz, California 95060

Hintz, M.
  Department of Biological Sciences
  Northwestern University
  Evanston, Illinois 60201

Hohberger, L. H.
  Brookhaven

Holt, B. R.
  Brookhaven

Hubby, J. L.
  Department of Biology
  University of Chicago
  1103 East 57th Street
  Chicago, Illinois 60637

Ino, I.
  Brookhaven

Jacob, R. J.
  Department of Biology
  Syracuse University
  Syracuse, New York 13210

Jahn, A.
  Brookhaven

Jelenkovic, G.
  Department of Horticulture and
    Forestry
  Rutgers University
  New Brunswick, New Jersey 08903

Kermicle, J. L.
  Laboratory of Genetics
  University of Wisconsin
  Madison, Wisconsin 53706

PARTICIPANTS

Klein, A.
Department of Biology
University of Rochester
Rochester, New York 14627

Klein, H.
Cornell University
Ithaca, New York 14850

Klein, J. R.
Brookhaven

Klug, E. E.
Brookhaven

Kohno, T.
Brookhaven

Kortt, A. A.
Brookhaven

Kosower, N. S.
Albert Einstein College of
Medicine
Yeshiva University
1300 Morris Park Avenue
Bronx, New York 10461

Kovacs, C. J.
Brookhaven

Kunz, W.
Department of Biology
Yale University
New Haven, Connecticut 06520

Lacy, A. M.
Department of Biological
Sciences
Goucher College
Towson, Maryland 21204

Lago, B.
Merck Sharp and Dohme Research
Laboratories
Rahway, New Jersey 07065

Ledbetter, M. C.
Brookhaven

Lederberg, S.
Division of Biomedical Sciences
Brown University
Providence, Rhode Island 02912

Lee, C. S.
Department of Biological
Chemistry
Harvard Medical School
Boston, Massachusetts 02115

Leichtling, B.
State University of New York
Stony Brook, New York 11790

Levine, R. P.
The Biological Laboratories
Harvard University
Cambridge, Massachusetts 02138

Liu, H. Z.
State University of New York
Plattsburgh, New York 12901

Lowenstein, R.
Cornell University
Ithaca, New York 14850

Lucov, Z.
Department of Biology
University of Rochester
Rochester, New York 14627

Lyman, H.
State University of New York
Stony Brook, New York 11790

MacIntyre, R. J.
Section of Genetics, Development
and Physiology
Cornell University
Ithaca, New York 14850

Mason, L. G.
Department of Biology
State University of New York
Albany, New York 12203

McCarthy, B. J.
Department of Microbiology
University of Washington
Seattle, Washington 98105

McGovern, G. E.
Brookhaven

Meiselman, N.
C. W. Post College
Greenvale, New York 11548

Melton, C. G., Jr.
Department of Biology
University of Pittsburgh
Pittsburgh, Pennsylvania 15213

Mericle, L. W.
Department of Botany and Plant
Pathology
Michigan State University
East Lansing, Michigan 48823

PARTICIPANTS

Mericle, R. P.
  Department of Botany and
    Plant Pathology
  Michigan State University
  East Lansing, Michigan 48823

Merriam, R. W.
  Biology Department
  State University of New York
  Stony Brook, New York 11790

Meyer, D. P.
  Brookhaven

Middleton, R. B.
  Department of Biology
  McGill University
  Montreal, Quebec, Canada

Milkman, R.
  Department of Zoology
  University of Iowa
  Iowa City, Iowa 52240

Moses, M. J.
  Department of Anatomy
  Duke University School of
    Medicine
  Durham, North Carolina 27706

Murray, M. J.
  A. M. Todd Company
  P. O. Box 711
  Kalamazoo, Michigan 49005

Nadolney, C. H.
  Brookhaven

Nakatani, H. Y.
  Brookhaven

Nasser, DeL.
  Department of Microbiology
  University of Florida
  Gainesville, Florida 32601

Nauman, A.
  Brookhaven

Nauman, C. H.
  Brookhaven

Nawrocky, M.
  Brookhaven

Nur, U.
  Department of Biology
  University of Rochester
  Rochester, New York 14627

O'Brien, S.
  Department of Genetics
  Cornell University
  Ithaca, New York 14850

Ohno, S.
  Department of Biology
  City of Hope Medical Center
  Duarte, California 91010

Ornston, L. N.
  Department of Biology
  Yale University
  New Haven, Connecticut 06520

Ornston, M. K.
  Department of Biology
  Yale University
  New Haven, Connecticut 06520

Osborne, T. S.
  Department of Biological Sciences
  Smith College
  Northampton, Massachusetts 01060

Ownbey, M.
  Department of Botany
  Washington State University
  Pullman, Washington 99163

Parent, J. B.
  Biology Department
  University of Virginia
  Charlottesville, Virginia 22904

Partridge, C.
  Department of Biology
  Yale University
  New Haven, Connecticut 06520

Pavan, C.
  Department of Zoology
  University of Texas
  Austin, Texas 78712

Pierce, D. A.
  Graduate Center
  Brown University
  Providence, Rhode Island 02912

Polan, M. L.
  Yale University
  New Haven, Connecticut 06520

Pond, V.
  Brookhaven

Price, H. J.
  Brookhaven

PARTICIPANTS

Prover, C. B.
Biology Department
Hofstra University
Hempstead, New York 11550

Ramirez, S. A.
Zoology Department
Indiana University
Bloomington, Indiana 47401

Rees, H.
Department of Agricultural
Botany
University College
Aberystwyth, Wales

Reincke, U.
Brookhaven

Rice, N. R.
Carnegie Institution of
Washington
5241 Broad Branch Road, N. W.
Washington, D. C. 20015

Rich, P. H.
Brookhaven

Richmond, J. Y.
U. S. Department of Agriculture
Plum Island Animal Disease
Laboratory
Greenport, New York 11944

Rossi, M.
Department of Biological Sciences
Stanford University
Stanford, California 94305

Rost, T. L.
Brookhaven

Rothstein, R.
University of Chicago
1103 East 57th Street
Chicago, Illinois 60637

Ruffing, R. N.
Brookhaven

Russell, P. J.
Section of Genetics, Development
and Physiology
Cornell University
Ithaca, New York 14850

Sager, R.
Department of Biological Sciences
Hunter College of the City
University of New York
New York, N. Y. 10021

Sarvella, P.
Science Research Division
U. S. Department of Agriculture
Beltsville, Maryland 20705

Saunders, G. F.
M. D. Anderson Hospital and Tumor
Institute
University of Texas
Houston, Texas 77025

Sautkulis, D. O.
Brookhaven

Sautkulis, R. C.
Brookhaven

Scandalios, J. G.
MSU/AEC Plant Research Laboratory
Michigan State University
East Lansing, Michigan 48823

Schairer, L. A.
Brookhaven

Schlagman, S.
Biology Department
Hofstra University
Hempstead, New York 11550

Schneiweiss, J. W.
Biology Department
Hofstra University
Hempstead, New York 11550

Schwemmer, S. S.
Brookhaven

Sciaky, D.
Program in Genetics
Washington State University
Pullman, Washington 99163

Serianni, R. W.
Brookhaven

Shifriss, O.
Department of Horticulture
Rutgers University
New Brunswick, New Jersey 08903

Shull, J. K., Jr.
Department of Biological Sciences
Florida State University
Tallahassee, Florida 32306

Siegelman, H. W.
Brookhaven

Simpson, M.
State University of New York
Stony Brook, New York 11790

Smith, H. H.
Brookhaven

PARTICIPANTS

Smithies, O.
  Department of Medical Genetics
  University of Wisconsin
  Madison, Wisconsin 53706

Smyth, D. R.
  University of California,
    San Diego
  La Jolla, California 92037

Somers, G. F.
  Department of Biological Sciences
  University of Delaware
  Newark, Delaware 19711

Sonea, S.
  Department of Microbiology and
    Immunology
  University of Montreal
  Montreal, Quebec, Canada

Sparrow, A. H.
  Brookhaven

Sparrow, R.
  Brookhaven

Spear, B. B.
  Kline Biology Tower
  Yale University
  New Haven, Connecticut 06520

Spofford, J. B.
  Department of Biology
  University of Chicago
  1103 East 57th Street
  Chicago, Illinois 60637

Squire, R. D.
  Long Island University
  385 Flatbush Avenue Extension
  Brooklyn, New York 11201

Stangby, J. G.
  Brookhaven

Stein, O. L.
  Department of Botany
  University of Massachusetts
  Amherst, Massachusetts 01002

Stevenson, H. Q.
  Southern Connecticut State
    College
  501 Crescent Street
  New Haven, Connecticut 06515

Stinson, H. T.
  237 Plant Science Building
  Cornell University
  Ithaca, New York 14850

Storck, R.
  Department of Biology
  Rice University
  Houston, Texas 77001

Storti, R. V.
  Zoology Department
  Indiana University
  Bloomington, Indiana 47401

Strahs, K. R.
  Biology Department
  Hofstra University
  Hempstead, New York 11550

Strickberger, M. W.
  Department of Biology
  University of Missouri
  8001 Natural Bridge Road
  St. Louis, Missouri 63121

Studier, F. W.
  Brookhaven

Swanson, R. F.
  Department of Biology
  University of Virginia
  Charlottesville, Virginia 22903

Tai, W.
  Department of Botany and Plant
    Pathology
  Michigan State University
  East Lansing, Michigan 48823

Taylor, J. H.
  Institute of Molecular Biophysics
  Florida State University
  Tallahassee, Florida 32306

Taylor, S.
  Institute of Molecular Biophysics
  Florida State University
  Tallahassee, Florida 32306

Tempel, N. R.
  Brookhaven

Thompson, W. F.
  The Biological Laboratories
  Harvard University
  Cambridge, Massachusetts 02138

Threlkeld, S. F. H.
  Biology Department
  McMaster University
  Hamilton, Ontario, Canada

Underbrink, A. G.
  Brookhaven

PARTICIPANTS

Van't Hof, J.
Brookhaven

Vickery, R. K., Jr.
Department of Biology
University of Utah
Salt Lake City, Utah 84112

Vincent, W. S.
Department of Biological Sciences
University of Delaware
Newark, Delaware 19711

Vodkin, M.
Section of Genetics, Development
and Physiology
Cornell University
Ithaca, New York 14850

Voeller, B.
The Rockefeller University
New York, N. Y. 10021

Walker, A. A.
Department of Genetics
Towson State College
Baltimore, Maryland 21204

Wall, D. A.
Section of Genetics, Development
and Physiology
Cornell University
Ithaca, New York 14850

Webster, P. L.
Department of Botany
University of Massachusetts
Amherst, Massachusetts 01002

Wells, R.
Department of Biological Sciences
Hunter College of the City
University of New York
New York, N. Y. 10021

Wemyss, C. T.
Department of Biology
Hofstra University
Hempstead, New York 11550

Wensink, P. C.
Department of Embryology
Carnegie Institution of Washington
115 West University Parkway
Baltimore, Maryland 21210

Whitehouse, H. L. K.
Botany School
University of Cambridge
Cambridge CB2 3EA, England

Wills, C.
Department of Biology
Wesleyan University
Middletown, Connecticut 06457

Woese, C. R.
Department of Microbiology
University of Illinois ·
Urbana, Illinois 61801

Woodwell, G. M.
Brookhaven

Wright, T. R. F.
Department of Biology
University of Virginia
Charlottesville, Virginia 22903

Wu, C-H.
Kline Biology Tower
Yale University
New Haven, Connecticut 06520

Wysocki, J. R.
Brookhaven

Yagi, S.
Brookhaven

Yep, D.
Section of Genetics, Development
and Physiology
Cornell University
Ithaca, New York 14850

Yourno, J. D.
Brookhaven

SUBJECT INDEX

Aberrant ratios, 293
Aberrant segregation, 296
ac-20 cells, 516
Acid phosphatase-1, 171, 180
    Dimer, 161
    Subunit reassociation, 164, 166
    Variation, 151
Adaptive basis, 220
Adaptive values, 127, 128, 131, 217
Adaptor, 330
Agnatha, genome size, 435
Algae, DNA content, 455
Alignment, 330
Alkaline phosphatase, 148
Alleles, multiple, 132
Allelic diversification, 123
Allium, pachytene analysis, 402
Allozymic variation, 221
Amino acid
    Recognition, 348
    Replacements, 205
    Sequences, 198
    Substitution, 15, 218
Amphibia, evolution of, 440
Ampholine electrofocusing, 256
Amplification, chromosome
    segments, 396
Amylase, 148
Angiospermae, DNA content, 461
Annuli, 239
Anitbiotics, 524
Anticodon arm, 342
Anura, genome size, 421
Ascobolus immersus, 293
Aspergillus nidulans, 299
Autoradiography, 272

B chromosomes, 403
Bacillus subtilis, 11
Bacterial genes, incorporation
    of, 547
Bacterial recombination, 547
Bar eye duplication, Drosophila
    melanogaster, 314
Base pairs, mismatched, 9, 11
Base substitutions, 6, 13, 17
β-galactosidase protomer, 95
Birds, genome size, 443
Bistable elements, 335
Bovine serum albumin, 97
Bryophyta, DNA contents, 459
Bufo genome, 426
Bufo marinus, 37

Carboxypeptidase A, 221
Cell(s)
    DNA values, 469
    Drosophila polytene, 285
    Size, 429
        And mass, 405
Cellular tape reading processes,
    326
Centromere, 236, 237
Centromeric heterochromatin, 282
Chimaeric polypeptides, evolution
    of multifunctional, 117
Chimp, 63
Chinese hamster, 57, 59, 277
Chironomus, polytene chromosomes,
    402
Chlamydomonas reinhardii, 308, 503
    Cytoplasmic genes, 495